METHODS IN
MICROBIOLOGY

Methods in Microbiology

Chairman
John Norris
University of Reading, UK

Advisory Board

A complete list of the books in this series is available from the publishers on request.

METHODS IN
MICROBIOLOGY

Volume 21
Plasmid Technology
2nd Edition

Edited by

J. GRINSTED

Department of Microbiology, University of Bristol, Medical School, Bristol, UK

and

P. M. BENNETT

Department of Microbiology, University of Bristol, Medical School, Bristol, UK

1988

ACADEMIC PRESS
Harcourt Brace Jovanovich, Publishers
London   San Diego   New York   Berkeley   Boston
Sydney   Tokyo   Toronto

ACADEMIC PRESS LIMITED
24/28 Oval Road
London NW1 7DX

United States Edition published by
ACADEMIC PRESS INC.
San Diego, CA 92101

British Library Cataloguing in Publication Data
Methods in microbiology
Vol. 21
1. Microbiology. Laboratory techniques
576'.028
ISBN 0-12-521521-5

Typeset by Page Bros (Norwich) Ltd
Printed in Great Britain by St Edmundsbury Press,
Bury St Edmunds, Suffolk

CONTRIBUTORS

P. M. Bennett, Department of Microbiology, University of Bristol, Medical School, University Walk, Bristol BS8 1TD, UK

N. L. Brown, Department of Biochemistry, University of Bristol, University Walk, Bristol BS8 1TD, UK

H. J. Burkhardt, Department of Microbiology and Biochemistry, University of Erlangen, FRG

G. Dougan, Microbiology Department, Trinity College, Dublin 2, Ireland (Present Address: Department of Molecular Biology, Wellcome Research Laboratories, Langley Court, Beckenham, Kent BR3 3BS, UK)

N. F. Fairweather, Department of Molecular Biology, Wellcome Research Laboratories, Langley Court, Beckenham, Kent BR3 3BS,UK

T. J. Foster, Microbiology Department, Trinity College, Dublin, Ireland

J. Grinsted, Department of Microbiology, University of Bristol, Medical School, University Walk, Bristol BS8 1TD

P. A. Lund, Department of Biochemistry, University of Bristol, Bristol BS8 1TD, UK (Present address: Advanced Genetic Sciences, Oakland, CA 94608, USA

A. Pühler, Department of Genetics, Faculty of Biology, University of Bielefeld, Postfach 8640, 4800 Bielfeld, FRG

J. R. Saunders, Department of Genetics and Microbiology, University of Liverpool, Liverpool L69 3BX, UK

V. A. Saunders, School of Natural Sciences, Liverpool Polytechnic, Liverpool, UK

V. Stanisich, Department of Microbiology, La Trobe University, Bundoora 3083, Australia

R. Thompson, Institute of Virology, University of Glasgow, Church Street, Glasgow G11 5JR, UK

N. Willetts, Department of Molecular Biology, University of Edinburgh, King's Building, Mayfield Road, Edinburgh EH9 3JR, UK (Present address: Biotechnology Australia Pty Ltd, 28 Barcoo Street, P.O. Box 20, Roseville, NSW 2069, Australia)

PREFACE TO 1ST EDITION

While initiating many students into the mysteries of plasmid biology during the last few years it has become increasingly obvious that there is a need for an introductory methodology text that deals with the techniques used to investigate bacterial plasmids. This volume is an attempt to fill that gap. The techniques covered range from those that are relatively simple and demand no sophisticated equipment to those now used routinely in molecular biology. We hope that both the novice and the experienced research worker will find something of use within the pages of this text.

P. M. Bennett
January 1984   J. Grinsted

PREFACE TO 2ND EDITION

Six years ago we believed that there was a need for a text that dealt with techniques used for studying plasmids; the result was the first edition of this book (*Methods in Microbiology, Vol. 17, 1984*). We still believe that there is such a need. But, as in most fields, the subject has moved a long way since that book was published, and no other text that covers a similar range of subjects has appeared since. So, a second edition seems appropriate. This is substantially different from the first: contributions have been thoroughly revised and updated, while new sections on cloning vectors and on gene expression have been included.

The ability to isolate and manipulate plasmids is at the heart of current progress in molecular genetics, and the current sophistication of the subject can mask the fact that many of the techniques employed are conceptually and technically simple. We hope that this edition will again emphasize this and show how much useful information can be obtained with a limited range of inexpensive equipment. We hope that the book will be both accessible to a novice and useful to a more experienced worker in the field.

J. Grinsted

July 1988   P. M. Bennett

CONTENTS

1

Introduction

J. GRINSTED AND P. M. BENNETT

Department of Microbiology and Unit of Molecular Genetics, University of Bristol, Medical School, Bristol, UK.

I. Chromosomes and plasmids

The genetic material that encodes the essential basic information that specifies an organism is carried in its chromosomes. In most organisms, this genetic material is double-stranded DNA, with the specific sequence of base pairs providing the information content. This use of double-stranded DNA (as opposed to single-stranded DNA or RNA) is an important factor in the maintenance of the genetic material. (DNA is more stable than RNA, and the double-stranded nature provides two copies of the information so that damage to one strand does not inevitably result in irretrievable loss of the information.)

Structurally, chromosomes can be more than just the genetic material: particularly in eukaryotic organisms, for instance, they are complicated structures that involve proteins as well. Genetically, however, they can simply be viewed as arrays of genetic material, which, at the molecular level translates into a single DNA molecule per chromosome. Although a growing bacterium can appear to have more than one nuclear body (up to four in fast-growing organisms), these are actually copies of a single chromosome that result from the manner in which the bacterium synchronizes chromosome replication with cellular division. Thus, a non-growing bacterium usually contains a single chromosome; this single molecule of DNA encodes all essential functions. In *Escherichia coli*, for instance, this molecule is 4 704 000 base pairs (4704 kb) and encodes about 2000 genes (Kohara *et al.*, 1987). Higher organisms distribute

METHODS IN MICROBIOLOGY
VOLUME 21   ISBN 0-12-521521-5

their genetic information between a number of chromosomes. The total genetic information within a cell is referred to as the genome. (This term is often restricted to chromosomally-encoded information.)

Plasmids are genetic elements that are independent of the chromosomes. They have been found in most groups of bacteria (see Chapter 2) and have also been reported in some lower eukaryotic organisms. (Yeasts, for example, have the '2 micron plasmid', which is often used for cloning purposes (Broach, 1981).) This book is solely concerned with bacterial plasmids. At the physical level, plasmids are separate molecules that replicate independently of and that can be isolated from the chromosomes. (It should be noted that certain intracellular forms of virus could be considered to be plasmids, and that some 'plasmids' are capable of inserting into the chromosome by recombination, so that they can exist in two states, a plasmid state and a chromosomal state — see Hayes (1968).) In principle, plasmids could consist of any suitable genetic material; in practice, in bacteria at least, they are generally circular double-stranded DNA molecules that are covalently closed in each strand (see Section III).

The independence of plasmids from the chromosome implies that they do not encode essential functions; indeed, bacteria often contain plasmids that encode no obvious characteristic, apart from their ability to be maintained independently of the chromosome (cryptic plasmids).

II. History

The first plasmid to be identified was the *E. coli* fertility factor F, which was discovered because of its ability to mediate transfer of chromosomal markers from one strain of *E. coli* to another (Hayes, 1968). These initial genetic experiments established that the fertility factor was physically separate from the bacterial chromosome, and some years after, it was shown that the factor is a circular, double-stranded DNA molecule (Hayes, 1968). The discovery of F was of fundamental importance for the development of bacterial genetics but generated little interest outside certain academic circles. However, the discovery of infectious resistance to antibiotics in epidemic strains of *Shigella* in the late 1950s in Japan, and the fact that this was encoded by plasmids (Watanabe, 1963) changed people's awareness. The tempo of plasmid research accelerated dramatically as the clinical importance of the elements was appreciated. Throughout the next two decades, much plasmid research centred on drug-resistance plasmids, particularly on their epidemiology and the mechanism of drug resistance. However, it also became apparent during this period that plasmids were responsible for conferring a much greater variety of phenotypes than just fertility or drug resistance and that most groups of bacteria contain plasmids (Chapter 2; the extent of the reference list in Chapter 2 attests to the enormous diversity of plasmids).

Plasmids that encode antibiotic resistance continue to be a scourge in hospitals. One of the reasons for this is that they can very easily acquire genes that encode different antibiotic resistances. The resistance to currently-used drugs is then transferred through the bacterial population by conjugation. The main route for this acquisition is transposition, in which the genes are part of a segment of DNA that can insert into other DNA sequences (see Chapter 9). Knowledge of this process has increased in step with knowledge of the plasmids themselves.

In recent years, plasmid research has taken an entirely new turn with the advent of gene cloning and the use of plasmids as cloning vectors (i.e. plasmids are important in genetic engineering). Thus, DNA sequences of interest can be isolated from one organism and then inserted *in vitro* into a bacterial plasmid, which is then introduced by transformation into a bacterial cell. The technique allows, in principle, any DNA to be introduced into a bacterial cell, and, because of the ease with which bacterial cultures can be manipulated, this allows molecular biologists to use a whole range of experimental techniques on the heterologous sequence. It also provides the opportunity of expressing the foreign gene in the bacterium; this, of course has led to the recent industrial interest in the technique. Therefore, now, the study of plasmids is not only of academic interest, but also of great practical importance, commercially, socially and ecologically.

III. Properties and categories of plasmids

Bacterial plasmids are usually double-stranded DNA molecules that are covalently closed in each of its single strands, so that they are circular. Indeed, it is this character that is the basis of the standard method for the isolation of pure plasmid DNA (see Chapter 5), and there is the danger that this method (ethidium bromide/CsCl gradient centrifugation) might be used alone to determine whether or not a strain contains plasmid. There are good reasons why plasmids should be double-stranded DNA (see Section I), and circular (see Section IV). But this does not rule out other forms, and, recently, linear plasmids have been described (see, for example, Kinashi *et al.*, 1987). Furthermore, covalently closed circular plasmids sometimes also exist as single-stranded circles, as a result of their mode of replication (Te Riele *et al.*, 1986). And, in principle, a plasmid could be RNA, if the necessary enzymes for its replication were present. In this volume, though, only circular double-stranded DNA plasmids will be considered.

Plasmids vary enormously in size: they can be as small as 1000 base pairs

(1 kb) or greater than 400 000 base pairs (400 kb). Obviously, the size of the plasmid will put a lower limit on the proportion of the total cell DNA that is plasmid. For instance, in a stationary-phase cell of *E. coli* (in which the chromosome is 4700 kb) containing the fertility factor F (which is 100 kb), the proportion would be just over 2%. The other factor that is relevant here is the copy number of the plasmid: that is, the number of copies of the plasmid compared with the amount of chromosomal DNA (usually quoted in 'chromosomal equivalents' — in the example just given, the copy number would be 1). There is a wide variation in copy number between various types of plasmids; in general, large plasmids have low copy number (1–2) and small plasmids have high copy number (sometimes hundreds). (Low-copy number plasmids are sometimes referred to as being under stringent control, and high-copy number as under relaxed control, since the copy number must reflect how the replication of the plasmid is controlled. However it should be appreciated that even high copy number plasmids have to maintain their number and that they also encode mechanisms to effect such control – see Section IV.)

The simplest operational way of categorizing plasmids is by the genetic markers that they carry (such as genes that encode antibiotic resistance — see Chapter 2), and this has often been the first approach in epidemiological studies of resistance plasmids. (Indeed, the very category of 'resistance plasmids' is an example of such a classification.) However, this is not very satisfactory because of the very rapid changes that can take place in plasmids due to, for instance, transposition. A more satisfactory basic classification would be by replication functions: it is, after all, the capacity for independent replication that defines plasmids. Thus, there are a number of well-defined groups that can be characterized by common patterns of functional organization of their replication apparatuses: for instance, Projan *et al.* (1985) define the following groups on this basis: ColE1, IncFII, F-P1, R6K, RK2, pT181/pC221, pUB110/pBC16 (the first five are from Gram negative bacteria and the other two from Gram positives). A consequence of two plasmids having the same replication control functions is that they are incompatible, that is, they cannot coexist stably in the same cell. Incompatibility is simply tested genetically and is the basis of Inc groups in a number of groups of bacteria (see Chapter 2).

Initially there was no systematic way of naming plasmids, and plasmids that were characterized before about 1976 have numerous different sorts of name (e.g. F, R1, RP4, R6K). In 1976, a proposal to standardize the naming of plasmids was put forward (Novick *et al.*, 1976). This has been generally accepted, so that nowadays, plasmid names start with a lower case 'p', followed with a two letter uppercase identifier for the research worker or laboratory (three letters are now becoming necessary to preserve uniqueness), and then a number for the particular plasmid. For example, plasmid pUB110 was discovered at the University of Bristol and is the one hundred and tenth plasmid on the list of plasmids there. The Plasmid Reference Center in the United States allocates identifiers, which minimises the possibility of duplication (Lederberg, 1986). In addition, typical plasmids are also available from the Plasmid Section of the National Collection of Type Cultures, U.K. (Central Public Health Laboratory, 61 Colindale Avenue, London NW9 5HT).

IV. Plasmid replication

The unit of replication is the replicon. This is a segment of DNA on which replication can be initiated and terminated and so is capable of self-duplication. In bacteria, replicons are unique molecules of DNA and the bacterial chromosome or a plasmid are examples of replicons. In higher organisms, the single DNA molecules that comprise chromosomes are composed of many replicons. Replication of double-stranded DNA involves a DNA polymerase that usually operates as part of a replicating fork that traverses the DNA, resulting in daughter molecules that are 'semi-conservative' (each molecule has one old strand and one newly-synthesized strand). The initiation of the replicating forks occurs only at a specific origin of replication, and can result in monodirectional replication, with one replicating fork proceeding along the DNA in a particular direction, or bidirectional replication, in which there are two forks, one going in each direction. There are potential problems in termination of replication because of the anti-parallel nature of the DNA and the characteristics of DNA polymerases. These factors would result in incomplete replication if replication forks simply ran off the end of linear molecules. Thus, where there are linear replicons there are special provisions to cope with the problem of replicating the ends. (In eukaryotic chromosomes, for instance, there are telomeres (see Blackburn (1984).) However, with circular molecules, these problems do not arise, which seems a very good reason why most replicons in bacteria (including the chromosome and plasmids) are circular. In this case, termination of replication occurs either at some point roughly opposite the origin where the forks meet (if replication is bidirectional), or when the fork encounters the origin again after replicating the whole molecule (if replication is monodirectional).

Replication of both the chromosome and of plasmids requires many common replication functions, in particular those that carry out the basic polymerization reaction. Some plasmids have other requirements too; for example, high-copy number plasmids in *E. coli* not only require DNA polymerase III, which is the normal replicating enzyme in this organism, but also require DNA polymerase I (encoded by the *pol*A gene), for early stages of replication. In contrast, replication of the chromosome does not require the polymerizing activity of DNA polymerase I; thus, *pol*A mutants are viable, but the high copy number plasmids cannot replicate in them. This characteristic can be used to select against such plasmids (see Chapter 9).

There are some plasmids, from Gram-positive bacteria, for example, which do not use normal replicating forks: where one of the strands is replicated first, so that the other is displaced, resulting in single-stranded copies of the plasmid in the cell (Te Riele et al., 1986). The complementary strand of this single-stranded copy is then synthesized to generate the second daughter molecule. However, in general, plasmid replication proceeds using replication forks. There is variation in the way the replicating forks are organized. Some plasmids (e.g. F) replicate bidirectionally, like the chromosome, others replicate mono-directionally (R1 or ColE1, for instance), and yet others replicate bidirectionally but sequentially, so that one fork traverses its half of the molecule before the other one starts (R6K, for instance). Some plasmids also possess more than one replication origin; any one of which might be used for a particular round of replication (F or R6K, for instance).

Obviously, replication has to be synchronized in some way with cell division, so that daughter cells end up with the same amount of DNA as their parent had. As with almost all macromolecular processes in the cell, control of synthesis is exerted at the level of initiation. With the chromosome, specific proteins are produced that are required for initiation, and it is the production of these that is controlled. With plasmids that are present at low-copy number, the situation is broadly similar and a plasmid-encoded protein performs the same task for the plasmid. Indeed, it is possible for such a plasmid to take over control of chromosome replication: a mutant bacterium which has a chromosomal initiator protein that is inactive at high temperature cannot grow at that temperature since replication of the chromosome cannot be initiated; but if a plasmid that encodes its own initiator integrates into the chromosome, the chromosome may now be able to replicate from the plasmid origin, under the control of the plasmid initiator. This is called integrative suppression.

With high-copy number plasmids, there is no protein positively involved in regulation of initiation. Thus, protein synthesis is not required for continuing plasmid synthesis. This is the basis of 'chloramphenicol amplification' of plasmids (see Chapter 5, section VI.D). In the presence of chloramphenicol (which inhibits protein synthesis), chromosome replication stops after one round, due to lack of initiator protein, but high-copy number plasmids continue to be replicated because they do not require such a protein. Control of replication of these plasmids is purely negative, with the particular regulatory element inhibiting the formation of the primer required at initiation. Presumably the positive regulatory system used by low-copy number plasmids are more precise than the negative, giving more stringent control. Interestingly, one of the ways in which both regulatory systems operates is by the action of counter-transcripts; molecules of RNA that are complementary to the normal transcripts, can bind to them and inhibit their function (Cesarini & Banner, 1985; Projan et al., 1985).

It is on the basis of unique strategies for the control of replication that plasmids can be characterized (Projan *et al.*, 1986 — see Section III above). And it is also the characteristic of the replication systems and their controls that determine the incompatibility groups (see Section III above). The actual structure of the origin of the plasmid will determine its host range, that is, the range of different types of bacteria that will support replication and maintenance of that plasmid. Generally, plasmids have a restricted host range, and for genetic engineering purposes it is often necessary to construct vector plasmids that contain two separate origins of replication so that they can replicate in different hosts; these are called shuttle vectors (see Chapter 8). In a few cases, however, it has been reported that some plasmids have extraordinarily wide host ranges. Goursot *et al.* (1984) reported that some staphylococcal plasmids can be maintained in Gram negative bacteria and even in yeasts.

The coupling of the control of plasmid replication with cell division is obviously of great importance, otherwise there may not be a copy of the plasmid to be passed on to each of the daughter cells. Efficient partitioning of the plasmids is of crucial importance in achieving a stable plasmid. (It should be noted that, in the context of plasmids, 'stability' usually refers to maintenance in the host, so that an 'unstable plasmid' would be one that is lost easily from a strain.) Partitioning is often facilitated by particular sequences on the plasmids, such as the *par* locus on pSC101 or the *inc*D locus on F; these appear to act like the centromeres on eukaryotic chromosomes. (See Saunders & Saunders (1987) for a discussion of plasmid stability.)

V. Plasmids and evolution

Any piece of DNA that acquires the ability to replicate independently of the chromosome becomes a plasmid. This ability is simply conferred by an origin of replication, so that it is not difficult to see how plasmids might arise. But why do host cells tolerate plasmids, since they must exert a metabolic load on the cell? It may be that this is the wrong question and that plasmids should be considered a class of 'selfish DNA' (Doolittle and Sapienza, 1980), and persist due to their ability to replicate. Alternatively, they may be maintained because they are potential repositories of genetic information that will be useful to the host. Thus, cryptic plasmids can acquire genes that encode potentially useful functions. In these cases, there is no question as to the utility of the plasmids for the host.

Why should it normally be dispensable genes that are put on plasmids? One reason might be that this allows the bulk of the population to dispense with such genes when they are not needed, so that the vast majority of the population does not have to bear the metabolic load. Thus, as long as a small proportion of the population retains the plasmid, the rest can be relieved of the burden of maintaining it during these times. When the conditions change, the plasmid-containing cells will outgrow the rest of the population. This arrangement allows the population to take full advantage of the good times, in spite of the metabolic load required to retain genes that will overcome the bad times. For example, if a cell can grow and survive in the presence of ampicillin because it contains a resistance plasmid encoding a β-lactamase, it follows that the plasmid is indispensable in the presence of the drug. However, the plasmid is not necessary for growth in a drug-free medium and can be lost from the cell without affecting cell survival. Thus, the independent replication of dispensable genes leads to an important role for plasmids in the genetic versatility displayed by bacteria, and in the ability of bacteria to adapt.

Another reason for tolerating plasmids is because many of them can transfer between bacterial cells. Thus, the genetic pool that is available in a population of bacteria may be considerably greater than the chromosomal complement of the individuals, and beneficial changes that have occurred in one cell can be rapidly spread to others. This flexibility is increased by the fact that transposable elements (Chapter 9) can use the transferable plasmids as vectors. So, the combination of transferable plasmids and transposable elements provides a rapid method for distributing genetic information and is thus a potent force in evolution. (See Campbell (1981) for further discussion of the role of accessory genetic elements in evolution.)

VI. Scope of this volume

The rest of this volume is not specifically concerned with the plasmids themselves but rather with how they are studied. For further basic information concerning plasmids and their biology we recommend Hardy (1986); this is a very slim volume that is a very satisfactory and reasonably up-to-date introduction to plasmids. As regards research papers in general, there is a journal called *Plasmid*, which contains papers on all aspects of the subject; such papers are also distributed widely in most journals that are devoted to molecular genetics. We should also like to draw attention to the various technical publications that suppliers of molecular biology materials circulate free of charge. These are often regular pamphlets that contain many useful, short articles about various techniques. For instance, BRL publish their bulletin, 'Focus', four times a year, and there are a number of references to it through this volume.

This volume, then, is concerned with techniques that can be used to study plasmids. These range from the very simple genetic techniques that could be immediately applied in almost any laboratory, to some of the most sophisticated techniques used in a molecular genetics laboratory. It might be argued that the latter are hardly appropriate to what, we hope, will be an introductory text for some people; but use of such techniques often does follow automatically from the simpler type of experiments. Furthermore, we hope that detailed discussion of such advanced techniques will show people that they really are not that complicated or difficult to perform.

Chapter 2 lists the types of bacteria that are known to contain plasmids, and phenotypic characteristics that can be plasmid-borne. It then discusses how plasmids can be genetically characterized, and describes many types of experiments that might be done on first suspecting the presence of, or discovering a plasmid. Chapter 3 describes in detail how conjugation, a particular plasmid characteristic, can be studied. Chapter 4 describes another method of genetic transfer. This is transformation, in which pure DNA is used to transform cells. Ways in which plasmid DNA can be prepared and immediately analysed are discussed in Chapter 5, and Chapter 6 describes how this DNA can be analysed with restriction endonucleases. These techniques are, of course, all essential parts of genetic engineering: the DNA has to be isolated, then manipulated *in vitro*, and then reintroduced into the cell. The basic role of plasmids in genetic engineering is their role as vectors, that is, as carriers of particular heterologous genes, permitting their maintenance in the cell. Chapter 8 describes the latest generation of plasmid vectors. This chapter is less practically-oriented than the others, being simply a description of the sort of vectors that are available. This is somewhat at odds with the rest of the book, but we feel that the subject is appropriate here because it describes what is available for design of experiments. Chapter 7 describes analysis of plasmids by electron microscopy. Chapter 9 describes the detection and analysis of transposable elements; these movable genetic elements are of great importance in evolution generally and specifically in the evolution of the structure of plasmids. Transposable elements, for instance, encode many of the antibiotic resistances that are found on plasmids. Furthermore, these elements are powerful tools for the genetic analysis of replicons in general, and plasmids in particular. Finally, Chapters 10 and 11 discuss somewhat more sophisticated types of experiments: the first describes how transcription and translation of the genes of a plasmid can be studied, and the second describes how DNA can be sequenced. Determination of the sequence of a plasmid would, in principle, describe it completely, and there can be no question that many problems are simply solved by knowing the sequence of particular segments of the DNA. We hope that this Chapter will encourage people at least to consider the technique when appropriate.

References

Blackburn, E. H. (1984). *Cell* **37**, 7–8.
Broach, J. R. (1981). *In* "The Molecular Biology of the Yeast *Saccharomyces*, Life Cycle and Inheritance" (J. N. Strathern, E. W. Jones and J. R. Broach, Eds), pp. 445–470. Cold Spring Harbor, New York.

Campbell, A. (1981). *Ann. Rev. Microbiol.* **35**, 55–83.

Cesarini, G. & Banner, D. W. (1985). *Trends in Bioch. Sci.* **10**, 303–306.

Doolittle, W. F. and Sapienza, C. (1980). *Nature* (London), **284**, 601–603.

Goursot, R., Goze, A., Niaudet, B. and Ehrlich, S. D. (1984). *Nature (London)* **298**, 488–490.

Hardy, K. (1986). "Bacterial Plasmids" 2nd edn. Van Nostrand Reinhold, U.K.

Hayes, W. (1968). "The Genetics of Bacteria and their Viruses", 2nd edn. Blackwell, Oxford.

Kinashi, H., Shimaji, M. and Sakai, A. (1987) *Nature (London)* **328**, 454–456.

Kohara, Y., Akiyama, K. & Isono, K. (1987). *Cell* **50**, 495–508.

Lederberg, E. M. (1986). *Plasmid* **15**, 57–92.

Novick, R. P., Clowes, R. C., Cohen, S. N., Curtiss, R., Datta, N. and Falkow, S. (1976). *Bact. Rev.* **40**, 168–189.

Projan, S. J., Kornblum, J., Moghazeh, S. L., Edelman, I., Gennaro, M. L. and Novick, R. P. (1985). *Mol. Gen. Genet.* **199**, 452–464.

Saunders, V. A. & Saunders, J. R. (1987). "Microbial Genetics applied to Biotechnology" Croom Helm, London.

Te Riele H., Michel, B. and Ehrlich, S. D. (1986). *Proc. Natl. Acad. Sci. U.S.A.* **83**, 2541–2545.

Watanabe T. (1963). *Bacteriol. Rev.* **27**, 87–115.

2
Identification and Analysis of Plasmids at the Genetic Level

VILMA A. STANISICH

Department of Microbiology, La Trobe University, Bundoora 3083, Australia

1. Introduction

Many bacteria of diverse type and habitat are now known to harbour plasmid DNA. This observation lends credence to the view that these elements are ubiquitous among prokaryotes and are likely to be detected in any species in which a thorough search for them is made. In recent years the demonstration of plasmids in an increasing variety of bacteria has been striking and directly the result of the development of techniques that allow the physical demonstration, isolation and molecular characterization of plasmid DNA. Thus, plasmids in bacteria in which genetic analysis is presently limited or impossible are as amenable to detailed molecular study as those from genetically well-characterized bacteria. Such investigations are satisfactory where the primary concern is the molecular characterization, comparison and *in vitro* manipulation of plasmids for specific purposes such as the construction of cloning vectors.

METHODS IN MICROBIOLOGY
VOLUME 21 ISBN 0-12-521521-5

They may, however, be less appropriate for studies directed towards an evaluation of the contribution made by plasmids to the properties and adaptability
of the host bacterium. Thus, a list of plasmid-carrying bacteria such as that
shown in Table I includes many species in which extrachromosomal DNA
has been detected by physical methods, but no functions or host-expressed
phenotypes ascribed to it. In most instances, such 'cryptic' DNAs will be typical
plasmids, but in others they may represent bacteriophage DNAs comparable to
the coliphages of the P1 type that exist in the cell as extrachromosomally
replicating molecules (Sternberg and Austin, 1981; Scott, 1984). The bacteria
screened for extrachromosomal DNA have often been chosen because of some

TABLE I

Genera of bacteria in which plasmids or extrachromosomal DNA have been detected

Cyanobacteria	**Gram-negative, facultatively**
Synechococcus[1]	**anaerobic rods**
Synechocystis[1]	*Escherichia*[19]
Oscillatoria[1]	*Citrobacter*[19]
Nostoc[1]	*Salmonella*[19]
Anabena[1]	*Shigella*[19]
Calothrix[1]	*Klebsiella*[19]
Microcystis[2]	*Enterobacter*[19]
	Serratia[19]
Phototrophic bacteria	*Proteus*[19]
Rhodopseudomonas[3]	*Yersinia*[20]
Chromatium[4]	*Erwinia*[21]
	Legionella[22]
Gliding bacteria	*Vibrio*[23]
Myxococcus[5]	*Aeromonas*[24]
	Haemophilus[25]
Spiral and curved bacteria	*Caedibacter*[26]
Azospirillum[6]	
Campylobacter[7]	**Gram-negative, anaerobic bacteria**
	Bacteroides[27]
Gram-negative, aerobic rods	*Desulfovibrio*[28]
and cocci	
Pseudomonas[8]	**Gram-negative cocci and**
Xanthomonas[9]	**coccobacilli (aerobes)**
Zymomonas[10]	*Neisseria*[29]
Azotobacter[11]	*Acinetobacter*[30]
Rhizobium[12]	*Paracoccus*[31]
Agrobacterium[13]	
Halobacterium[14]	**Gram-negative chemolithotrophic**
Alcaligenes[15]	**bacteria**
Acetobacter[16]	*Thiobacillus*[32]
Bordetella[17]	
Thermus[18]	

TABLE I—*continued*

Methane producing bacteria
Methanobacterium[33]

Gram positive cocci
Deinococcus[34]
Staphylococcus[35]
Streptococcus[36]
Enterococcus[37]
Leuconostoc[38]
Pediococcus[39]

**Endospore forming rods
  and cocci**
Bacillus[40]
Clostridium[41]

**Gram positive, nonsporing,
  rod-shaped bacteria**
Lactobacillus[42]
Listeria[43]

Actinomyces and related organisms
Corynebacterium[44]
Arthrobacter[45]
Brevibacterium[46]
Mycobacterium[47]
Frankia[48]
Streptosporangium[49]
Nocardia[50]
Streptomyces[51]
Micromonospora[52]

Chlamydiales
Chlamydia[53]

Mycoplasmas
Mycoplasma[54]
Acholeplasma[54]
Spiroplasma[54]

[1] Herdman (1982); [2] Vakeria *et al.* (1985); [3] Fornari *et al.* (1984); [4] Saunders (1978); [5] Kaiser and Manoil (1979); [6] Elmerich (1983); [7] Taylor *et al.* (1987); [8] Jacoby (1986); [9] Kado and Liu (1981); [10] Walia *et al.* (1984); [11] Kennedy and Toukdarian (1987); [12] Dénarié *et al.* (1981); [13] Hooykaas and Schilperoort (1984); [14] Kushner (1985); [15] Bowien *et al.* (1984); [16] Valla *et al.* (1986); [17] Lax and Walker (1986); [18] Munster *et al.* (1985); [19] Jacob *et al.* (1977); [20] Portnoy and Martinez (1985); [21] Coplin *et al.* (1981); [22] Brown *et al.* (1982); [23] Guidolin and Manning (1987); [24] Aoki *et al.* (1986); [25] Chen and Clowes (1987); [26] Quackenbush (1983); [27] Odelson *et al.* (1987); [28] Postgate *et al.* (1984); [29] Cannon and Sparling (1984); [30] Goldstein *et al.* (1983); [31] Hogrefe and Friedrich (1984); [32] Rawlings and Woods (1985); [33] Whitman (1985); [34] Mackay *et al.* (1985); [35] Lyon and Skurray (1987); [36] Brunton (1984); [37] Colmar and Horaud (1987); [38] Janse *et al.* (1987); [39] Gonzales and Kunka (1986); [40] Carlton and Gonzáles (1985); [41] Rogers (1986); [42] Muriana and Klaenhammer (1987); [43] Péréz-Diaz *et al.* (1982); [44] Schiller *et al.* (1980); [45] Brandsch *et al.* (1982); [46] Santamaría *et al.* (1984); [47] Mizugucchi *et al.* (1981); [48] Simonet *et al.* (1985); [49] Fare *et al.* (1983); [50] Moretti *et al.* (1985); [51] Hopwood *et al.* (1986); [52] Oshida *et al.* (1986); [53] Palmer and Falkow (1986); [54] Razin (1985).

distinctive property that they display. This may be a pathogenic capability, an unusual metabolic capability or the ability to survive in an extreme environment. In these instances it has been usual to attempt to correlate plasmid carriage with the particular distinctive property of interest. Thus, while the physical characterization and manipulation of plasmid DNA is an essential aspect of their study, this cannot be divorced from the microbiological and genetic

TABLE II

Properties determined by bacterial plasmids

1. Resistance properties
   (a) Resistance to antibiotics[1]
      Aminoglycosides (e.g. streptomycin, gentamicin, amikacin)
      Chloramphenicol
      Fusidic acid
      β-lactam antibiotics (e.g. benzyl penicillin, ampicillin, carbenicillin)
      Sulphonamides, trimethoprim
      Tetracyclines
      Macrolides (MLS) (e.g. erythromycin)
   (b) Resistance to heavy metal cations[1]
      Mercuric ions and organomercurials
      Nickel, cobalt, lead, cadmium, bismuth, antimony, zinc, silver
   (c) Resistance to anions[1]
      Arsenate, arsenite, tellurite, borate, chromate
   (d) Other resistances
      To intercalating agents (e.g. acridines, ethidium)[1]
      To radiation (e.g. U.V., X-rays)[2]
      To phage and bacteriocins[3,4]
      To plasmid-specified Restriction/Modification (Hsd) systems[4,5,6]

2. Metabolic properties
   Antibiotic and bacteriocin production[3,7]
   Metabolism of simple carbohydrates (e.g. lactose[8], sucrose[9], raffinose[10])
   Metabolism of complex carbon compounds (e.g. octane, toluene, camphor, nicotine, aniline)[11] and halogenated compounds (e.g. 2,6-dichlorotoluene, 2,4-dichlorophenoxyacetic acid[11,12]
   Metabolism of proteins (e.g. casein[13], gelatin[14])
   Metabolism of opines (by Ti$^+$ *Agrobacterium*)[15]
   Nitrogen fixation (by Nif$^+$ *Rhizobium*)[15]
   Other properties
      Citrate utilization[14,16]
      Phosphoribulokinase activity by *Alcaligenes*[17]
      Thiamine synthesis by *Erwinia*[18] and *Rhizobium*[19]
      Denitrification activity by *Alcaligenes*[20]
      Proline biosynthesis by Ti$^+$ *Agrobacterium*[21]
      Pigmentation in *Erwinia*[22]
      H_2S production[23]
      Extracellular DNase[24]

3. Properties contributing to pathogenicity or symbiosis
   Antibiotic resistance and bacteriocin production[6]
   Toxin production
      Enterotoxins of *Escherichia coli*[25]
      Exfoliative toxin of *Staphylococcus aureus*[26]
      Exotoxin of *Bacillus anthracis*[27]
      δ-endotoxin of *Bacillus thuringiensis*[28]
      Neurotoxin of *Clostridium tetani*[29]
   Colonization antigens of *Escherichia coli* (e.g. K88, K99, CFAI, CFAII)[26]

TABLE II—*continued*

Haemolysin synthesis (e.g. in *Escherichia coli*[30,31] and *Streptococcus*[14])
Serum resistance of enterobacteria[32]
Virulence of *Yersinia* species[33]
Capsule production of *Bacillus anthracis*[27]
Crown gall and hairy root disease of plants (by Ti[+] and Ri[+]*Agrobacterium*)[34]
Infection and nodulation of legumes (by Sym[+] *Rhizobium*)[35]
Iron transport (e.g. in *Escherichia coli* and *Vibrio anguillarum*)[36]

4. Conjugal properties
 Sex pili and associated sensitivity to pilus-specific phages[37]
 Surface exclusion[38]
 Fertility inhibition[39]
 Primase activity[40]
 Mobilization functions of nonconjugative plasmids[41,42]
 Response to and inhibition of pheromones (in *Streptococcus* and *Staphylococcus*)[43]

5. Replication-maintenance properties
 Sensitivity to curing agents[44]
 Incompatibility[45,46]
 Host range[47,48]
 Copy number[49]

6. Other properties
 Gas vacuole formation in *Halobacterium*[50]
 Pock formation (lethal zygosis) in *Streptomyces*[7]
 Killing of *Klebsiella pneumoniae* by Kik[+] IncN plasmids[51]
 Sensitivity to bacteriocins (in *Agrobacterium*)[52]
 Translucent/opaque colony variation in *Mycobacterium*[53]
 Rhizosphere protein by Nod[+] Fix[+] *Rhizobium leguminosarum*[54]
 R-inclusion body production in *Caedibacter*[55]
 Endopeptidase activity (by *Staphylococcus*)[56]
 Chemotaxis towards acetosyringone (by Ti[+] *Agrobacterium*)[57]

[1] Foster (1983); [2] Strike and Lodwick (1987); [3] Hardy (1975); [4] Jacoby (1986); [5] Bannister and Glover (1968); [6] Jacob et al. (1977); [7] Hopwood et al. (1986); [8] Cornelis (1981); [9] Schmid et al. (1982); [10] Schmid et al. (1979); [11] Franz and Chakrabarty (1986); [12] Ghosal et al. (1985); [13] Blaschek and Solberg (1981); [14] Clewell (1981); [15] Hooykaas (1983); [16] Hirato et al. (1986); [17] Klintworth et al. (1985); [18] Ganotti et al. (1982); [19] Finan et al. (1986); [20] Romermann and Friedrich (1985); [21] Farrand and Dessaux (1986); [22] Ganotti and Beer (1982); [23] Ørskov and Ørskov (1973); [24] Matsumoto et al. (1978); [25] Betley et al. (1986); [26] Elwell and Shipley (1980); [27] Ivins et al. (1986); [28] Carlton and González (1985); [29] Finn et al. (1984); [30] Cavalieri et al. (1984); [31] Hacker and Hughes (1985); [32] Montenegro et al. (1985); [33] Skurnik et al. (1984); [34] Bevan and Chilton (1982); [35] Brewin et al. (1983); [36] Crosa (1984); [37] Bradley (1981); [38] Willetts and Skurray (1987); [39] Willetts and Skurray (1980); [40] Wilkins et al. (1985); [41] Willetts and Wilkins (1984); [42] Veltkamp and Stuitje (1980); [43] Clewell et al. (1985); [44] Wechsler and Kline (1980); [45] Timmis (1979); [46] Novick (1987); [47] Thomas and Smith (1987); [48] Scholz et al. (1984); [49] Scott (1984); [50] Kushner (1985); [51] Thatte et al. (1985); [52] Van Montagu and Schell (1979); [53] Mizuguchi et al (1981); [54] Dibb et al. (1984); [55] Quackenbush et al. (1986); [56] Heath et al. (1987); [57] Ashby et al. (1987).

investigations that provide additional information regarding the plasmid's contribution to host phenotype. This chapter is concerned with analyses of the latter type and deals with the primarily genetic procedures that can be used, first, to identify a plasmid-encoded phenotype, and then to characterize further the element involved.

II. Evidence for plasmid-determined phenotypes

Bacterial plasmids were first identified among members of the Enterobacteriaceae (Falkow, 1975) where they were found to display the following characteristics: they were responsible for particular properties exhibited by the bacterium, their replication was independent of that of the bacterial chromosome and they could be lost from the bacterium in an irreversible manner. The properties conferred on the host bacterium were those of chromosome mobilization, conjugal transmissibility (and the associated transfer of other fertility properties), antibiotic resistance and colicin production. As can be seen from Table II, this list has now expanded dramatically to include a wide array of properties in different bacterial species and many examples exist of plasmids that determine combinations of these properties.

The observation of an unusual phenotype in an isolate of an otherwise well-characterized bacterial species is often the first indicator that the property is plasmid-determined. Many examples of this have been found involving the properties used for the rapid diagnostic identification of clinically significant bacteria. Thus, plasmid-mediated lactose fermentation (Walia et al., 1987), H_2S production (Ørskov and Ørskov, 1973) and citrate utilization (Hirato et al., 1986) have all been found among atypically responding members of the Enterobacteriaceae. Some of the properties used to classify bacteria at the species or even the genus level may also be properties determined by plasmids. For example, members of the Rhizobiaceae are assigned to the genus *Rhizobium* or *Agrobacterium* largely on the basis of their nitrogen-fixing or plant-pathogenic capabilities. Both properties have been found to be plasmid-determined in the respective bacteria (Hooykaas, 1983). It would seem, therefore, that literally any bacterial property could be plasmid-encoded. Such an assumption flows fairly naturally from the broad variety of plasmid-determined properties already defined (Table II), and from the knowledge that, at least in theory, some of the products of recombinational interchanges between DNAs in a single bacterium have the possibility of being transferred to other species of the same or different genus of bacteria by the variety of gene transfer processes known (Starlinger, 1977; Brooks Low and Porter, 1978). Table III summarizes the way that the experimental approaches that are described in this section might be used to identify plasmid-encoded functions. Procedures for the isolation and examination of DNA are detailed in Chapters 5 and 6.

TABLE III

Criteria for determining whether the property A⁺ in bacterium B(A⁺) is plasmid mediated

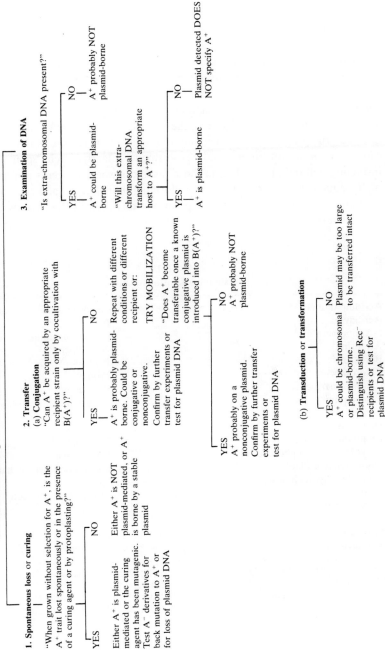

1. Spontaneous loss or curing

"When grown without selection for A⁺, is the A⁺ trait lost spontaneously or in the presence of a curing agent or by protoplasting?"

YES

Either A⁺ is plasmid-mediated or the curing agent has been mutagenic. Test A⁻ derivatives for back mutation to A⁺ or for loss of plasmid DNA

NO

Either A⁺ is NOT plasmid-mediated, or A⁺ is borne by a stable plasmid

2. Transfer

(a) Conjugation

"Can A⁺ be acquired by an appropriate recipient strain only by cocultivation with B(A⁺)?"

YES

A⁺ is probably plasmid-borne. Could be conjugative or nonconjugative. Confirm by further transfer experiments or test for plasmid DNA

NO

Repeat with different conditions or different recipient or: TRY MOBILIZATION "Does A⁺ become transferable once a known conjugative plasmid is introduced into B(A⁺)?"

YES

A⁺ probably on a nonconjugative plasmid. Confirm by further transfer experiments or test for plasmid DNA

NO

A⁺ probably NOT plasmid-borne

(b) Transduction or transformation

YES

A⁺ could be chromosomal or plasmid-borne. Distinguish using Rec⁻ recipients or test for plasmid DNA

NO

Plasmid may be too large to be transferred intact

3. Examination of DNA

"Is extra-chromosomal DNA present?"

YES

A⁺ could be plasmid-borne

NO

A⁺ probably NOT plasmid-borne

"Will this extra-chromosomal DNA transform an appropriate host to A⁺?"

YES

A⁺ is plasmid-borne

NO

Plasmid detected DOES NOT specify A⁺

A. Plasmid instability and curing

Plasmids are defined as extrachromosomally replicating molecules of DNA. It is expected, therefore, that in any growing population of plasmid-carrying bacteria, plasmidless segregants will occasionally be produced as the result of an error in the process of plasmid replication or partitioning to daughter bacteria. Such bacteria will survive provided that the plasmid lost does not encode functions vital for growth under the prevailing environmental conditions. Furthermore, the bacteria can only regain the lost functions by acquiring, from an external source, the necessary genes. Thus, the instability of a bacterial property, either a common or uncommon feature of that species, can be an indicator of plasmid involvement. Clearly, such 'spontaneous' instability will only be evident in those instances where it occurs at relatively high frequency, say $>10^{-3}$ per colony forming units (c.f.u.) and is associated with an easily-recognizable phenotype or one that is easily detected by testing individual colonies. The former would include changes in colony morphology

TABLE IV
Example of curing agents

Acridine orange[1]	Nalidixic Acid[16]
Acriflavin[2]	Ni^{2+} and Co^{2+} [17]
5-Amino acridine[3]	Nitrosoguanidine[10]
Attabrine[4]	Novobiocin[18]
Berberine[5]	Penicillin[19]
Chlorobiocin[6]	Proflavin[1]
Chlorpromazine[7]	Propidium[7]
Chloroquin[7]	Quinacrine[7]
Coumermycin[8]	Quinine[7]
Daunomycin[7]	4-Quinolones[16]
Ethidium bromide[9]	Rifampicin[20]
Ethyl methane sulphonate[10]	Sarkomycin[21]
Fluorodeoxyuridine[11]	Sodium dodecyl sulphate[22]
Guanidine hydrochloride[12]	Spermine[7]
Heat[13]	Thymidine starvation[23]
Imipramine[14]	Tilorone[7]
Methylene blue[7]	Trimethoprim[24]
Miracil D[7]	Urea[25]
Mitomycin C[15]	U.V. radiation[10]

[1] Hirota (1960); [2] Mitsuhashi et al. (1961); [3] Meynell (1972); [4] Yoshikawa and Sevag (1967); [5] Hahn and Ciak (1971); [6] Cejka et al. (1982); [7] Hahn and Ciak (1976); [8] Danislevakaya and Gragerov (1980); [9] Bouanchaud et al. (1969); [10] Willetts (1967); [11] Pinney et al. (1974); [12] Costa et al. (1980); [13] Stadler and Adelberg (1972); [14] Molnar et al. (1978); [15] Rheinwald et al. (1973); [16] Weisser and Wiedemann (1985); [17] Hirota (1956); [18] Wolfson et al. (1983); [19] Lacey (1975); [20] Riva et al. (1973); [21] Ikeda et al. (1967); [22] Salisbury et al. (1972); [23] Pinney and Smith (1971); [24] Pinney and Smith (1973); [25] Tomoeda et al. (1970).

or pigmentation, and the latter, properties such as resistance to antimicrobial agents, antibiotic production or various metabolic capabilities.

In those instances where the plasmid is stable or the loss of the property difficult to determine, the bacteria can be treated with 'curing' agents (Table IV). These include chemical and physical agents, some of which can mutate DNA, interfere specifically with its replication, or affect particular structural components or enzymes of the bacterial cell. Although all of these agents have been used to enhance the recovery of plasmidless derivatives of various bacteria, they are individually effective only against some plasmids and their likely response is unpredictable. The efficiency of curing can also vary widely (<0.1–100%) depending on the plasmid and the particular bacterial host carrying it (Bastarrachea and Willetts, 1968; Riva et al., 1973). In most instances, the underlying mechanism(s) of curing is not known. The agent may interfere directly with plasmid replication as occurs with the heat-induced curing of certain temperature-sensitive plasmids (Terawaki et al., 1967) or the curing of F by acridines or ethidium bromide (Hohn and Kohn, 1969; Wechsler and Kline, 1980). The coumarine antibiotics (coumermycin, novobiocin and clorobiocin), which inhibit subunit B of DNA gyrase, presumably affect plasmid replication by preferentially inhibiting gyrase-dependent supercoiling of the plasmid molecule (Gellert et al., 1976). Alternatively, curing may interfere with the growth of plasmid-carrying bacteria thereby allowing spontaneously arising plasmidless segregants to become predominant. This occurs in certain instances of curing by acridines, sodium dodecyl sulphate and urea (Tomoeda et al., 1968; Yoshikawa, 1971; Salisbury et al., 1972).

Since no universally effective curing agent has been identified, curing experiments tend to be conducted on a trial-and-error basis both with respect to the choice of agent and the conditions used. Some general comments can, nevertheless, be made. The DNA intercalating compounds such as acridine orange and ethidium bromide are the most commonly employed because they have been found to be effective against plasmids in a wide variety of genera. This includes species of the Enterobacteriaceae (Falkow, 1975), *Clostridium* (Rood et al., 1978), *Bacillus* (Bernhard et al., 1978), *Staphylococcus* (Bouanchaud et al., 1969), *Streptococcus* (Clewell et al., 1974) and *Rhizobium* (Higashi, 1967). Treatment with these agents can be combined with growth of the bacteria at an elevated temperature (Meynell, 1972), although heat alone can also be an effective curing agent (Della Latta et al., 1978; Zurkowski and Lorkiewcz 1978; Hershfield, 1979; Blaschek and Solberg, 1981). Curing by sodium dodecyl sulphate seems also be be broadly effective having been used successfully with such genera as *Clostridium* (Rood et al., 1978), *Bacillus* (Gonzales et al., 1981), *Staphylococcus* (Lacey, 1975) and *Pseudomonas* (Ingram et al., 1972). Mitomycin C is the agent of choice for curing plasmids in *P. putida* and related species (Chakrabarty, 1972; Whellis, 1975) and is also effective in *Streptomyces* (Hopwood *et al.*, 1986). In those cases where plasmids are refractory to curing by agents of the above type, strong mutagens such as ethyl methane sulphonate or nitrosoguanidine can be used in an attempt to eliminate the plasmid by specific mutation of its replication functions (Willetts, 1967).

Curing experiments are usually performed under conditions similar to those used for the routine culture of the bacteria unless the limitation of an essential component is specifically intended (e.g. thymine starvation). When treatment is to be for a continuous period, the agent is used at a concentration just less than that required to inhibit growth, and the number of bacteria initially exposed is small (10^2–10^4 cells ml^{-1}). If acridines are used, the culture is maintained at pH 7.6 and incubated in the dark (Hirota, 1960; Silver *et al.*, 1968). If heat-induced curing is used, the temperature chosen is appropriate to the bacterium and is generally 5–10°C above that used for its routine culture. The time allowed for curing depends on the efficiency of the curing agent and the rate of bacterial growth. Where curing is inefficient, the bacteria can be subcultured several times in the presence of the agent to increase the chances of plasmid loss. When more potent mutagens are used (e.g. ethyl methane sulphonate, nitrosoguanidine, U.V. radiation), treatment with the agent is likely to be limited to a single exposure that reduces cell viability by several orders of magnitude (Willetts, 1967). The treated bacteria are then subcultured to medium lacking the agent and growth allowed so that plasmidless segregants can be produced. A precaution that should be taken with any curing procedure that may involve a conjugative plasmid is to limit cell density to $<5 \times 10^7$ cells ml^{-1}. This will reduce the chances of conjugal reinfection of any plasmidless derivatives that may be produced.

In some cases, it may be possible to use the curing procedures outlined above in conjunction with a method which enriches for plasmidless cells. For example, if the plasmid of interest specifies resistance to an antibiotic whose action is bacteriostatic (e.g. tetracycline), growth of the culture in the presence of this drug and a second bacteriocidal drug (e.g. penicillin, ampicillin or cycloserine) to which the cells are sensitive will result in enrichment for tetracycline-sensitive (non-growing) cells (Maloy and Nunn, 1981). Cycling the culture in rich medium with and without both drugs results in selection against tetracycline-resistant, plasmid-carrying cells in the population. The method is identical to that often employed in enrichment of auxotrophic mutations (Miller, 1972).

In those instances where a curing experiment has resulted in a significant increase in the number of bacteria lacking the property under study, plasmid involvement is indicated. Confirmation of this will require supportive evidence that can be of several kinds. In the final analysis, however, it will involve the demonstration of a strict correlation between the expression of the particular property and the presence of a specific species of plasmid DNA. Where a low frequency of curing has been obtained, the possibility that loss of the property is due to a specific gene mutation must be excluded. In the first instance, attempts can be made to isolate revertants that have regained the lost property. Such bacteria could only be recovered where loss of the phenotype was due to mutation of a plasmid-associated or chromosomally-associated gene rather than to loss of an entire plasmid. This approach is only practical if revertants for the property are easily detectable, for example, those that can be isolated on an appropriate selective medium. Failure to recover revertants is consistent with plasmid loss but is not conclusive evidence for this because of the stability of some mutations. Direct evidence of curing may therefore be required by comparing the plasmid content of the parent and potentially cured derivatives.

Thus, the inability to cure a phenotypic property is almost never used as evidence against plasmid involvement. The plasmid may simply be refractory to the curing agent used, or appear so because it carries functions vital for cell viability. This latter possibility is perhaps most likely for very large plasmids whose contribution to host function may be considerable. Where curing cannot be demonstrated, transfer of the property to other bacteria can be attempted.

B. Plasmid loss by incompatibility

The incompatibility response exhibited by certain pairs of conjugative and non-conjugative plasmids (see Section III.D) can be used as a means of plasmid elimination (Toh-E and Wickner, 1981; Hynes et al., 1985). Though potentially having a wide application, the use of incompatibility is limited by a number of factors. Firstly, a large number of incompatibility groups is likely to occur in any given species or group of bacteria. The choice of the plasmid to serve as the superinfecting element may not, therefore, be a simple one. Secondly, displacement of the resident plasmid by the superinfecting one essentially reestablishes the *status quo* unless the latter can subsequently also be removed. This would be possible if the superinfecting plasmid is naturally unstable or can be made so by mutation (e.g. by the introduction of a mutation for temperature-sensitive replication). Thirdly, the superinfecting plasmid must carry a selective marker and the resident plasmid a discriminatory phenotype to allow inheritance or loss of the plasmids to be ascertained. Fourthly, a suitable system of gene transfer must be available.

C. Plasmid loss by protoplasting

Techniques that remove the cell wall from bacteria provide an alternative method to that of standard curing experiments for the identification of plasmid-associated phenotypes. This was first suggested by the observation that up to 80% of regenerated protoplasts from *Staph. aureus* (Novick et al., 1980) and

25% of those from *Strep. coelicolor* (Hopwood, 1981) did not carry plasmids present in the untreated bacteria. Curing by protoplast regeneration has subsequently been applied successfully to a variety of *Streptomyces* (Hopwood *et et al.*, 1986) and *Micromonospora* species (Oshida *et al.*, 1986). The degree of success achieved, however, varies with the plasmid and strain used, the age of the protoplasts (fresh or frozen for several days) and the osmolarity of the regeneration medium. Other species in which protoplast curing has occurred are *B. subtilis* (Edger *et al.*, 1981), Group B *Streptococcus* (Schmitt-Slomska *et al.*, 1979) and *S. lactis* (Gasson, 1983). Where strains carry multiple plasmids including small multicopy plasmids, the individual plasmid species can be lost independently of each other (Gruss and Novick, 1986) and the chances of total loss enhanced by subjecting incompletely cured strains to sequential rounds of protoplast regeneration (Gasson, 1983). While the mechanism of protoplast curing remains to be fully elucidated, the method would appear to be generally suitable for Gram-positive bacteria given the range in which protoplast formation and regeneration has been demonstrated. For example, protoplasting by lysozyme is effective for *Streptococcus* (Schöller *et al.*, 1983; Smith 1985), *Listeria* (Vincente *et al.*, 1987) and *Clostridium* (Minton and Morris, 1983) as is lysostaphin for *Staphylococcus* (Novick *et al.*, 1980). Of perhaps wider applicability is polyethylene glycol which is effective for several *Bacillus* species (Chang and Cohen, 1979; Brown and Carlton, 1980; Martin *et al.*, 1981; Imanaka *et al.*, 1982) and for *Corynebacterium* (Katsumata *et al.*, 1984), *Brevibacterium* (Santamaría *et al.*, 1985) and *Lactobacillus* (Morelli *et al.*, 1987). More complex protoplasting procedures have also been developed for several Gram-negative bacteria (Hopwood, 1981; Klebe *et al.*, 1983).

D. Additional criteria

Several additional criteria may also be appropriate for the primary identification of a plasmid or plasmid-associated phenotype in individual instances. These are the U.V. inactivation kinetics of the property (Section II.E.2), and the response of the bacterium to donor specific bacteriophages (Section III.E). These aspects are discussed later as indicated.

E. Plasmid transfer

The ability of a property to be transferred from one bacterium to another provides good presumptive evidence of plasmid involvement, particularly if the frequency of transfer is high. Attempts to demonstrate such transfer can, therefore, be used as an initial means of identifying a plasmid-encoded property or of providing evidence to support this if plasmid involvement has already been indicated by curing experiments or DNA examination. In many instances the frequency of gene transfer will be low ($<10^{-3}$ per donor) so that experiments to detect this are only feasible where a positive selection for inheritance of the plasmid-encoded property can be made. Transfer experiments that are conducted simply by mixing cultures of the test bacterium with a suitable recipient strain can give rise to progeny by conjugation, transduction or transformation, or, indeed, by a variety of other less well defined processes (Brooks Low and Porter, 1978). Sometimes, the interpretation of data obtained from transfer experiments can be facilitated by a knowledge of the type of process involved. This may be roughly gauged as follows: if treatment of the bacterial mixture with DNase reduces the yield of progeny, transformation is indicated; if mixtures of the recipient with a culture supernatant of the test bacterium yields progeny, transduction is indicated and the presence of bacteriophage should be demonstrable; if neither transformation nor transduction can be demonstrated, transfer via cell-to-cell contact (i.e. conjugation) is indicated. Table V presents examples of the main gene transfer processes that have been detected in various bacterial genera.

1. Conjugation

Many plasmids encode functions that allow their transfer from cell-to-cell by direct contact. This process, termed conjugation, is controlled by conjugative plasmids whose individual sizes vary over a wide range. Such plasmids of only 17×10^6 daltons have been found in *Streptococcus* (Clewell, 1981) and the Enterobacteriaceae (Jacob *et al.*, 1977), whereas others of more than 200×10^6 daltons occur in *Pseudomonas* (Handen and Olsen, 1978) and the Rhizobiaceae (Hooykaas, 1983). Conjugation has been detected in many members of the Enterobacteriaceae and in other Gram-negative and Gram-variable bacteria (Table V). It also occurs in certain Gram-positive species, for example, *Bacillus* (Carlton and Gonzáles, 1985) and *Bacteroides* (Odelson *et al.*, 1987) (Table V) and includes interspecies (e.g. Battisti *et al.*, 1985) and intergeneric transfer events (e.g. Clewell, 1981). Studies with F and F-like plasmids in *E. coli* K12 have shown that the conjugation process is complex (Willetts and Skurray, 1987). Several features of these systems are common to other plasmids of the Enterobacteriaceae and the Pseudomonadaceae despite the fact that at the detailed genetic level, each is unique. These plasmids all control the formation of a filamentous surface appendage, the sex pilus, that is required for the initial mating contact (Bradley *et al.*, 1980; Bradley, 1981; Bradley, 1983). They have a special system of replication and transfer of plasmid DNA (Willetts and Wilkins, 1984), and they have a mechanism of 'surface exclusion' that reduces conjugal efficiency between bacteria carrying the same plasmid type (Willetts and Skurray, 1987). While these features may or may not apply widely among Gram-negative bacteria, more obvious differences exist in the conjugation systems of Gram-positive bacteria. For example, sex pili have not been identified and, while 'typical' conjugative plasmids are recognized, an additional cell-to-cell transfer process mediated by transposons occurs in *Streptococcus* (Clewell and Gawron-Burke, 1986) compared to one mediated by bacteriophages in *Staphylococcus* (Lyon and Skurray, 1987). Moreover, in *Streptococcus*, the conjugal efficiency of one class of plasmids is strongly influenced by extracellular pheromones (Clewell *et al.*, 1985). In *Streptomyces*, conjugation appears to occur by simple hyphal fusion, but the mode of DNA transfer (plasmid or chromosomal) may be fundamentally different from that in Enterobacteriaceae, i.e. it is neither unidirectional nor single-stranded (Hopwood *et al.*, 1986).

For details of conjugation, see Chapter 3.

2. Transduction

Transduction is the transfer of bacterial DNA from one cell (the donor) to another (the recipient) by a bacteriophage capsid. It proceeds because the entity involved, the transducing particle, retains the adsorption and DNA-infection mechanisms typical of the normal phage. Once the DNA has entered the recipient cell three pathways are possible: if the DNA is a replicon, for example a plasmid, it can be inherited intact by the recipient — the process is termed repliconation (Clark and Warren, 1979); if it is only a fragment of a chromosome or plasmid, stable inheritance of any part of it will require its recombination with DNA already carried by the recipient; if neither of these events occur the DNA can survive for a period without replication, but will eventually be degraded. In this last instance, the functions carried by the DNA can, however, be expressed during the interim period and the altered phenotype of the cells (abortive transduction) observed under special circumstances (Ozeki and Ikeda, 1968). The transductional processes that can give rise to these events fall broadly into two classes termed specialized and generalized transduction.

Specialized transduction occurs when the transducing particle carries the bacterial DNA as an insertion or substitution within the viral genome. The covalent linkage of the phage and bacterial DNAs occur during the establishment of the prophage state in the donor bacterium. Subsequently, the induction of vegetative phage replication involves the detachment of the prophage from the bacterial DNA by an excision or 'looping out' process. If this event is imprecise, bacterial DNA that adjoins the prophage can become part of the excised unit and thence enclosed in the phage capsid during the viral maturation process. Clearly, only bacterial DNA that occurs in the 'special' regions flanking the prophage can be transduced in this way. For λ phages, these are usually the *gal* and *bio* regions of the *E. coli* K12 chromosome adjacent to the λ-attachment site (Franklin, 1971). In contrast, generalized transduction appears to occur by the accidental packaging of bacterial DNA inside a phage capsid — the transducing particles contain only, or almost only, bacterial DNA and any segment of the bacterial gemone can be encapsulated (Ozeki and Ikeda, 1968; Schmieger and Buch, 1975). While most transducing phages studied fall clearly into one or other category, the two forms of transduction are not mutually exclusive. This is seen in the ability of the classical transducing phages P22 and P1 to promote specialized transduction in their respective hosts, *S. typhimurium* (Smith-Keary, 1960) and *E. coli* (Luria *et al.*, 1960), and of the specialized transducing λ to promote generalized transduction in *E. coli* under certain conditions (Sternberg and Weisberg, 1975).

Most examples of generalized transduction involve temperate phages that are able to associate with their host bacteria to establish a lysogenic or pseudo-lysogenic state. This is not, however, an essential requirement, as can be seen from the transductional capabilities of typical virulent phages such as T1 in *E. coli* (Drexler, 1970), SPP1 in *B. subtilis* (Yasbin and Young, 1974), E79 in *P. aeruginosa* (Morgan, 1979) and RL38 in *R. leguminosarum* (Buchanan-Wollaston, 1979), and in the common use of clear plaque mutants of P1 for transduction experiments in *E. coli* K12 (Lennox, 1955). This is not to suggest that all phages have transducing capability. In the first instance, DNA from the donor bacterium must be packaged, either by an accidental or specific process, to produce the transducing particle. This may not occur or may not be possible in cases where phage infection leads to rapid and extensive degradation of the host DNA (Sadowski and Kerr, 1970). Secondly, the transducing particle must be able to adsorb to the recipient bacterium and inject its DNA without otherwise interfering with cell viability. With some phages, adsorption is sufficient to cause cell death (Duckworth *et al.*, 1981). Thirdly, the injected DNA must survive the barrier posed by nuclease degradation in the recipient and become established, via repliconation or recombination, to produce stable transductants. Bearing in mind that the phage lysates used for generalized transduction experiments contain a minority of transducing particles, the conditions used must be such as to prevent excessive killing of potential transductants by lytic growth of the accompanying normal phage particles.

Phages containing double-stranded DNA range in size from 12×10^6 to 480×10^6 daltons and have been detected in a wide variety of bacteria (Reanney and Ackermann, 1981). Those capable of transduction have been detected in a more limited number of genera (Table V) and, with a few exceptions, such as PBS1 (170×10^6 daltons) in *B. subtilis*, they have a smaller size range (about 20×10^6 to 70×10^6 daltons). Gene transfer by generalized transduction occurs at a frequency in the range 10^{-5}–10^{-7} per plaque forming units (p.f.u.), depending on the phage, and has been used to detect plasmid-associated properties, to separate individual plasmids from cells harbouring several different plasmids and to define the range of properties associated with a particular plasmid. Such studies are limited by the fact that the distribution of phages among bacterial species is far from even. Transductional studies may, therefore, be precluded, not only because of the unavailability of documented transducing phages, but also because of a lack of virulent or temperate phages whose transductional capabilities might be examined. In cases where phages are available, their individual host ranges determine the number of bacteria that can serve as recipients. Some phages such as PBS1 in *Bacillus* (Bramucci and Lovett, 1976), pf20 in *Pseudomonas* (Stanisich and Richmond, 1975), P1323mo in *Streptococcus* (Skjold *et al.*, 1979) and RL38 in *Rhizobium* (Buchanan-Wollaston, 1979) can infect or transduce several species. More usually, transduction is confined to the same species but may be extendable to a larger number of strains if the effects of restriction can be overcome. For example a temporary restriction-deficient phenotype may be inducible by heat treatment as occurs in *Pseudomonas* (Rolfe and Holloway, 1966), *Staphylococcus* (Lacey, 1975) and *Streptomyces* (Bailey and Winstanley, 1986). Finally, the size of the DNA that can be transduced is limited by the DNA capacity of the phage. Plasmids of similar or smaller size than the phage can be transduced intact and give rise to stable transductants. In contrast, larger plasmid are transduced as fragments and will not form stable transductants unless appropriate recipients are used that allow recombinational 'rescue' of properties from the DNA (Stanisich *et al.*, 1976). The phenomenon of 'transductional shortening' is sometimes observed with large plasmids (Falkow, 1975) and probably represents the packaging of spontaneously arising deleted derivatives of the plasmid that occur at low frequency within the population. Successful and frequent transduction of a plasmid-associated property, therefore, provides an upper limit on the size of the plasmid concerned.

Attempts to transduce a bacterial property will involve propagation of an appropriate generalized transducing phage on the donor or induction of a prophage already carried by it. The former is best carried out by cocultivation of the phage and bacterium in soft agar overlays on nutrient medium, since this generally yields higher titres than does propagation in liquid. Bacteria (about 10^7 cells) and phage (10^4–10^5 p.f.u.) in about 1 ml are mixed with 2–3 ml of molten (45–48°C) soft agar, which is then poured on to the surface of a dried nutrient agar plate to set as a 1–2 mm layer. After appropriate incubation, near-confluent lysis of the bacterial lawn should be evident. The layer is scraped off, emulsified in about 5 ml of broth and the mixture centrifuged to remove debris. The supernatant is then freed of viable bacteria by the addition of a few drops of chloroform or by membrane filtration. Induction of prophage is usually by treatment with U.V. radiation of wavelength about 260 nm. Growing bacteria are suspended in saline or some other inorganic medium at density of about 10^8 cell ml^{-1}, then irradiated as a thin layer (1–2 mm) in a glass Petri dish. The dose used (between 10–100 joules mm^{-2}) will depend on the bacteria but should be sufficient to obtain 10–15% survival. After irradiaton, the bacteria are concentrated, transferred to broth and allowed to grow for 1–2 h until lysis occurs. Alternatively, mitomycin C can be added to growing broth cultures (about 10^8 cells ml^{-1}) at a concentration of 0.5 μg ml^{-1}, then incubation continued in the dark until lysis occurs. In both procedures, residual contaminating bacteria are removed after lysis is completed. Additional details concerning these and other methods of phage propagation are described by Billing (1969), Kay (1972), Meynell and Meynell (1965) and Eisenstark (1966).

The transduction procedure involves mixing the transducing lysate with growing broth cultures of the recipient (about 10^9 cells ml^{-1}) at the normal incubation temperature, allowing a period for phage adsorption (about 30 min), then concentrating the bacteria and plating on medium containing the selective agent in order to recover transductants. If a period of expression seems appropriate for the particular property under study, the bacteria can be reincubated in fresh broth after the centrifugation stage and selection imposed later. A multiplicity of infection of between one to ten is usually used for transduction experiments with temperate phages. This leads to the recovery of lysogenic transductants that have arisen by simultaneous infection of the bacterium with both a transducing and normal phage particle or, more probably, by superinfection of the transduced cell during the course of its growth. When virulent phages are used, infections by normal phage particles must be avoided, since such events will kill the transductants. This is usually achieved by U.V. irradiating the transducing lysate to reduce plaque-forming (i.e. phage viability), using a phage to bacterium ratio of <1 and removing unadsorbed phage by thorough washing of the bacteria at the concentration stage or treating them with appropriate phage antiserum. An alternative method used for successful transduction by the virulent phage E79tv-1 in *P. aeruginosa* involved a plasmid-carrying recipient, where the plasmid concerned could interfere with phage propagation but not with either phage adsorption or DNA injection (Morgan, 1979).

The distinction between transduction of a plasmid associated or a chromosomally associated property can be made in a number of ways: by examining the transductants for the inheritance of a unique species of plasmid DNA, by comparing the transducability of the property to recombination-proficient (Rec$^+$) and recombination-deficient (Rec$^-$) bacteria, if the latter are available, or by studying the effects of U.V. irradiation of the lysate on the frequency of transduction. The latter approach is based on the observation of Arber (1960) and others (Benzinger and Hartman, 1962; Asheshov, 1966) that the transduction frequency of certain properties can be increased up to ten-fold or more by irradiation of the lysate with small doses of U.V., whereas for other properties, irradiation causes only a progressive decline in both transduction frequency and plaque-forming ability. The former is characteristic of chromo-

TABLE V

Gene transfer in bacteria[a]

Family or Genus	Conjugation[b]	Transduction	Transformation
Enterobacteriaceae	Jacob et al. (1977)	Brooks Low and Porter (1978)	Smith et al. (1981)
Acinetobacter	Hinchliffe et al. (1980)	Herman and Juni (1974)	Stewart and Carlson (1986)
Agrobacterium	Dellaporta and Pesano (1981)		Holsters et al. (1978)
Amycolatopsis			Matsushima et al. (1987)
Anacystis	Herdman (1982)		Golden and Sherman (1984)
Azospirillum	Fani et al. (1986)		Fani et al. (1986)
Azotobacter	Kennedy and Toukdarian (1987)	Kennedy and Toukdarian (1987)	Kennedy and Toukdarian (1987)
Bacillus	Carlton and Gonzáles (1985)	Carlton and Gonzáles (1985)	Carlton and Gonzáles (1985)
Bacteroides	Odelson et al. (1987)		Odelson et al. (1987)
Brevibacterium			Santamaría et al. (1985)
Campylobacter	Taylor et al. (1987)		
Caulobacter		Poindexter (1981)	
Clostridium	Odelson et al. (1987)		Odelson et al. (1987)
Corynebacterium		Hirai and Yanagawa (1970)	
Erwinia	Chatterjee et al. (1979)		Hinton et al. (1985)
Haemophilus	Laufs et al. (1981)		Stewart and Carlson (1986)
Halobacterium	Mevarech and Werczberger (1985)		Cline and Doolittle (1987)
Lactobacillus			Chassy and Flickinger (1987)
Methanococcus			Bertani and Baresi (1987)

Methylobacterium	Holloway (1984)		Holloway (1984)
Methylococcus			Holloway (1984)
Micrococcus			Tirgari and Moseley (1980)
Micromonospora	Beretta et al. (1971)		
Moraxella			
Mycobacterium	Grange (1982)	Grange (1982)	Stewart and Carlson (1986)
Neisseria	Cannon and Sparling (1984)		Grange (1982)
Nostoc			Biswas et al. (1985)
			Trehan and Sinha (1981)
Pasteurella	Jacob et al. (1977)		Tyeryar and Lawton (1969)
Pseudomonas	Holloway (1986)	Holloway (1986)	Holloway (1986)
Rhizobium	Kondorosi and Johnson (1981)	Finan et al. (1984)	Kondorosi and Johnston (1981)
Rhodopseudomonas	Scolnik and Marrs (1987)	Scolnik and Marrs (1987)	Scolnik and Marrs (1987)
Staphylococcus	Lyon and Skurray (1987)	Lyon and Skurray (1987)	Lyon and Skurray (1987)
Streptococcus	Clewell and Gawron-Burke (1986)	Clewell (1981)	Stewart and Carlson (1986)
Streptomyces	Hopwood et al. (1986)	Chater (1986)	Hopwood et al. (1986)
Synechococcus			Chauvat et al. (1983)
Thermus			Koyama et al. (1986)
Vibrio	Guidolin and Manning (1987)	Guidolin and Manning (1987)	
Xanthobacter	Wilke (1980)	Wilke and Schlegel (1979)	
Xanthomonas			Corey and Starr (1957)
Zymomonas			Browne et al. (1984)

[a] Other systems of gene transfer are described by Brooks Low and Porter (1978).

[b] Transfer of IncP plasmids to various genera and conjugation mediated by these plasmids is described by Holloway (1979), Thomas and Smith (1987) and Holloway (1986).

somal properties and is due to the stimulation of recombination events by irradiation-damaged DNA. The latter is characteristic of plasmid properties whose stable inheritance is not dependent on recombination. Irradiation, therefore, only leads to mutation of the property or of other plasmid functions required for survival in the recipient. This U.V. inactivation test has been used frequently to identify plasmids in *Staphylococcus* (Lacey, 1975) and should be applicable to other bacteria in which plasmids can be transduced intact.

Two additional observations related to transduction are worthy of note. First, transduction can be used to separate plasmids from a bacterium carrying several different plasmids, since the transducing particle packages only a single segment of DNA. Although this applies generally, there have been observations of the joint transduction of plasmids arising from a single transduction event (Grubb and O'Reilly, 1971; Stiffler *et al.*, 1974; Iordanescu *et al*, 1978; Novick *et al.*, 1981; Lyon and Skurray, 1987). Such joint transduction probably reflects an efficient cointegration type interaction mediated by the host and phage recombination systems (Novick *et al.*, 1984). Packaging of a recombinant molecule could, therefore, occur and subsequently persist stably in the transductant or dissociate into its constituent replicons if the recombination event is a reversible one. Under these circumstances, transduction of a particular property could give rise to the unexpected finding that the transductants are heterogeneous with respect to the number and size of the plasmids that they carry.

The second observation concerns the unusual phage-mediated conjugation system in *Staph. aureus* (Lacey, 1975; Lyon and Skurray, 1987). Phages are implicated since gene transfer requires the presence of calcium ions (presumably for adsorption), and a prophage in either the donor or recipient bacterium. Typical transduction is not involved since cell-to-cell contact of the bacteria is essential and no transfer occurs if the recipient is mixed only with culture supernatants of the donor. Indeed, in many instances, such supernatants contain little plaque-forming activity, suggesting that the phage may be cell bound, defective or highly unstable, so that cell apposition is required for adsorption and DNA transfer. Unusual gene transfer processes that similarly appear to involve phage-like particles have been described in *Rhodopseudomonas capsulata* (Saunders, 1978) and *Strep. lactis* (McKay *et al.*, 1973; Klaenhammer and McKay, 1976). It would seem, therefore, that transfer experiments involving species in which phages are common should be attempted both by the procedures typically used for transduction and by the mixed culture method more usually associated with conjugation.

3. Transformation

Transformation is the process by which DNA from one cell (the donor) is taken up by another (the recipient) directly from the surrounding medium. Table V shows examples of genera in which transformation has been demonstrated. Although 'competence' (the ability to be transformed) can arise naturally with some genera (Stewart and Carlson, 1986), of much greater value in attempts to demonstrate a plasmid-associated property *via* transformation are the artificial systems of competence induction. These favour transformation by plasmid rather than by chromosomal DNA and can be applied to bacteria irrespective of their natural transformability.

Confirmation of the plasmid association of a transformed property can be obtained by a comparison of the efficiency of transformation to Rec^+ and Rec^- recipients, or by the identification of a unique plasmid species among the transformants. Bearing in mind that a competent bacterium can take up more than one molecule of DNA, the concentration of DNA used in the experiment should be adjusted to reduce this possibility. Alternatively, species of plasmid DNA can be individually recovered from the transformants and retested in an appropriate recipient to determine association with the property under study.

Transformation is discussed in detail in Chapter 4.

III. Phenotypic characterization of plasmids

Identification of a plasmid by physical or genetic means is usually accompanied by efforts to define its contribution to cell phenotype and to compare its properties and interactions with those of other plasmids identified in the same or related species. Such studies contribute to a broader understanding of the epidemiology and evolution of these molecules and of their control of many of the unique attributes displayed by bacteria. As far as possible, such genetic characterization should be undertaken by transfer of the plasmid to a bacterial host whose properties are well defined and which carries no other plasmids. Table II lists examples of the properties that have been found associated with plasmids in various bacteria. Combinations of several of these properties can occur on a single plasmid (Jacob *et al.*, 1977; Clewell, 1981; Jacoby, 1986; Lyon and Skurray, 1987) and, in theory, all combinations may occur in natural isolates. In practice, only a limited number of tests are applied to characterize a new plasmid, the choice of these depending on the origin of the plasmid, the ease of the test, whether it is applicable in the host strain and the particular interests of the investigator. In Section III some of the common genetic criteria used in plasmid characterization and classification are discussed.

A. Resistance properties

Plasmids that determine resistance to antibiotics (R-plasmids) are most common in bacteria from clinical and veterinary sources, where exposure to antibiotics is likely to be high (Linton, 1977; Smith, 1977). They also occur at varying frequencies in bacteria from water and other sources (Gonzal et al., 1979; Murray et al., 1984; Baya et al., 1986; Hermansson et al., 1987) and from healthy individuals living in urban or isolated communities (Falkow, 1975). R-plasmids have been detected in many members of the Enterobacteriaceae and in a variety of other clinically significant bacteria including species of *Bacteroides*, *Haemophilus*, *Neisseria*, *Pseudomonas* and *Streptococcus* (Table I for references). The antibiotic resistances conferred by plasmids broadly reflect the agents commonly used against their bacterial hosts. Thus, erythromycin resistance occurs on plasmids from the Gram-positive cocci, *Clostridium* and *Corynebacterium*, whereas resistance to tetracycline and to certain penicillins and aminoglycosides occurs on plasmids from bacteria of either Gram type. Plasmids may also confer resistance to other antimicrobial agents. These include heavy metal ions such as mercury(II) in *Staphylococcus* and *Pseudomonas* (Robinson and Tuovinen, 1984), various anions such as arsenate, chromate and tellurite in *Staphylococcus*, *Streptococcus* and *Pseudomonas* (Foster, 1983) and U.V. radiation in *Streptococcus*, *Pseudomonas* and the Enterobacteriaceae (Strike and Lodwick, 1987). The particular resistance properties conferred by a plasmid are of little value as an aid to establishing the relatedness of independent isolates, although in a few instances they may serve as a guide to plasmid type — for example, tellurite resistance and U.V. radiation resistance are common attributes of the Inc P2 and Inc N plasmids in *Pseudomonas* and *E. coli*, respectively (Jacoby, 1986; Strike and Lodwick, 1987).

The characterization of plasmids with respect to these properties generally poses no special difficulties, and requires only the determination of the minimum inhibitory concentration of the agent against the R^- control and its R^+ derivative. The range of concentrations that are tested will depend on the medium used, the natural level of susceptibility of the R^- strain and whether an attempt is made to induce resistance by growth of the R^+ strain in subinhibitory concentrations of the agent. Most basal media should be suitable although certain agents, such as the sulphonamides, trimethoprim and mercury(II), should be tested using diagnostic sensitivity testing agar or defined minimal medium, perhaps supplemented with about 1% casamino acids. The addition of chemicals to test for anion or cation resistance can result in marked changes in the pH of the medium; this should be appropriately adjusted to avoid an inhibition of bacterial growth that could mask the presence of a resistance property. Resistance to U.V. radiation can be determined by exposing bacteria on nutrient agar to varying doses of U.V. radiation. This can be achieved by preparing a series of similar plates inoculated with a dilution series of a culture of the bacterium. The individual plates are then exposed to increasingly higher doses of U.V. radiation (from about 10 to 100 joules mm^{-2}) and, after incubation in the dark, the effect of U.V. radiation quantitated from the number of colony forming bacteria.

B. Metabolic properties

Plasmids can confer a wide variety of metabolic properties on their host bacterium. This includes properties that may interfere with the rapid diagnostic identification of certain bacteria; for example plasmid-determined H_2S production (Ørskov and Ørskov, 1973) or utilization of lactose (Walia et al., 1987), sucrose (Schmid et al., 1982), raffinose (Schmid et al., 1979) or citrate (Hirato et al., 1986) occurs among members of the Enteriobacteriaceae. Other properties are significant in the species designation of bacteria: for example, plasmid-determined β-haemolysin production by Streptococcus zymogenes (Jacob et al., 1975) and citrate utilization by S. diacetylactis (Kempler and McKay, 1979) accounts for the distinction between these bacteria and S. faecalis and S. lactis respectively. Metabolic versatility enables bacteria to occupy ecological niches not available to those with more stringent nutritional requirements. Many species of Pseudomonas can use a wide variety of compounds as sole carbon and energy sources (Stanier et al., 1966) and, in P. putida especially, this can be associated with the presence of 'degradative' plasmids. Depending on the catabolic pathway encoded by the plasmids, compounds such as toluates, xylenes, naphthalene and salicylate can be degraded via the intermediate, catechol, and thence further by plasmid-encoded (meta-cleavage) or chromosomally encoded (ortho-cleavage) pathways (Whelis, 1975; Franz and Chakrabarty, 1986). Other unusual compounds metabolized via plasmid-encoded pathways include opines (α-N-substituted amino acid derivatives) by Ti-containing Agrobacterium (Dellaporta and Pesano, 1981) and the herbicides 2-methyl-4-chlorophenoxyacetic acid and bromoxynil by Alcaligenes (Franz and Chakrabarty, 1986) and Klebsiella (Stalker and McBride, 1987) respectively. In the former situation the presence of Ti DNA stimulates crown gall tissue of the infected plant to produce one of the three major types of opine (octopine, nopaline or agropine). This is then available as a nutrient source for the bacteria since the Ti plasmid also encodes the pathway required for utilization of the specific opine induced by it. The properties indicated here are all readily tested on the appropriate diagnostic media or on chemically defined media in which the substance to be degraded is provided as the sole carbon source.

An additional plasmid determined metabolic property is that of antibiotic production. Two groups of substances are involved: these are bacteriocins, which are usually active only against bacteria related to the producer strain (Hardy, 1975; Tagg et al., 1976; Konisky, 1982), and true antibiotics that are active against diverse microbial genera (Hopwood, 1978). The former occur among members of the Enterobacteriaceae (Hardy, 1975), Pseudomonas (Kageyama, 1975) and various other Gram positive and Gram negative bacteria (Tagg et al., 1976). Plasmid-determined antibiotic production includes the microcins from E. coli and P. aeruginosa (Bacquero et al., 1978), methylenomycin A from *Strep. coelicolor* (Hopwood *et al.*, 1986) and possibly nisin from *Strep. lactis* (Kozak *et al.*, 1974; Fuchs *et al.*, 1975). Production of these types of substances can be demonstrated by exposure of a sensitive bacterium to the producer strain or culture supernatants of the producer strain. For example, the producer strain can be grown as single colonies or as a streak on nutrient medium and then overlayed with agar containing the sensitive bacterium. Alternatively, bacteria-free culture supernatants of the producer strain can be spotted on lawns of the sensitive bacteria. After appropriate incubation, inhibition of growth of the sensitive strain would indicate antibiotic production. In some cases the amounts of bacteriocin produced can be increased by induction of the strain with mitomycin C or U.V. radiation.

C. Pathogenic and symbiotic properties

Various aspects of the host-bacterium relationship are controlled by plasmids. Properties such as bacteriocin production, antibiotic resistance and haemolysin production are likely to contribute to the establishment of the organism, its epidemic spread and the severity of the disease caused by it. Other properties, however, are more clearly implicated in the pathogenic state. This is seen in the case of plant infections by *Agrobacterium* (Bevan and Chilton, 1982) and toxin production by both Gram-positive and Gram-negative bacteria. The latter includes the heat-stable (ST) and heat-labile (LT) enterotoxins produced by *E. coli* (Betley *et al.*, 1986) and the exotoxin and neurotoxin of *B. anthracis* (Ivins *et al.*, 1986) and *Clos. tetani* (Finn *et al.*, 1984), respectively. The host specificity and colonization ability of bacteria may also be determined by plasmids. This is seen in the control of host range and nodulation specificity of *Rhizobium* (Dénarié *et al.*, 1981) and *Agrobacterium* (Dellaporta and Pesano, 1981), respectively, and in the adhesion of enteropathogenic *E. coli* to the intestinal mucose brought about by specific colonization antigens that occur on their surface (Elwell and Shipley, 1980).

Evaluation of most of the properties indicated here depend on *in vivo* animal or plant tests, or *in vitro* serological or tissue culture tests. Details of these procedures are given in the references cited here and by Cabello and Timmis (1979) and Macrina (1984).

D. Incompatibility and exclusion properties

Certain plasmids are unable to coexist stably in the same cell line, whereas others are able to do so. The former are referred to as incompatible plasmids and where several exhibit this property (i.e. A is incompatible with B, B with

C and C with A) they are assigned to the same incompatibility (Inc) group. Incompatibility is an inherent characteristic of all plasmids and can be demonstrated between homologous plasmids when experiments with appropriately marked derivatives are undertaken. It is generally accepted, therefore, that incompatibility between independently isolated heterologous plasmids reflects their similarity or evolutionary relatedness. This view is strengthened by the finding that incompatible plasmids usually share more sequence homology (>60%) than do compatible plasmids (<15%) (Falkow, 1975; Gorai et al., 1979; de la Cruz et al., 1980; Shalita et al., 1980) and, in the case of conjugative plasmids, usually specify similar transfer systems (Willetts and Skurray, 1980; Bradley, 1981). The genetic control of incompatibility is complex and several distinct interactions are involved (Timmis, 1979; Cowan and Scott, 1981; Lane, 1981; Novick, 1987). For the purpose of plasmid characterization, however, it is sufficient that incompatibility provides a feature of distinction that applies both to conjugative and nonconjugative plasmids (Datta, 1979) and to extrachromosomally maintained phages (Cowan and Scott, 1981). Some 24 incompatibility groups have been defined in *E. coli* K12 for plasmids from the Enterobacteriaceae (Datta, 1979), 13 in *Staph. aureus* (Iordanescu and Surdeanu, 1980), 10 in *P. aeruginosa* PAO (Jacoby, 1986) and 3 in *Strep. faecalis* (Romero et al., 1979).

The incompatibility of two plasmids, A and B, can be determined if each controls a distinctive phenotype that allows its presence to be inferred. A plasmid can then be introduced into cells containing B (by conjugation, transduction or transformation) by imposing a selection only for A. The progeny so derived is purified once or twice in the presence of the selective agent, then tested for the distinctive property determined by B. If this is absent, displacement is indicated, and the two plasmids are considered to be incompatible. If the polarity of the cross is reversed so that B is transferred to A⁺ cells, incompatibility should again be observed. This pattern of behaviour, in which a superinfecting plasmid invariably displaces a resident one, is the basis of definition of the Inc groups indicated above. An alternative, and formally more rigorous, test for incompatibility is to establish both plasmids in the same cell (usually a Rec⁻ strain, to limit recombination between the plasmids), and then to examine segregation. The cell containing both plasmids is constructed and purified with selection for both plasmids. It is then grown in nonselective conditions and plated to give single colonies. These colonies are then tested for the markers carried by the two plasmids; loss of one set or other of these markers indicates plasmid incompatibility.

Unfortunately, the situation is often less clear-cut than implied above, and many reports of anomolous incompatibility interactions exist. These include unidirectional incompatibility, where plasmid A can displace plasmid B but the two are compatible when B is the superinfecting element; dual incompatibility, where a plasmid is incompatible with members of more than one Inc group, and 'dislodgement' where low level incompatibility occurs with members of one or several groups (Datta, 1979). The basis of these interactions is not known but they will often prevent the unambiguous assignment of a new plasmid to a defined Inc group. Despite this, the particular interaction observed will be a reproducible property of the plasmid and hence serve to identify it if it is isolated again from independent sources. In those instances where incompatibility is observed in some but not all progeny of the cross, the latter should be repurified several times on nonselective medium and again tested for the distinctive phenotypes. If both are still present, the bacteria may be carrying the plasmids as a compatible pair or as a recombinant that has retained both the phenotypes. These possibilities can be distinguished by a further transfer and linkage analysis study, or by examining the DNA for one or two plasmid species.

The most commonly encountered limitation in incompatibility studies is the similarity of properties determined by the plasmids under study. This can be overcome by the tedious task of isolating specific plasmid mutants following mutagenesis, or screening cultures for spontaneously arising plasmid segregants. Alternatively, plasmids can be 'labelled' with appropriate transposons (Finger and Krishnapillai, 1980) or suitable recombinants constructed in other ways. In the latter regard, Iordanescu and Surdeanu (1980) have developed an efficient method of labelling *Staph. aureus* plasmids by cointegrating them with a temperature-sensitive Tc^r plasmid. When these are subsequently used in incompatibility experiments, they displace plasmids homologous to or related to the Tc^s component of the cointegrate. In *E. coli* K12, Sasakawa *et al.* (1980) have simplified the task of incompatibility testing by isolating a series of strains each carrying a different plasmid integrated into the chromosome. Eleven Inc groups are represented (Inc FI, FII, FIV, Iα, H1, H2, L, M, N, T and X). Conjugation is used to introduce the test plasmid into a control R^- strain and into the series of R^+ derivatives. Incompatibility is observed as a decreased yield of progeny ($<10^{-2}$) in matings involving the incompatible plasmid pair. In a similar vein, Davey *et al.* (1984) have constructed a set of mini-*gal* plasmids carrying the incompatibility regions of six plasmid types (Inc 9, B, FII, I, M and N). The conjugative R-plasmid under test is transferred to each mini-*gal*-containing recipient, then a replica plating system to galactose-tetrazolium indicator agar is used to determine maintenance or loss of the *gal* plasmids on the basis of a colour change. Feilberg Jørgensen *et al.* (1982) have exploited the serological differences in the cell surface of piliated bacteria to identify the conjugative plasmids responsible. Indirect haemagglutination was found to distinguish unequivocally between plasmids of different incompatibility groups (Inc F, I and N) but was less appropriate for those of the same Inc complex (FI and FII). Finally, where the plasmids under study differ in only a single selective property, incompatibility can be detected by purifying the progeny and directly examining their DNA for the two plasmid species (Palomares and Perea, 1980).

Surface exclusion is a property that is often observed when tests for incompatibility are carried out by conjugation (Chapter 3). The two phenomena are, however, quite distinct. Surface exclusion refers to the decrease in transconjugant yield that occurs when a donor is mated with a plasmid-carrying recipient compared to a plasmidless recipient. In the case of F, exclusion is controlled by two cistrons in the major transfer operon which exert their effects by reducing the formation of stable mating aggregates and DNA transfer (Willetts and Skurray, 1987). Exclusion of about 400-fold occurs in matings between two F-carrying bacteria and is bidirectional in that each bacterium used as the recipient will exclude the superinfecting plasmid. Since plasmids of the same Inc group usually specify similar transfer systems, it is not surprising to find both exclusion and incompatibility when certain heterologous combinations are studied. Unfortunately, there are many examples of exclusion (unidirectional and bidirectional) between compatible plasmids, and of incompatible plasmids that do not exhibit exclusion. Thus, while the observation of exclusion serves to define this characteristic of a plasmid, it cannot be used alone to infer the relatedness of the plasmids involved. This may not apply in *P. aeruginosa* where good correlation was observed when FP plasmids were grouped by their exclusion or incompatibility response (Finger and Krishnapillai, 1980).

Exclusion can interfere with incompatibility testing if it abolishes completely or almost completely the yield of transconjugants. This effect can sometimes be overcome if stationary phase rather than log phase cultures of the recipient are used. If not, the required progeny will have to be obtained by transduction or transformation. Exclusion can also lead to spurious results in instances where the plasmid carried by the recipient is unstable. Any plasmidless segregants arising in the population will have a greater chance of mating with the donor than will their exclusion-proficient sibs. Particularly in instances where the effect of exclusion is strong, even the low frequency occurrence of such segregants can lead to a significant proportion (if not all) of the transconjugants arising from such matings. These progeny will appear to have resulted from displacement of the resident plasmid by the superinfecting one, and hence to the erroneous conclusion that the two plasmids under study are incompatible. Thus, when incompatibility is observed in combination with a strong exclusion effect, it should be confirmed by demonstrating that it also occurs if the polarity of the conjugational cross is reversed or if the progeny are constructed by transduction or transformation. (It should be noted that if incompatibility was tested by first establishing both plasmids in the same cell and then examining segregation, exclusion will not present a problem).

E. Interactions with phages

Conjugative plasmids carried by Gram-negative bacteria determine a fila-mentous surface appendage, the sex pilus. The pili determined by plasmids of a particular Inc group are serologically related and usually distinguishable from those of other groups (Bradley, 1980; Feilberg Jørgernsen *et al.*, 1982). These differences in structure and morphology are also reflected in the adsorptional specificity of certain phages, which are termed sex specific or donor specific, and whose receptors occur on the sex pilus. Such phages can, therefore, be used as a primary means of detecting the presence of a conjugative plasmid in a wild-type isolate and of establishing its probable incompatibility type. Sex pilus specific phages and procedures for their use are described in Chapter 3.

Plasmids may also interfere with phage propagation. This phenomenon has been described in many bacteria, and a variety of mechanisms have been implicated (Duckworth *et al.*, 1981). These include plasmid-mediated restriction/modification (Boyer, 1971), effects on phage transcription (Moyer *et al.*, 1972) or translation (Blumberg *et al.*, 1976), or more general effects on eell functions (Britton and Haselkorn, 1975; Condit, 1976). In *P. aeruginosa* PAO and *E. coli* K12 such plasmid-phage interactions have been used for the rapid identification of Inc P2 and Inc N plasmids respectively (Jacob *et al.*, 1977; Jacoby, 1986).

Acknowledgements

I gratefully acknowledge the use of material dealing with plasmid curing taken from the contribution of Caro *et al.* from the 1st edition of this book.

I am grateful to S. T. Fong for assistance with the manuscript and preparation of the typescript. Research in the author's laboratory is supported by the Australian Research Grants Scheme.

References

Aoki, T., Mitoma, Y. and Crosa, J. H. (1986). *Plasmid* **16**, 213–218.
Arber, W. (1960). *Virology* **11**, 273–288.
Ashby, A. M., Watson, M. D. and Shaw, C. H. (1987). *FEMS Microbiol. Lett.* **41**, 189–192.
Asheshov, E. H. (1966). *Nature (London)* **210**, 804–806.
Bacquero, F., Bouanchaud, D., Martinez-Perez, M. C. and Fernandez, C. (1978). *J. Bacteriol.* **135**, 342–347.
Bailey, C. R. and Winstanley, D. J. (1986). *J. Gen. Microbiol.* **132**, 2945–2947.

Bannister, D. E. and Glover, S. W. (1968). *Biochem. Biophys. Res. Commun.* **30**, 735–738.
Bastarrachea, F. and Willetts, N. S. (1968). *Genet. Res.* **59**, 153–166.
Battisti, L., Green, B. D. and Thorne, C. B. (1985). *J. Bacteriol.* **162**, 543–550.
Baya, A. M., Brayton, P. R., Brown V. L., Grimes, D. J., Russek-Cohen, E. and Colwell, R. R. (1986). *Appl. Environ. Microbiol.* **51**, 1285–1292.
Benzinger, R. and Hartman, P. E. (1962). *Virology* **18**, 614–626.
Beretta, M., Betti, M. and Polsinelli, M. (1971). *J. Bacteriol.* **107**, 415–419.
Bernhard, K., Schrempf, H. and Goebel, W. (1978). *J. Bacteriol.* **133**, 897–903.
Bertani, G. and Baresi, L. (1987). *J. Bacteriol.* **169**, 2730–2738.
Betley, M. L., Miller, V. L. and Mekalanos, J. J. (1986). *Annu. Rev. Microbiol.* **40**, 577–605.
Bevan, M. W. and Chilton, M. D. (1982). *Annu. Rev. Genet.* **16**, 357–359.
Billing, E. (1969). *In* "Methods in Microbiology" (J. R. Norris and D. W. Ribbons, Eds), Vol. 3B, pp. 315–329. Academic Press, London.
Biswas, G. D., Burnstein, K. and Sparling, P. F. (1985). *In* "The Pathogenic Gonorrhoeae" (G. A. Schoolnick, Ed.), pp. 204–208. Amer. Soc. Microbiol., Washington, DC.
Blaschek, H. P. and Solberg, M. (1981). *J. Bacteriol.* **147**, 262–266.
Blumberg, D. D., Mabie, C. T. and Malamy, M. H. (1976). *J. Virol.* **17**, 94–105.
Bouanchaud, D. H., Scavizzi, M. R. and Chabbert, Y. A. (1969). *J. Gen. Microbiol.* **54**, 417–425.
Bowien, B., Friedrich, B. and Friedrich, C. (1984). *Arch. Microbiol.* **139**, 305–310.
Boyer, H. W. (1971). *Annu. Rev. Microbiol.* **25**, 153–176.
Bradley, D. E. (1980). *Plasmid* **4**, 155–169.
Bradley, D. E. (1981). *In* "Molecular Biology, Pathogenicity and Ecology of Bacterial Plasmids" (S. Levy, R. C. Clowes and E. L. Koenig, Eds), pp. 217–226. Plenum Press, New York.
Bradley, D. E. (1983). *J. Gen. Microbiol.* **29**, 2545–2556.
Bradley, D. E., Taylor, D. E. and Cohen, D. R. (1980). *J. Bacteriol.* **143**, 1466–1470.
Bramucci, M. G. and Lovett, P. S. (1976). *J. Bacteriol.* **127**, 829–831.
Brandsch, R., Hinkkanen, A. E. and Decker, K. (1982). *Arch. Microbiol.* **132**, 26–30.
Brewin, N. J., Wood, E. A. and Young, J. P. W. (1983). *J. Gen. Microbiol.* **129**, 2973–2977.
Britton, J. R. and Haselkorn, R. (1975). *Proc. Natl. Acad. Sci. U.S.A.* **72**, 2222–2226.
Brooks Low, K. and Porter, D. D. (1978). *Annu. Rev. Genet.* **12**, 249–287.
Brown, A., Vickers, R. M., Elder, E. M., Lema, M. and Garrity, G. M. (1982). *J. Clin. Microbiol.* **16**, 230–235.
Brown, B. J. and Carlton, B. C. (1980). *J. Bacteriol.* **142**, 508–512.
Browne, G. M., Skotnicki, M. L., Goodman, A. E. and Rogers, P. L. (1984). *Plasmid* **12**, 211–214.
Brunton, J. (1984). *In* "Antimicrobial Drug Resistance" (L. E. Bryan, Ed.), pp. 529–565. New York: Academic Press.
Buchanan-Wollaston, V. (1979). *J. Gen. Microbiol.* **112**, 135–142.
Cabello, F. and Timmis, K. N. (1979). *In* "Plasmids of Medical, Environmental and Commercial Importance" (K. N. Timmis and A. Pühler, Eds), pp. 55–69. Elsevier, Amsterdam.
Cannon, J. G. and Sparling, P. F. (1984). *Annu. Rev. Microbiol.* **38**, 111–133.
Carlton, B. C. and Gonzáles, J. M. (1985). *In* "The Molecular Biology of the Bacilli" (D. A. Dubnau, Ed.), pp. 211–249. Academic Press, New York.

Caro, L. Churchward, G. and Chandler, M. (1984). *In* "Methods in Microbiology" (P. M. Bennett and J. Grinsted, Eds), Vol. 17, pp. 97–122. Academic Press.

Cavalieri, S. J., Bohach, G. A. and Snyder, I. S. (1984). *Microbiol. Rev.* **48**, 326–343.

Cejka, K., Holubova, I. and Hubacek, J. (1982). *Molec. Gen. Genet.* **186**, 153–155.

Chakrabarty, A. M. (1972). *J. Bacteriol.* **112**, 815–823.

Chang, A. and Cohen, S. N. (1979). *Molec. Gen. Genet.* **168**, 111–115.

Chassy, B. M. and Flickinger, J. L. (1987). *FEMS Microbiol. Lett.* **44**, 173–177.

Chater, K. F. (1986). *In* "The Bacteria: A Treatise on Structure and Function" (S. W. Queener and L. E. Day, Eds), Vol. IX, pp. 119–158. Academic Press, New York.

Chatterjee, A. M., Behrens, M. K. and Starr, M. P. (1979). *Proc. Int. Conf. Plant. Pathol. Bacteriol.* **4**, 75–79.

Chauvat, F., Astier, C., Vedel, F. and Joset-Espardellier, F. (1983). *Molec. Gen. Genet.* **191**, 39–45.

Chen, S-T. and Clowes, R. C. (1987). *J. Bacteriol.* **169**, 3124–3130.

Clark, A. J. and Warren, G. J. (1979). *Annu. Rev. Microbiol.* **13**, 99–125.

Clewell, D. B. (1981), *Microbiol. Rev.* **45**, 409–436.

Clewell, D. B. and Gawron-Burke, C. (1986). *Annu. Rev. Microbiol.* **40**, 635–659.

Clewell, D. B., An, F. Y., White, B. A. and Gawron-Burke, C. (1985). *J. Bacteriol.* **162**, 1212–1220.

Clewell, D. B., Yagi, Y., Dunny, G. M. and Schultz, S. K. (1974). *J. Bacteriol.* **117**, 283–289.

Cline, S. W. and Doolittle, F. W. (1987). *J. Bacteriol.* **169**, 1341–1344.

Colmar, I. and Horaud, T. (1987). *Appl. Environ. Microbiol.* **53**, 567–570.

Condit, R. (1976). *Nature (London)* **260**, 287–288.

Corey, R. R. and Starr, M. P. (1957). *J. Bacteriol.* **74**, 146–150.

Cornelis, G. (1981). *J. Gen. Microbiol.* **124**, 91–97.

Coplin, D. L., Rowan, R. G., Chisholm, D. A. and Whitmoyer, R. E. (1981). *Appl. Environ. Microbiol.* **42**, 599–604.

Costa, M. L. P., Penido, E. and Costa, S. O. P. (1980). *J. Gen. Microbiol.* **118**, 543–547.

Cowan, J. A. and Scott, J. R. (1981). *Plasmid* **6**, 202–222.

Crosa, J. (1984). *Annu. Rev. Microbiol.* **38**, 69–89.

Danislevskaya, O. N. and Gragerov, A. I. (1980). *Molec. Gen. Genet.* **178**, 233–235.

Datta, N. (1979). *In* "Plasmids of Medical, Environmental and Commercial Importance" (K. N. Timmis and A. Pühler, Eds), pp. 3–12. Elsevier, Amsterdam.

Davey, R. B., Bird, P. I., Nikoletti, S. M., Praszkier, J. and Pittard, J. (1984). *Plasmid* **11**, 234–242.

de la Cruz, F., Zabala, J. C. and Ortiz, J. M. (1980). *Plasmid* **4**, 76–81.

Della Latta, P., Bouanchaud, D. and Novick, R. P. (1978). *Plasmid* **1**, 366–375.

Dellaporta, S. I. and Pesano, R. I. (1981). *In* "Biology of the Rhizobiaceae" (K. L. Giles and A. G. Atherly, Eds), pp. 83–104. *Int. Rev. Cytol.* Suppl. 13.

Dénarié, J., Boistard, P. and Casse-Delbert, F. (1981). *In* "Biology of the Rhizobiaceae" (K. L. Giles and A. G. Atherly, Eds), pp. 225–246. *Int. Rev. Cytol.* Suppl. 13.

Dibb, N. J., Downie, J. A. and Brewin, N. J. (1984). *J. Bacteriol.* **158**, 621–627.

Drexler, H. (1970). *Proc. Natl. Acad. Sci. U.S.A.* **66**, 1083–1088.

Duckworth, D. H., Glenn, J. and McCorquodale, D. J. (1981). *Microbiol. Rev.* **45**, 52–71.

Edger, F., Orzech McDonald, N. K. and Burke, W. F. (1981). *FEMS Microbiol. Lett.* **12**, 131–133.

Eisenstark, A. (1966). *In* "Methods in Virology" (K. Maramorosch and H. Koprowski, Eds), pp. 449–524. Academic Press, New York and London.

Elmerich, C. (1983). *In* "Molecular Genetics of the Bacterial-Plant Interaction" (A. Pühler, Ed.), pp. 367–372. Springer-Verlag, Berlin.

Elwell, L. P. and Shipley, P. L. (1980). *Annu. Rev. Microbiol.* **34**, 465–496.

Ersfeld-Dressen H., Sahl, H-G. and Brandis, H. (1984). *J. Gen. Microbiol.* **130**, 3029–3035.

Falkow, S. (1975). "Infectious Multiple Drug Resistance" Pion, London.

Fani, R., Bazzicalupo, M., Coianiz, P. and Polsinelli, M. (1986). *FEMS Microbiol. Lett.* **35**, 23–27.

Fare, L. R., Taylor, D. P., Toth, M. J. and Nash, C. H. (1983). *Plasmid* **9**, 240–246.

Farrand, S. K. and Dessaux, Y. (1986). *J. Bacteriol.* **167**, 732–734.

Feilberg-Jørgensen, N. H., Møller, J. K. and Stenderup, A. (1982). *J. Gen. Microbiol.* **128**, 2991–2996.

Finan, T. M., Kunkel, B., De Vos, G. F. and Signer, E. R. (1986). *J. Bacteriol.* **167**, 66–72.

Finan, T. M., Hartwieg, E., Lemieux, K., Bergman, K., Walker, G. C. and Signer, E. R. (1984). *J. Bacteriol.* **159**, 120–124.

Finger, J. and Krishnapillai, V. (1980). *Plasmid* **3**, 332–342.

Finn, C. W., Silver, R. P., Habig, W. H., Hardegree, M. C., Zon, G. and Garon, C. F. (1984). *Science* **224**, 881–884.

Fornari, C. S., Watkins, M. and Kaplan, S. (1984). *Plasmid* **11**, 39–47.

Foster, T. J. (1983). *Microbiol. Rev.* **47**, 361–409.

Franklin, N. C. (1971). *In* "The Bacteriophage Lambda" (A. D. Hershey, Ed.), pp. 175–194. Cold Spring Harbor, New York.

Franz, B. and Chakrabarty, A. M. (1986). *In* "The Bacteria: A Treatise on Structure and Function" (J. R. Sokatch and L. N. Ornston, Eds), Vol. X, pp. 295–317. Academic Press, New York.

Fuchs, P. G., Zajdel, J. Dobrzanski, W. T. (1975). *J. Gen. Microbiol.* **88**, 189–192.

Ganotti, B. V. and Beer, S. V. (1982). *J. Bacteriol.* **151**, 1627–1629.

Gasson, M. J. (1983). *J. Bacteriol.* **154**, 1–9.

Gellert, M., O'Dea, M. H., Itoh, T. and Tomizawa, J. I. (1976). *Proc. Natl. Acad. Sci. U.S.A.* **73**, 4474–4478.

Ghosal, D., You, I. S., Chatterjee, D. K. and Chakrabarty, A. M. (1985). *Science* **228**, 135–142.

Golden, S. S. and Sherman, L. A. (1984). *J. Bacteriol.* **158**, 36–42.

Goldstein, F. W., Labigne-Roussel, A., Gerbaud, G., Carlier, C., Collatz, E. and Courvalin, P. (1983). *Plasmid* **10**, 138–147.

Gonzal, S. M., Gerba, C. P. and Melnick, J. L. (1979). *Water Res.* **13**, 349–356.

Gonzalez, C. F. and Kunka, B. S. (1986). *Appl. Environ. Microbiol.* **51**, 105–109.

Gonzáles, J. M., Dulmage, H. T. and Carlton, B. C. (1981). *Plasmid* **5**, 351–365.

Gorai, A. P., Heffron, F., Falkow, S., Hedges, R. W. and Datta, N. (1979). *Plasmid* **2**, 485–492.

Grange, S. M. (1982). *In* "The Biology of the Mycobacteria" (C. Ratledge and J. Stanford, Eds), Vol. I, pp. 309–351. Academic Press, New York.

Grubb, W. B. and O'Reilly, R. J. (1971). *Biochem. Biophys. Res. Commun.* **42**, 377–383.

Gruss, A. and Novick, R. (1986). *J. Bacteriol.* **165**, 878–883.

Guidolin, A. and Manning, P. A. (1987). *Microbiol. Rev.* **51**, 285–298.

Hacker, J. and Hughes, C. (1985). *Curr. Topics Microbiol. Immunol.* **118**, 139–162.

Hahn, F. E. and Ciak, J. (1971). *Ann. N.Y. Acad. Sci.* **182**, 295–304.
Hahn, F. E. and Ciak, J. (1976). *Antimicrob. Agents Chemother.* **9**, 77–80.
Hansen, J. B. and Olsen, R. H. (1978). *J. Bacteriol.* **135**, 227–238.
Hardy, K. G. (1975). *Bacteriol. Rev.* **39**, 464–515.
Heath, L. S., Heath, H. E. and Sloan, G. L. (1987). *FEMS Microbiol. Lett.* **44**, 129–133.
Herdman, M. (1982). *In* "The Biology of Cyanobacteria" (N. G. Carr and B. A. Whitton, Eds), Vol. 19, pp. 263–305. Botanical Monographs. University of California Press, Berkeley.
Herman, J. J. and Juni, E. (1974). *J. Virol.* **13**, 46–52.
Hermansson, M., Jones, G. W. and Kjelleberg, S. (1987). *Appl. Environ. Microbiol.* **53**, 2338–2342.
Hershfield, V. (1979). *Plasmid* **2**, 137–149.
Higashi, S. (1967). *J. Gen. Appl. Microbiol.* **13**, 391–403.
Hinchliffe, E., Nugent, M. E. and Vivian, A. (1980). *J. Gen. Microbiol.* **121**, 411–418.
Hinton, J. C. D., Perombelon, M. C. M. and Salmond, G. P. C. (1985). *J. Bacteriol.* **161**, 786–788.
Hirai, K. and Yanagawa, R. (1970). *J. Bacteriol.* **101**, 1086–1087.
Hirato, T., Ishiguro, N., Shinagawa, M. and Sato, G. (1986). *J. Bacteriol.* **165**, 324–327.
Hirota, Y. (1956). *Nature (London)* **178**, 92.
Hirota, Y. (1960). *Proc. Natl. Acad. Sci. U.S.A.* **46**, 57–64.
Hogrefe, C. and Friedrich, B. (1984). *Plasmid* **12**, 161–169.
Hohn, B. and Kohn, D. (1969). *J. Mol. Biol.* **45**, 385–395.
Holloway, B. W. (1979). *Plasmid* **2**, 1–19.
Holloway, B. W. (1984). *In* "Methylotrophs: Microbiology, Biochemistry and Genetics" (C. T. Hou, Ed.), pp. 87–105. CRC Press, Florida.
Holloway, B. W. (1986). *In* "The Bacteria: A Treatise on Structure and Function" (J. R. Sokatch, Ed.), Vol. X, pp. 217–249. Academic Press, New York.
Holsters, M., de Waele, D., Depicker, A., Messens, E., van Montagu, M. and Schell, L. (1978). *Molec. Gen. Genet.* **163**, 181–187.
Hooykaas P. J. J. (1983). *In* "Molecular Genetics of the Bacteria-Plant Interaction" (A. Pühler, Ed.), pp. 229–239. Springer-Verlag, Berlin.
Hooykaas, P. J. J. and Schilperoort, R. A. (1984). *Adv. Genet.* **22**, 209–283.
Hopwood, D. A. (1978). *Annu. Rev. Microbiol.* **32**, 373–392.
Hopwood, D. A. (1981). *Annu. Rev. Microbiol.* **35**, 237–272.
Hopwood, D. A., Kieser, T., Lydiate, D. J. and Bibb, M. J. (1986). *In* "The Bacteria: A Treatise on Structure and Function" (S. W. Queener and L. E. Day, Eds), Vol. IX, pp. 159–229. Academic Press, New York.
Hynes, M. F., Simon, R. and Pühler, A. (1985). *Plasmid* **13**, 99–105.
Ikeda, Y., Iijima, T. and Tajima, K. (1967). *J. Gen. Appl. Microbiol.* **13**, 247–254.
Imanaka, T., Fujii, M., Aramori, I. and Aiba, S. (1982). *J. Bacteriol.* **149**, 824–830.
Ingram, L., Sykes, R. B., Grinsted, J., Saunders, J. R. and Richmond, M. H. (1972). *J. Gen. Microbiol.* **72**, 269–279.
Iordanescu, S. and Surdeanu, M. (1980). *Plasmid* **4**, 256–260.
Iordanescu, S., Surdeanu, M., Della Latta, P. and Novick, R. (1978). *Plasmid* **1**, 468–479.
Ivins, B. E., Ezzell, J. W., Jemski, J., Hedlund, K. W., Ristroph, J. D. and Leppla, S. H. (1986). *Infect. Immun.* **52**, 454–458.
Jacob, A. E., Douglas, G. L. and Hobbs, S. J. (1975). *J. Bacteriol.* **121**, 863–872.

Jacob, A. E., Shapiro, J. A., Yamamoto, L., Smith, D. I., Cohen, S. N. and Berg, D. (1977). *In* "DNA Insertion Elements, Plasmids and Episomes" (A. I. Bukhari, J. A. Shapiro and S. Adhya, Eds), pp. 607–638. Cold Spring Harbor, New York.

Jacoby, G. A. (1986). *In* "The Bacteria: A Treatise on Structure and Function" (J. R. Sokatch, Ed.), Vol. X, pp. 265–293. Academic Press, New York.

Janse, B. J. H., Wingfield, B. D., Pretorius, I. S. and van Vuuren, H. J. J. (1987). *Plasmid* **17**, 173–175.

Kado, C. I. and Liu, S-T. (1981). *J. Bacteriol.* **145**, 1365–1373.

Kageyama, M. (1975). *In* "Microbial Drug Resistance" (S. Mitsuhashi and H. Hashimoto, Eds), pp. 291–305. Tokyo: Univ. Tokyo.

Kaiser, D. and Manoil, C. (1979). *Annu. Rev. Microbiol.* **33**, 595–639.

Katsumata, R., Ozaki, A., Oka, T. and Furuya, A. (1984). *J. Bacteriol.* **159**, 306–311.

Kay, D. (1972). *In* "Methods in Microbiology" (J. R. Norris and D. W. Ribbons, Eds), Vol. 7A, pp. 191–313, Academic Press, London.

Kempler, G. M. and McKay, L. L. (1979). *Appl. Environ. Microbiol.* **37**, 316–323.

Kennedy, C. and Toukdarian, A. (1987). *Annu. Rev. Microbiol.* **41**, 227–258.

Klaenhammer, T. R. and McKay, L. L. (1976). *J. Dairy Sci.* **59**, 396–404.

Klebe, R. J., Harriss, J. V., Sharp, Z. D. and Douglas, M. G. (1983). *Gene* **25**, 333–341.

Klintworth, R., Husemann, M., Salnikow, J. and Bowien, B. (1985) *J. Bacteriol.* **164**, 954–956.

Kondorosi, A. and Johnston, A. W. B. (1981). *In* "Biology of the Rhizobiaceae" (K. L. Giles and A. G. Atherly, Eds), pp. 191–223. *Int. Rev. Cytol.* Suppl. 13.

Konisky, J. (1982). *Annu. Rev. Microbiol.* **36**, 125–144.

Koyama, Y., Hoshino, T., Tomizuka, N. and Furukawa, K. (1986). *J. Bacteriol.* **166**, 338–340.

Kozak, W., Rajchert-Trzpil, M. and Dobrzanski, W. T. (1974). *J. Gen. Microbiol.* **83**, 295–302.

Kushner, D. J. (1985). *In* "The Bacteria: A Treatise on Structure and Function" (C. R. Woese and R. S. Wolfe, Eds), Vol. VIII, pp. 171–214. Academic Press, New York.

Lacey, R. W. (1975). *Bacteriol. Rev.* **39**, 1–32.

Lane, H. E. D. (1981). *Plasmid* **5**, 110–126.

Laufs, R., Riess, F. C., Jahn, G., Fock, R. and Kaulfers, P. (1981). *J. Bacteriol.* **147**, 563–567.

Lax, A. J. and Walker, C. A. (1986). *Plasmid* **15**, 210–216.

Lennox, E. S. (1955). *Virology* **1**, 190–206.

Linton, A. H. (1977). *In* "Antibiotics and Antibiosis in Agriculture" (M. Woodbine, Ed.), pp. 315–343. Butterworths, London.

Luria, S. E., Adams, J. N. and Ting, R. C. (1960). *Virology* **12**, 348–390.

Lyon, B. R. and Skurray, R. (1987). *Microbiol. Rev.* **51**, 88–134.

Mackay, M. W., Al-Bakri, G. H. and Moseley, B. E. B. (1985). *Arch. Microbiol.* **141**, 91–94.

Macrina, F. L. (1984). *Annu. Rev. Microbiol.* **38**, 193–219.

Maloy, S. R. and Nunn, W. D. (1981). *J. Bacteriol.* **145**, 1110–1112.

Martin, P. A. W., Lohr, J. R. and Dean, D. H. (1981). *J. Bacteriol.* **145**, 980–983.

Matsumoto, H., Kamio, Y., Kobayashi, R. and Terawaki, Y. (1978). *J. Bacteriol.* **133**, 387–389.

Matsushima, P., McHenney, M. A., and Baltz, R. H. (1987). *J. Bacteriol.* **169**, 2298–2300.

McKay, L. L., Cords, B. R. and Baldwin, K. A. (1973). *J. Bacteriol.* **115**, 801–815.

Mevarech, M. and Werczberger, R. (1985). *J. Bacteriol.* **162**, 461–462.

Meynell, G. G. (1972). "Bacterial Plasmids". Macmillan, London and New York.

Meynell, G. G. and Meynell, E. (1965). "Theory and Practice in Experimental Bacteriology". Cambridge Univ. Press, London and New York.

Miller, J. H. (1972). "Experiments in Molecular Genetics". Cold Spring Harbor, New York.

Minton, N. P. and Morris, J. G. (1983). *J. Bacteriol.* **155**, 432–434.

Mitsuhashi, S., Harada, K. and Kameda, M. (1961). *Nature (London)* **189**, 947.

Mizuguchi, Y., Fukunaga, M. and Taniguchi, H. (1981). *J. Bacteriol.* **146**, 656–659.

Molnar, J., Beladi, I. and Hollanb, I. B. (1978). *Genet. Res.* **31**, 197–201.

Montenegro, M. A., Bitter-Suermann, D., Timmis, J. K., Agüero, M. E., Cabello, F. C., Sanyal, S. C. and Timmis, K. N. (1985). *J. Gen. Microbiol.* **131**, 1511–1521.

Morelli, L., Cocconcelli, P. S., Bottazzi, V., Damiani, G., Ferrett, L. and Sgaramella, V. (1987). *Plasmid* **17**, 73–75.

Moretti, P., Hintermann, G. and Hütter, R. (1985). *Plasmid* **14**, 126–133.

Morgan, A. F. (1979). *J. Bacteriol.* **139**, 137–140.

Moyer, R. W., Fu, A. S. and Szabo, C. (1972). *J. Virol.* **9**, 804–812.

Munster, M. J., Munster, A. P. and Sharp, R. J. (1985). *Appl. Environ. Microbiol.* **50**, 1325–1327.

Muriana, P. M. and Klaenhammer, T. R. (1987). *Appl. Environ. Microbiol.* **53**, 553–560.

Murray, G. E., Tobin, R. S., Junkins, B. and Kushner, D. J. (1984). *Appl. Environ. Microbiol.* **48**, 73–77.

Novick, R. P. (1987). *Microbiol. Rev.* **51**, 381–395.

Novick, R. P., Iordanescu, S., Surdeanu, M. and Edelman, I. (1981). *Plasmid* **6**, 159–172.

Novick, R. P., Sanchez-Rivas, C., Gruss, A. and Edelman, I. (1980). *Plasmid* **3**, 348–358.

Novick, R. P., Projan, S. J., Rosenblum, W. and Edelman, I. (1984). *Molec. Gen. Genet.* **195**, 357–377.

Odelson, D. A., Rasmussen, J. L., Smith, C. J. and Macrina, F. L. (1987). *Plasmid* **17**, 87–109.

Ørskov, I. and Ørskov, F. (1973). *J. Gen. Microbiol.* **77**, 487–499.

Oshida, T., Takeda, K., Yamaguchi, T., Ohshima, S. and Ito, Y. (1986). *Plasmid* **16**, 74–76.

Ozeki, H. and Ikeda, H. (1968). *Annu. Rev. Genet.* **2**, 245–278.

Palmer, L. and Falkow, S. (1986). *Plasmid* **16**, 52–62.

Palomares, J. C. and Perea, E. J. (1980). *Plasmid* **4**, 352–353.

Pérez-Díaz, J. C., Vicente, M. F. and Baquero, F. (1982). *Plasmid* **8**, 112–118.

Pinney, R. J. and Smith, J. T. (1971). *Genet. Res.* **18**, 173–177.

Pinney, R. J. and Smith, J. T. (1973). *Antimicrob. Agents Chemother.* **3**, 670–676.

Pinney, R. J., Bremer, K. and Smith, J. T. (1974). *Molec. Gen. Genet.* **133**, 163–174.

Poindexter, S. J. (1981). *Microbiol. Rev.* **45**, 123–179.

Portnoy, D. A. and Martinez, R. J. (1985). *Curr. Topics Microbiol. Immunol.* **118**, 29–51.

Postgate, J. R., Kent, H. M., Robson, R. L. and Chesshyre, J. A. (1984). *J. Gen. Microbiol.* **130**, 1597–1601.

Quackenbush, R. L. (1983). *Plasmid* **9**, 298–306.

Quackenbush, R. L., Dilts, J. A. and Cox, B. J. (1986). *J. Bacteriol.* **166**, 349–352.

Rawlings, D. E. and Woods, D. R. (1985). *Appl. Environ. Microbiol.* **49**, 1323–1325.

Razin, S. (1985). *Microbiol. Rev.* **49**, 419–455.
Reanney, D. C. and Ackermann, H-W. (1981). *Intervirology* **15**, 190–197.
Rheinwald, J. G., Chakrabarty, A. M. and Gunsalus, I. C. (1973). *Proc. Natl. Acad. Sci. U.S.A.* **70**, 885–889.
Riva, S., Fietta, A., Berti, M., Silverstri, L. G. and Romero, E. (1973). *Antimicrob. Agents Chemother.* **3**, 456–462.
Robinson, J. B. and Tuovinen, O. H. (1984). *Microbiol. Rev.* **48**, 95–124.
Rogers, P. (1986). *Adv. Appl. Microbiol.* **31**, 1–60.
Rolfe, B. and Holloway, B. W. (1966). *J. Bacteriol.* **92**, 43–48.
Romermann, D. and Friedrich, B. (1985). *J. Bacteriol.* **162**, 852–854.
Romero, E., Perduca, M. and Pagani, L. (1979). *Microbiologica* **2**, 421–424.
Rood, J. I., Scott, V. N. and Duncan, C. L. (1978). *Plasmid* **1**, 563–570.
Sadowski, P. D. and Kerr, C. (1970). *J. Virol.* **6**, 149–155.
Salisbury, V., Hedges, R. W. and Datta, N. (1972). *J. Gen. Microbiol.* **70**, 443–452.
Santamaría, R. I., Gil, J. A. and Martin, J. F. (1985). *J. Bacteriol.* **162**, 463–467.
Santamaría, R. I., Gil, J. A., Mesas, J. M. and Martin, J. F. (1984). *J. Gen. Microbiol.* **130**, 2237–2246.
Sasakawa, C., Takamatsu, N., Danbara, H. and Yoshikawa, M. (1980). *Plasmid* **3**, 116–127.
Saunders, V. A. (1978). *Microbiol. Rev.* **42**, 357–384.
Schiller, J., Groman, N. and Coyle, M. (1980). *Antimicrob. Agents Chemother.* **18**, 814–821.
Schmid, K., Ritschewald, S. and Schmitt, R. (1979). *J. Gen. Microbiol.* **114**, 477–481.
Schmid, K., Schupfner, M. and Schmitt, R. (1982). *J. Bacteriol.* **151**, 68–76.
Schmieger, H. and Buch, U. (1975). *Molec. Gen. Genet.* **140**, 111–122.
Schmitt-Slomska, J., Caravano, R. and El-Solh, N. (1979). *Ann. Microbiol. (Paris)* **130A**, 23–27.
Schöller, M., Klein, J. P., Sommer, P. and Frank, R. (1983). *J. Gen. Microbiol.* **129**, 3271–3279.
Scholz, P., Haring, V., Scherzinger, E., Lurz, R., Bagdasarian, M. M., Schuster, H. and Bagdasarian, M. (1984). *In* "Plasmids in Bacteria" (D. R. Helinski, S. N. Cohen, D. B. Clewell, D. A. Jackson and A. Hollaender, Eds), pp. 243–259. Plenum Press, New York.
Scolnik, P. A. and Marrs, B. L. (1987). *Annu. Rev. Microbiol.* **41**, 703–726.
Scott, J. R. (1984). *Microbiol. Rev.* **48**, 1–23.
Shalita, Z., Murphy, E. and Novick, R. P. (1980). *Plasmid* **3**, 291–311.
Silver, S., Levine, E. and Spielman, P. M. (1968). *J. Bacteriol.* **95**, 333–339.
Simonet, P., Normand, P., Moiroud, A. and Lalondem, M. (1985). *Plant and Soil* **87**, 49–60.
Skjold, S. A., Malke, H. and Wannamaker, L. K. (1979). *In* "Pathogenic Streptococci" (M. T. Parker, Ed.), pp. 274–275. Reedbooks, Chertsey.
Skurnik, M., Bölin, I., Heikkinen, H., Piha, S. and Wolf-Watz, H. (1984). *J. Bacteriol.* **158**, 1033–1036.
Smith, H. O., Danner, D. B. and Deich, R. A. (1981). *Annu. Rev. Biochem.* **50**, 41–68.
Smith, H. W. (1977). *In* "Antibiotics and Antibiosis in Agriculture" (M. Woodbine, Ed.), pp. 344–357. Butterworths, London.
Smith, M. D. (1985). *J. Bacteriol.* **162**, 92–97.
Smith-Keary, P. F. (1960). *Heredity* **14**, 61–71.
Stadler, J. and Adelberg, E. A. (1972). *J. Bacteriol.* **109**, 447–449.

46   V. A. STANISICH

Stalker, D. M. and McBride, K. E. (1987). *J. Bacteriol.* **169**, 955–960.
Stanier, R. Y., Palleroni, N. J. and Doudoroff, M. (1966). *J. Gen. Microbiol.* **43**, 159–273.
Stanisich, V. A. and Richmond, M. H. (1975). In "Genetics and Biochemistry of *Pseudomonas*" (P. H. Clarke and M. H. Richmond, Eds), pp. 163–190. Wiley, London.
Stanisich, V. A., Bennett, P. M. and Ortiz, J. M. (1976). *Molec. Gen. Genet.* **143**, 333–337.
Starlinger, P. (1977). *Annu. Rev. Genet.* **11**, 103–126.
Sternberg, N. and Austin, S. (1981). *Plasmid* **5**, 20–31.
Sternberg, N. and Weisberg, R. (1975). *Nature (London)* **256**, 97–103.
Stiffler, P. W., Sweeney, H. M. and Cohen, S. (1974). *J. Bacteriol.* **120**, 934–944.
Stewart, G. J. and Carlson, C. A. (1986). *Annu. Rev. Microbiol.* **40**, 211–235.
Strike, P. and Lodwick, D. (1987). *J. Cell Sci.* Suppl. **6**, 303–321.
Tagg, J. R., Dajani, A. S. and Wannamaker, L. W. (1976). *Bacteriol. Rev.* **40**, 722–756.
Taylor, D. E., Hiratsuka, K., Ray, H. and Manavathu, E. K. (1987). *J. Bacteriol.* **169**, 2984–2989.
Terawaki, Y., Takatasu, H. and Akiba, T. (1967). *J. Bacteriol.* **94**, 687–690.
Thatte, V., Gill, S. and Iyer, V. N. (1985). *J. Bacteriol.* **163**, 1296–1299.
Thomas, C. M. and Smith, C. A. (1987). *Annu. Rev. Microbiol.* **41**, 77–101.
Timmis, K. N. (1979). In "Plasmids of Medical, Environmental and Commercial Importance" (K. N. Timmis and A. Pühler, Eds), pp. 13–22. Elsevier/North Holland Biomedical Press, New York.
Tirgari, S. and Moseley, B. E. B. (1980). *J. Gen. Microbiol.* **119**, 287–296.
Toh-E, A. and Wickner, R. B. (1981). *J. Bacteriol.* **145**, 1421–1424.
Tomoeda, M., Inuzuka, M., Kubo, N. and Nakamura, S. (1968). *J. Bacteriol.* **95**, 1078–1089.
Tomoeda, M., Kokubu, M., Nabata, H. and Minamikawa, S. (1970). *J. Bacteriol.* **104**, 864–870.
Trehan, K. and Sinha, U. (1981). *J. Gen. Microbiol.* **124**, 349–352.
Tyeryar, F. J. and Lawton, W. D. (1969). *J. Bacteriol.* **110**, 1112–1113.
Vakeria, D., Codd, G. A., Bell, S. G., Beattie, K. A. and Priestly, I. M. (1985). *FEMS Microbiol. Lett.* **29**, 69–72.
Valla, S., Coucheron, D. H. and Kjosbakken, J. (1986). *J. Bacteriol.* **165**, 336–339.
Van Montagu, M. and Schell, J. (1979). In "Plasmids of Medical, Environmental and Commercial Importance" (K. N. Timmis and A. Pühler, Eds), pp. 55–95. Elsevier/North Holland Biomedical Press, New York.
Veltkamp, E. and Stuitje, A. R. (1980). *Plasmid* **5**, 76–99.
Vincente, M. F., Baquero, F. and Péréz-Diaz, J. C. (1987). *Plasmid* **18**, 89–92.
Walia, S. K., Carey, V. C., All, B. P. and Ingram, L. O. (1984). *Appl. Environ. Microbiol.* **47**, 198–200.
Walia, S. K., Madhavan, T., Chugh, T. D. and Sharma, K. B. (1987). *Plasmid* **17**, 3–12.
Wechsler, J. and Kline, B. C. (1980). *Plasmid* **4**, 276–280.
Weisser, J. and Wiedemann, B. (1985). *Antimicrob. Agents Chemother.* **28**, 700–702.
Whellis, M. L. (1975). *Annu. Rev. Microbiol.* **29**, 505–524.
Whitman, W. B. (1985). In "The Bacteria: A Treatise on Structure and Function" (C. R. Woese and R. S. Wolfe, Eds), Vol. VIII, pp. 3–84. Academic Press, New York.
Wilke, D. (1980). *J. Gen. Microbiol.* **117**, 431–436.

Wilke, D. and Schlegel, H. G. (1979). *J. Gen. Microbiol.* **115**, 403–410.

Wilkins, B. M., Chatfield, L. K., Wymbs, C. C. and Merryweather, A. (1985). *In* "Plasmids in Bacteria" (D. R. Helinski, S. N. Cohen, D. B. Clewell, D. A. Jackson and A. Hollaender, Eds), pp. 585–603. Plenum, New York.

Willetts, N. S. (1967). *Biochem. Biophys. Res. Commun.* **27**, 112–117.

Willetts, N. S. and Skurray, R. (1980). *Annu. Rev. Genet.* **14**, 41–76.

Willetts, N. S. and Skurray, R. (1987). *In* "*Escherichia coli* and *Salmonella typhimurium*: Cellular and Molecular Biology" (F. C. Neidhardt, Ed.), Vol. 2, pp. 1110–1133. Amer. Soc. Microbiol. Washington DC.

Willetts, N. S. and Wilkins, B. (1984). *Microbiol. Rev.* **48**, 24–41.

Wolfson, J. S., Hooper, D. C., Swartz, M. N., Swartz, M. D. and McHugh, G. L. (1983). *J. Bacteriol.* **156**, 1165–1170.

Yasbin, R. E. and Young, F. E. (1974). *J. Virol.* **14**, 1343–1348.

Yoshikawa, M. (1971). *Genet. Res.* **17**, 9–16.

Yoshikawa, M. and Sevag, M. G. (1967). *J. Bacteriol.* **93**, 245–253.

Zurkowski, W. and Loriewcz, Z. (1978). *Genet. Res.* **32**, 311–314.

3

Conjugation

N. WILLETTS

Biotechnology Australia Pty. Ltd., P.O. Box 20, Roseville, NSW 2069, Australia

I. General introduction

Conjugation is the process whereby DNA is transferred from one bacterial cell to another by a mechanism that requires cell-to-cell contact. Requirement for the physical presence of the donor organism, together with insensitivity to deoxyribonuclease (DNase), allow conjugation to be readily distinguished from transduction or transformation. Conjugation is determined almost invariably by plasmids and not by bacterial chromosomes, and gives the subset of gene types usually located on plasmids, an evolutionary advantage. Firstly, since conjugation is essentially a replication process, it allows plasmids to replicate more frequently than chromosomal genes, and secondly, it enables them to transfer to alternative bacterial hosts. It is therefore not surprising that a large proportion of plasmids of various types, isolated from many different bacterial genera, are conjugative.

The specification of a conjugation system requires a relatively large amount of DNA and typically takes up about one-third of the plasmid genome (Fig. 1). In the case of the plasmid F this is equivalent to 33 kilobases (kb), and this large input of genetic information can perhaps be correlated with the efficiency of conjugative transfer which can be 100% even in short term matings. Genetic analysis of conjugation by F (briefly described in Section VIII) shows that 25 or more genes are involved. Both for F and other plasmids where more rudimentary investigations have been carried out, conjugation can be divided on both physiological and genetic bases into two parts: the recognition of recipient cells by donor cells that leads to mating pair formation, and the subsequent physical transfer of plasmid DNA (Fig. 2). Initiation of DNA transfer takes place at a specific site on the DNA called the origin of transfer, *oriT* (Willetts, 1972): this site has been sequenced in some cases (Guiney and Yakobson, 1983; Thompson *et al.*, 1984; Willetts and Wilkins, 1984; Coupland *et al.*, 1987). In addition, plasmids usually specify a surface (or entry) exclusion system that prevents the unproductive transfer of the plasmid to a cell that already possesses a copy.

Despite these overall similarities, numerous genetically distinct and noninteracting conjugation systems have been identified. Plasmids can therefore be classified on this basis, and there is a notable correlation between the incompatibility group (Chapter 2) and conjugation group; plasmids with a given conjugation system usually belong to a single incompatibility group, or to a small number of such groups (e.g. the IncFI, IncFII, etc. complex). This relationship probably indicates co-evolution of incompatibility and conjugation systems, since these function entirely independently of each other.

II. The conjugation pilus

Conjugative plasmids from Gram-negative (but not Gram-positive, see Section VII) bacteria synthesize an extracellular organelle called a pilus that is essential for recognition of recipient cells and formation of mating pairs with them. In addition, pili provide the sites for adsorption of certain bacteriophages, which can therefore be described as plasmid- or pilus-specific (Fig. 3). Since pili, and the pilus-specific phages to which they are sensitive, vary between different conjugation systems, determination of precisely which of these phages will infect a plasmid-containing cell provides a relatively simple method for the preliminary identification of the conjugation system (Table I).

TABLE I
Pilus types and pilus-specific phages[a]

Plasmid Inc group	Representative plasmid	Pilus type	Ratio of plate to broth transfer frequencies	Phages[b] Isometric RNA	Filamentous single-stranded DNA	Double stranded DNA[c]
IncI complex	R64-11	Thin & thick flexible[c]	0.9	φIα	If1	—
IncF complex	R100-1	Thick flexible	0.7	f2	f1	—
IncS complex	EDP208	Thick flexible	0.9	UA6, SR	f1, SF	—
J[d]	R391	Thick flexible	0.9	—	—	φJ[e,f]
C[d]	RA1	Thick flexible	45	φC-1	φC-2	φJ
P-9	TOL	Thick flexible	18	—	—	PR4[f]
T	Rts1	Thick flexible	260	φt	tf-1[g]	—
X	R6K	Thick flexible	250	—	φX[h]	—
N	N3	Rigid	1×10^4	—	Ike, (φX)	PR4
P	RP1	Rigid	2×10^4	PRR1	Pf3	PR4
P-10	R91-5	Rigid	47	—	—	PR4
W	Sa	Rigid	4×10^4	—	(φX)	PR4

[a] Data are taken from Bradley (1981, 1984) and Coetzee et al. (1986).
[b] Other related phages are as follows: If1, If2, PR64FS; f2, R17, M12, MS2, Qβ; f1, fd, M13; UA6, φF$_0$lac; PR4, PRD1. Parentheses indicate that a positive titre-increase, but not spot test, was obtained.
[c] Interestingly, plasmids of the IncI complex all encode two morphologically distinct types of pili (Bradley, 1984).
[d] The pili of IncJ and IncC plasmids are serologically related: other pili in Table I are not.
[e] φJ also infects cells carrying IncD plasmids and φX infects cells carrying IncM and IncU plasmids.
[f] φJ is similar in morphology to T3, whereas PR4 is a lipid-containing phage with head and tail structures.
[g] Phage tf-1 grows on strains carrying the IncT plasmid pIN25, not Rts1 (Coetzee et al., 1987).
[h] φX is unrelated to φX174.

Pili can also be differentiated by their morphology as visualized under the electron microscope, and the three major morphological groups (Fig. 4) can be further subdivided by serological comparisons (Bradley, 1980). Pili encoded by *Pseudomonas* plasmids have been similarly classified (Bradley, 1983). However, these techniques are beyond the scope of this chapter.

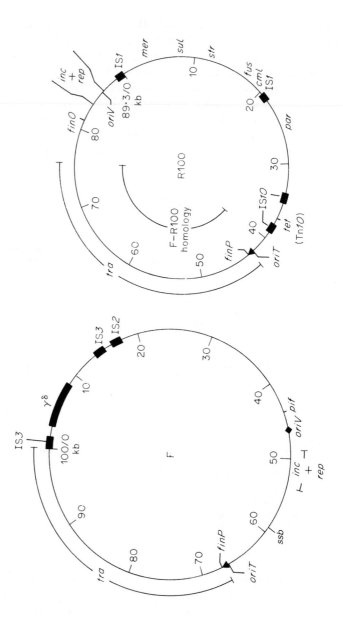

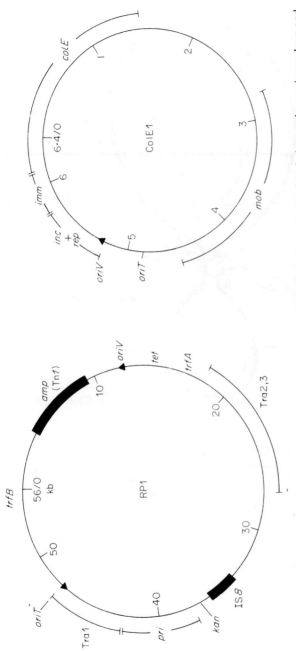

Fig. 1. Genetic and physical maps of four plasmids. Kilobase coordinates are shown inside the circles, and genetic markers and insertion sequences outside. Genetic nomenclature is as follows: *inc*, incompatibility; *rep*, replication; *tra* and Tra, transfer region; *finO*, fertility inhibition; *pif*, phage inhibition; resistances to mercuric ions (*mer*), sulphonamides (*sul*), streptomycin (*str*), fusidic acid (*fus*), chloramphenicol (*cml*), tetracycline (*tet*), ampicillin (*amp*), kanamycin (*kan*); *pri*, DNA primase; *ssb*, single-strand binding protein; *trfA*, B replication functions; *colE*, colicin E; *imm*, immunity to colicin E; *mob*, mobilization; *oriT*, origin of transfer replication; *oriV*, origin of plasmid replication.

DONOR

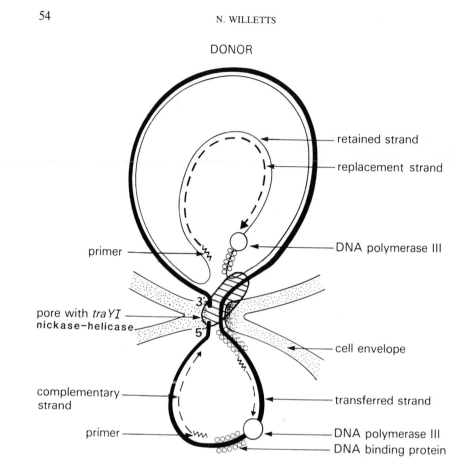

Fig. 2. A model for DNA transfer during conjugation. In this model the termini of the transferred strand are attached to the cell membrane, and DNA helicase I is bound to the membrane near to the conjugation 'pore'; migration of helicase I along the transferred strand provides the motive force displacing this into the recipient cell. The model assumes that a primer is required, and that single-stranded DNA-binding protein coats the single-stranded DNA (from Willetts and Skurray, 1987, with permission).

A. Preparation of phage lysates

Add about 10^6 plaque-forming units (p.f.u.) and 0.1 ml of an overnight L broth (10 g Difco tryptone, 5 g yeast extract, 10 g NaCl per litre, pH 6.8) culture of a sensitive bacterial strain to 2.5 ml molten (46°C) LC top agar (L broth plus 7 g per litre of Difco agar plus 5 mM $CaCl_2$, which is necessary for infection by these phages), and pour onto a nutrient plate. Incubate at 37°C overnight.

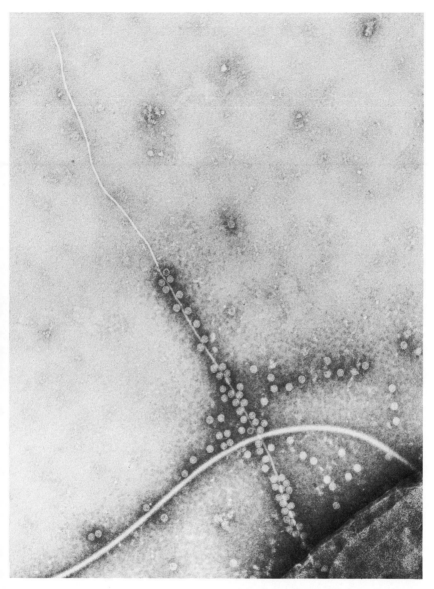

Fig. 3. Adsorption of F-specific phages MS2 and f1 to the sides and tip of an F pilus on an *Escherichia coli* K12 cell. The thicker organelle is a flagellum (courtesy of E. Boy de la Tour and L. Caro, previously published by Broda, 1979).

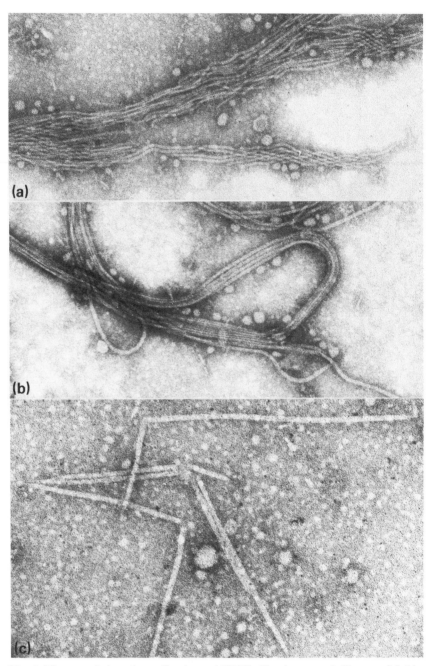

Fig. 4. Pilus morphology (magnification × 140 000). The three panels illustrate (a) thin flexible (of R64-11; 6 nm diameter), (b) thick flexible (of the unclassified plasmid pIN32; 9 nm diameter) and (c) rigid (of N3; 9.5 nm diameter, though rigid pili vary from 8 to 12 nm depending upon the plasmid) pili (courtesy of D. E. Bradley).

Scrape off the top agar layer, and emulsify with 5 ml of broth. Centrifuge, remove the supernatant and sterilize this by adding a few drops of chloroform or by membrane filtration. The latter technique must be used for lipid-containing phages, such as PR4, and for filamentous single-strand DNA phages. Phage lysates for IncP-specific RNA phages (e.g. PRR1) or filamentous phages (e.g. Pf3) are best grown in *Pseudomonas aeruginosa* rather than *Escherichia coli*.

B. Titring the lysates

Mix 0.1 ml aliquots of appropriate dilutions in phage dilution buffer (3 g KH_2PO_4, 7 g Na_2HPO_4, 5 g NaCl per litre, plus 1 mM $MgSO_4$, 0.1 mM $CaCl_2$, 0.001% gelatin) and 0.1 ml of an overnight L broth culture of the host strain with 2.5 ml of molten LC top agar and pour onto a nutrient plate. After overnight incubation, count the plaques.

C. Sensitivity by spot test

This is the simplest qualitative test for phage sensitivity. Add 0.1 ml of an overnight culture of the bacterial strain to 2.5 ml of molten LC top agar and pour onto a nutrient plate. After the top agar has set, mark the bottom of the plate and place 10 μl spots of the phage lysates, diluted to 10^5–10^6 p.f.u. per millilitre, on the surface. Four to eight spots can be accommodated on a single plate. Let the spots dry in, and then incubate the plate at 37°C overnight. Record the turbidity of the spot with respect to a control (a standard plasmid-containing phage-sensitive bacterial host), as this can give some idea of the level of sensitivity. It is also wise to include a plasmid-free strain as a negative control.

It should be noted that apparently unrelated plasmids in the same cell can sometimes inhibit pilus formation and transfer by a coexisting conjugative plasmid. For example, IncN plasmids reduce IncP plasmid transfer by about 10^4-fold (Brown, 1981). It is therefore preferable to test strains which carry only a single plasmid. The nature of the bacterial host can also markedly affect the response to these spot tests (Bradley *et al.*, 1981). For example, when enterobacterial hosts are tested using IncP-specific phages grown in *P. aeruginosa*, it is necessary to use an increased phage inoculum (e.g. 10^5 for PRR1 and 10^8 for Pf3).

D. Efficiency of plating

This parameter is sometimes worth measuring since it can vary from plasmid to plasmid: for example, amongst IncF plasmids, the efficiency of plating (e.o.p.) of RNA phages varies over a more than 100-fold range (Alfaro and

Willetts, 1972). This is a consequence of variations in the sequence of the pilin subunit proteins (Frost *et al.*, 1984, 1985).

Mix about 2×10^2 p.f.u. of the phage with 0.1 ml of an overnight broth culture of the test strain and pour onto a nutrient plate to determine the titre as described in Section II.B. Simultaneously, repeat this titre using the standard plasmid-containing host. The ratio of the two titres gives the e.o.p.

E. Measurement of titre increase

The conjugation systems of many naturally occurring plasmids are repressed (Section VI), and resistance to a pilus-specific phage may simply be due to this. In general terms, this is true for IncF and IncI plasmids, but not for IncN, IncP or IncW plasmids. In order to test the phage sensitivities of plasmids that transfer at low frequencies (<0.1) and so might be repressed, a titre increase assay can be carried out on a transiently derepressed culture prepared as described in Section VI. 10^4 p.f.u. of the phage are added to 1 ml of the culture, and shaken in a test-tube for 2 h. After chloroforming, the phage titre is measured as described in Section II.B. The ratio of this titre to the phage inoculum titre gives the titre increase. This typically ranges from 10^3 to 10^5 for RNA phage infection of transiently derepressed cultures containing IncF plasmids (Meynell and Data, 1966).

An alternative approach is to determine the sensitivity by spot test of a genetically derepressed plasmid mutant, obtained as described in Section VI.

III. Conjugative plasmid transfer

Detection or measurement of plasmid transfer is straightforward when the plasmid carries a selectable marker such as antibiotic resistance. Of the commonly used antibiotics, care should be taken with chloramphenicol and β-lactams such as ampicillin, since these become ineffective if too many cells (10^6–10^8) are plated; in addition, ampicillin is unstable both in solution and in plates. Trimethoprim is inactivated in many nutrient media, and minimal medium is preferable for this antibiotic and also for sulphonamides. Dilute solutions and plates containing mercuric ions are notoriously unstable. Note, furthermore, that a high level of antibiotic resistance can sometimes be determined by chromosomal markers, for example in *Pseudomonas* strains (Holloway *et al.*, 1979).

In essence, a conjugation experiment is carried out by mixing broth cultures of the donor and recipient strains under appropriate conditions, and selecting plasmid-containing transconjugants of the recipient strain. Selection against the donor strain is often achieved by using a recipient strain resistant to an antibiotic such as streptomycin, spectinomycin, nalidixic acid or rifamycin to which the donor is sensitive. Alternatively, a (multi)auxotrophic donor strain can be crossed with a prototrophic (or differently auxotrophic) recipient strain, and transconjugants selected on the appropriate minimal medium.

The following notes apply to all quantitative matings. Cultures of donor and recipient strains (obtained by diluting overnight cultures 1:20 with fresh L broth) should be grown separately, with shaking, to 2×10^8 cells per millilitre. These cultures can then be stored on ice (for up to 1 h) until required. Culture density is most easily measured during growth by using side-arm flasks as the culture vessels, and a Klett–Summerson colorimeter with a red filter, into which the side-arms can be placed. Alternatively, absorption at 620 nm can be followed. Whenever culture dilutions are made in saline or phosphate buffer rather than broth, and plating is on minimal medium, 0.1 ml broth per plate should be included to prevent step-down conditions that can lead to a five-fold reduction in transconjugant recovery. Usually a 0.1 ml sample of an appropriate dilution is plated. Plating can be done either by a top layer overlay technique, mixing the sample with 2.5 ml molten water agar (0.7% Difco agar), or by spreading the sample with an alcohol-sterilized glass rod spreader. Incubate plates at 37°C for one (nutrient medium) or two (minimal medium) days, or as otherwise appropriate. The efficiency of transfer can then be calculated as the number of transconjugants per donor cell. Plasmids differ in their transfer frequencies over a wide range (10^{-8}–1) so that some preliminary experiments may be necessary to establish the appropriate dilutions for selecting trans-conjugants of a new plasmid. It may also be worthwhile to test fresh overnight cultures instead of exponential ones, since for many plasmids these give almost equally efficient transfer. It is wise to check about 25 colonies from the donor viable count plates for continued presence of the plasmid, by a patching (see below) and replica-plate technique, since some plasmids are unstable.

Plasmids synthesizing rigid pili and some synthesizing thick flexible pili, transfer efficiently only when the mating cells are supported on a solid surface, while plasmids that determine thin flexible pili also transfer efficiently in broth matings (Table I; Bradley et al., 1980). For an unknown plasmid, then, it is best to choose a surface mating technique (Section III.A or III.B), or to compare the two methods.

A. Membrane filter matings

This is perhaps the most efficient technique, although it is tedious if a large number of matings has to be carried out. Mix 0.1 ml of the donor culture and 1.8 ml of the recipient culture and filter through a 2.5 cm diameter 0.45 μm pore-size filter. A Millipore Swinnex assembly and disposable syringe provides the simplest means for this. Place the filter (using sterile forceps) on a prewarmed nutrient agar plate, cells uppermost, and incubate at 37°C for $\frac{1}{2}$ to 1 h or longer if required. Then immerse the filter in 2 ml of broth, resuspend the cells by vortexing and plate appropriate dilutions on selective plates. Because of the variable efficiency of resuspensions, it is best to measure the donor viable count at this stage by plating dilutions on plates selective for the donor strain.

B. Plate mating

This technique is simpler, but may be up to ten times less efficient than membrane filter mating. Spread 0.1 ml of an appropriate dilution (found in a preliminary experiment) of an exponential broth culture of the donor strain with 0.1 ml of an exponential (or overnight) culture of the recipient strain together on a selective plate, and incubate at 37°C for one to two days. Other steps are as described in Section III.A, except that the donor culture viable count can be determined on nutrient medium.

In these plate matings, the contraselective marker should be Str^r or Spc^r, not Nal^r since nalidixic acid inhibits conjugation (Barbour, 1967). If nalidixic acid must be used, or if a delay prior to selection is necessary to allow expression of the plasmid marker in the recipient cells, then the mating should be carried out for the desired time on a nutrient plate, before resuspending the cells in 1 ml of broth and plating dilutions on selective plates.

C. Broth mating

Mix 0.2 ml of the donor strain culture with 1.8 ml of the recipient strain culture in a large (18 mm × 150 mm) test-tube. This ensures adequate aeration without shaking, which might disrupt mating pairs. Incubate at 37°C for 30 min (or longer if required), and in the meantime plate dilutions to measure the viable count of the donor culture. Stop the mating by placing the tubes in ice, and plate dilutions on selective plates as described.

D. Transfer of plasmids without selectable markers

Transfer of such plasmids can only be measured if it is relatively frequent, such that when a reasonable number (10–1000) of exconjugant recipient colonies are screened, a proportion carrying the plasmid will be found. After mating as described in Section III.A, III.B or III.C (but using a 1:1 donor to recipient ratio), dilutions of the mating mix are plated on medium selective for the recipient cells only. Individual colonies are then 'patched' in a geometric array of up to 100 areas per nutrient plate, using flat-ended sterile toothpicks. After growing up, the colonies can be replica-plated onto whatever medium is appropriate for detecting plasmid-containing cells. If the plasmid has no detectable marker, colonies can be screened using a small-scale technique

(Chapter 5) for the physical presence of the plasmid: however, it is probably simpler to construct a transposon-carrying derivative in such cases (Chapter 9).

To measure transfer of Col (colicin-producing) plasmids (to a colicin-resistant recipient strain), the mating mixture is plated on recipient-selective nutrient medium using a top agar overlay technique, this layer when set being overlayed with a further 2.5 ml of agar to prevent the colonies from growing through the agar. After incubation overnight, the transconjugant colonies are killed by brief exposure to chloroform vapour (introduced on a piece of filter paper on the lid of the inverted plate) and a third overlay applied, containing 0.1 ml of an overnight culture of a colicin-sensitive strain able to grow on the recipient-selective nutrient medium. After further overnight growth, clear areas (lacunae) will be observed around Col$^+$ colonies (Monk and Clowes, 1964).

E. Qualitative transfer techniques

These are useful for screening for transfer ability, or for constructing plasmid-containing strains. 10^2–10^3 colonies are most easily screened by patching them in a geometric array on nutrient agar plates (50–100 patches per plate). After incubation at 37°C for about 6 h, to give exponentially growing cells, this array is replica-plated onto a selective plate that has been spread with 0.1 ml of an overnight culture of the recipient strain concentrated ten times in buffer. After one to two days incubation, transconjugant patches will appear where the donor colonies carried a conjugation-proficient plasmid.

One to ten colonies can be tested for transfer by a cross-streak mating, which has the advantage that it includes control areas showing that neither donor nor recipient strain alone will grow on the selective plate. Streak two loopfuls of an overnight broth culture of the recipient strain across the centre of a selective plate, and allow to dry in. Then streak loopfuls of broth cultures of the donor strains perpendicularly across this in marked positions. After incubation, transconjugant growth beyond where the donor and recipient streaks cross identifies transfer-proficient donor cells.

To construct plasmid-containing derivatives, perhaps the simplest and most rapid technique is to mix loopfuls of the donor and recipient cultures in a small area on a selective plate. After incubation for 1–2 h, cells can be streaked out from this area, and further incubation will give single transconjugant colonies, ready for a second round of purification on nutrient plates.

F. Measurement of surface exclusion

The presence of a plasmid commonly reduces the ability of a cell to function as a recipient in matings with donor cells containing the same or a related plasmid. The quantitative level of this property can be measured by comparing the frequency of transconjugant formation by plasmid-free and plasmid-containing cells, and the ratio is called the surface exclusion index. Any of the mating techniques described in Section III.A, III.B or III.C can be used for the measurement. The donor plasmid must carry an extra selectable marker and if necessary this is most easily achieved by constructing a transposon derivative, or by deleting a marker from the plasmid in the recipient.

Surface exclusion indices usually range from 100 to 1000, and surface exclusion systems are highly specific; for example, four different ones have been detected amongst IncF plasmids (Willetts and Maule, 1974, 1985).

IV. Mobilization of nonconjugative plasmids

Many naturally occurring plasmids are nonconjugative but nonetheless have an *oriT* site and genes that allow their efficient transfer, or mobilization, if the appropriate conjugation system is provided by a coexisting conjugative plasmid (Willetts and Wilkins, 1984; Chan *et al.*, 1985; Derbyshire and Willetts, 1987; Derbyshire *et al.*, 1987). The most common nonconjugative plasmids are relatively small (2–10 kb). Examples of such plasmids are listed in Table II, together with their frequencies of mobilization by some conjugative plasmids. All these plasmids have mobilization genes with products that probably initiate transfer from their *oriT* sites, and substitute for similar functions encoded by the conjugative plasmid that are specific for its own *oriT*. However, only some conjugative plasmids will mobilize a given nonconjugative plasmid. In a practical sense, this means that several different conjugative plasmids must be

TABLE II

Mobilization of nonconjugative plasmids[a]

Conjugative plasmid	Incompatibility group	Transfer (fraction of conjugative)			
		ColE1	ColE2	pSC101	RSF1010
R64-11	Iα	0.7	0.8	5×10^{-3}	0.06
F*lac*	FI	1	$<10^{-4}$	10^{-7}	10^{-5}
R100-1	FII	10^{-3}	$<10^{-4}$	5×10^{-5}	10^{-5}
R91-5	P-10	—	—	—	0.6
R6K	X	5×10^{-4}	—	10^{-3}	0.1
RP1	P	0.5	—	0.13	1
R46	N	10^{-6}	—	$<10^{-5}$	10^{-7}
R388	W	10^{-2}	—	$<4 \times 10^{-7}$	10^{-4}

[a] Data are taken from Willetts (1981), and the unpublished data of K. Derbyshire and N. S. Willetts.

tested in any search for ability to mobilize a new nonconjugative plasmid. Experience suggests that these should include (derepressed) IncP, IncI and IncF plasmids since these have the broadest mobilization abilities; if all give negative results, a wider search is necessary.

A. Mobilization from double-plasmid strains

The simplest technique is to transfer various conjugative plasmids into the strain carrying the nonconjugative plasmid, as described in Section III.E, and then to use these strains as donors with a suitable recipient strain. Use of a RecA⁻ host will prevent mobilization via possible homologous recombination with the conjugative plasmid.

A low frequency of mobilization may indicate not that this is taking place via an *ori*T site on the nonconjugative plasmid, but rather that one of the two plasmids carries a transposable DNA sequence that allows cointegrate formation, and hence mobilization (even in a RecA⁻ host). This is in fact a simple method for the detection of such a sequence, in systems where *ori*T mobilization does not occur (Chapter 9).

B. Triparental matings

This technique is not so sensitive nor is it quantitative, but it avoids strain construction when many pairs of plasmids have to be tested, and will give transient double-plasmid cells (Section VIII) even when the plasmids are incompatible. The simplest variation is to mix 0.05 ml spots of overnight broth cultures of the donor strain carrying the conjugative plasmid and the intermediate strain carrying the nonconjugative plasmid on an appropriate selective plate that has been spread with 0.1 ml of an overnight culture of a recipient strain resistant to some antibiotic contraselective to both donor and intermediate strains. If necessary, this technique can be modified by using exponential cultures, membrane filter or broth matings, or by carrying out the donor and intermediate, and intermediate and recipient matings separately.

V. Mobilization of the bacterial chromosome

An important use of conjugative plasmids is for transfer, by conjugation, of segments of the host bacterial chromosome, thus providing a genetic system for its analysis. Such transfer is achieved by covalently linking the chromosome to the origin of transfer sequence of a conjugative plasmid, so that it is transferred as an 'extension' of the plasmid DNA. This covalent linkage can be established by a variety of recombinational mechanisms, and may be transient or reversible rather than essentially permanent as in the classical system for transfer of the *E. coli* K12 chromosome by integration of F to give Hfr strains (Hayes, 1953, Davidson *et al.*, 1975).

A further important feature of mobilization of the bacterial chromosome by its covalent linkage to a conjugative plasmid is that by a further recombination event between sites different from those used for plasmid integration, plasmid primes can be formed. These carry a segment of the chromosome integrated into the plasmid genome in a relatively stable fashion (although it is often preferable to use RecA⁻ hosts to prevent or reduce their breakdown). Plasmid primes are useful for constructing cells that are diploid for that region of the chromosome or as a source of that segment of chromosomal DNA.

The review by Bachmann (1983) gives the most up to date linkage map of *E. coli*, and those of Bachmann and Low (1980), Low (1987) and Holloway and Low (1987) are excellent sources of information about Hfr and F or R prime strains, and genetic techniques useful for mapping in this organism. Given below are details of techniques for measuring the frequency of transfer of chromosomal markers and for the classical interrupted mating experiment, and brief descriptions of the various methods for mobilizing the bacterial chromosome and for constructing plasmid primes.

A. Measurement of chromosome transfer

The techniques are similar to those given in Section III, except that mating should be carried out for 2 h to allow transfer of long DNA segments. Recipient strains are usually (multi)auxotrophic or unable to ferment one or more sugars, so that selection for inheritance of the wild type allele can be performed on the appropriate minimal selective medium. Depending upon the nature of the donor strain, markers may be transferred at frequencies varying from 10^{-8} to 10^{-1}. The recipient strain is also usually resistant to an antibiotic to allow contraselection of the donor strain. Care should be taken that the antibiotic-sensitive allele is not transferred as an early marker by the donor strain, especially if it is dominant (e.g. *str*ˢ, *spc*ˢ, *rif*ˢ). Antibiotic resistances can be introduced at alternative chromosomal locations if required, by inserting a transposon. Note also that strains containing an R plasmid with an Smʳ/Spʳ marker give rise, at a relatively high frequency (about 10^{-5}), to moderate to high level Smʳ mutants (Pearce and Meynell, 1968) so that such strains should be contraselected with a high level (1–2 mg ml⁻¹) of streptomycin, or with some other antibiotic.

B. The interrupted mating experiment

The following technique is one variant for carrying out an interrupted broth mating between an *E. coli* K12 Hfr strain and a multiauxotrophic recipient strain. It can be improved by taking more samples (every 15 s) at critical periods if an efficient interrupting device (Low and Wood, 1965) is available, or if mating is interrupted by treating the samples for 10 min with a concentrated T6 lysate (using a T6^s donor and T6^r recipient). It is also possible to carry out an interrupted plate mating by using Nal^s donor and Nal^r recipient strains, and adding nalidixic acid (which prevents further DNA transfer) at various intervals (Haas and Holloway, 1978).

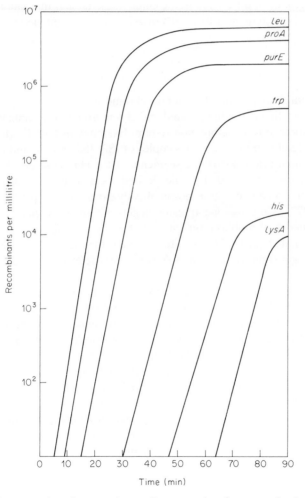

Fig. 5. The interrupted mating experiment. Data are taken from a mating between Hfr Hayes and a (multi)auxotrophic recipient strain.

Mix 1 ml of an exponential broth culture of the Hfr strain with 9 ml of a similar culture of the recipient strain in a 100-ml conical flask, and incubate in a water bath at 37°C without agitation. Leaving the flask in the bath, remove 0.5 ml samples to test-tubes containing 4.5 ml of phosphate buffer or saline at 0, 5, 10, 15, 20, 25, 30, 40, 50, 60, 75 and 90 min. Vortex for 1 min to disrupt mating pairs and plate appropriate dilutions (found by trial and error) on plates selective for each marker in turn. Count the colonies after incubation at 37°C for two days, and plot on semi-logarithmic graph paper (Fig. 5). The curves obtained demonstrate the differing times of entry (obtained by extrapolation to the abscissa) of different chromosomal markers, and that late markers are transferred less frequently than early ones because of spontaneous chromosome breakage.

C. Chromosome transfer by autonomous plasmids

Some conjugative plasmids transfer the chromosome because they carry regions of homology that allow the plasmid and chromosome to recombine, using the host's homologous recombination system. Examples include F, which, like the *E. coli* K12 chromosome, carries copies of IS2, IS3 and $\gamma\delta$, and F primes (or other plasmid primes) carrying segments of the chromosome. Alternatively, homology can be provided at the desired location by introducing into the chromosome a copy of a transposon already carried by the plasmid (Kleckner *et al.*, 1977). Chromosome transfer begins within the region of homology. The frequency of mobilization varies with the extent of homology being, for example, quite high for F primes (10^{-2}) but low for F itself (10^{-5}) and essentially zero if the insertion sequence homology region is deleted (Willetts and Johnson, 1981).

Other conjugative plasmids mobilize the chromosome because they carry a transposable DNA sequence, since a cointegrate between the two is formed as an intermediate in the transposition of this sequence to the chromosome (Chapter 9). Mobilization by R68.45 (Haas and Holloway, 1978) [which carries the highly transposable IS21 (Willetts *et al.*, 1981)] and by plasmids with an inserted Mu (Faelen and Toussaint, 1976; Denarié *et al.*, 1977) or mini-Mu (Toussaint *et al.*, 1981; Van Gijsegem and Toussaint, 1982) prophage is by this mechanism; these plasmids are of particular value since they have a broad host range and will mobilize the chromosomes of a variety of Gram-negative genera (Holloway, 1979). The frequency of chromosome transfer is relatively low (10^{-4}–10^{-8}) and is initiated at a large number of chromosome sites owing to the lack of specificity of the transposition process.

Finally, yet other plasmids mobilize the chromosome by mechanisms that have not been determined. These include the FP plasmids in *Pseudomonas* (Holloway, 1979) and *Streptococcus* plasmids (Franke *et al.*, 1978).

D. Chromosome transfer by integrated plasmids

To study chromosomal genetics by conjugation, it is clearly advantageous to construct an Hfr strain with a conjugative plasmid stably integrated into the chromosome, since transfer then occurs at a high frequency from a defined point. F-derived *E. coli* K12 Hfr strains can be isolated, and are relatively stable, because the bacterial homologous recombination system is inefficient in carrying out recombination between the short IS regions of homology. For the same reason, Hfr strains arise at low frequencies (10^{-6}) in F^+ cells, and the frequency may be even lower for other plasmids with perhaps even shorter regions of homology or yet more inefficient plasmid-chromosome recombination. However, it is possible to select for such integration events if the plasmid carries a suitable marker and its replication (or that of the chromosome, for *E. coli* K12 *dnaA* hosts) is temperature-sensitive; temperature-resistant survivors carrying the marker frequently have the plasmid integrated into the chromosome, allowing both molecules to replicate (Broda and Meacock, 1971; Tresguerres *et al.*, 1975; Harayama *et al.*, 1980).

An alternative method for constructing enterobacterial F-derived Hfr strains is to transpose Tn*2301*, a transposon construction in which the entire F transfer region (without the replication and incompatibility genes) has been cloned into an ampicillin-resistant transposon, from a plasmid to the bacterial chromosome (Johnson and Willetts, 1980). A related technique applicable to most Gram-negative bacteria is to transpose into the chromosome a Tn*5* derivative into which the *oriT* sequence of the broad host-range plasmid RP4 has been cloned; in the presence of RP4 to provide the transfer functions, chromosome transfer is initiated from this *oriT* sequence (Yakobson and Guiney, 1984; Simon, 1984).

E. Construction of plasmid primes

There are two general techniques available for plasmid prime construction, both of which depend upon preventing transferred chromosome segments from recombining into the recipient chromosome. One technique uses a recombination deficient (*RecA*) recipient (Low, 1973), whereas the other uses a heterospecific recipient in which the donor genes are expressed, but DNA homology is not sufficient for recombination (Van Gijsegem and Toussaint, 1982; Morgan, 1982). Any of the donor strains transferring the bacterial chromosome and described in Section V.C or V.D can be used, and the rare transconjugants that arise in such matings include a proportion that are plasmid prime partial diploids. These can be identified in further qualitative matings as described in Section III.E.

VI. Regulation of conjugation

Expression of some, but not all, plasmid conjugation systems is subject to control by fertility inhibition (*fin*) genes also present on the plasmid. For example, most IncF plasmids have two genes, *finO* and *finP*, the products of which act in concert to prevent transcription of *traJ*, a positive control gene necessary for transcription of all the other transfer genes (Finnegan and Willetts, 1973; Willetts, 1977; Cuozzo and Silverman, 1986). Consequently, established cells carrying wild-type IncF plasmids transfer these at about 0.1% of maximal frequency, and are resistant to pilus-specific phages (Meynell *et al.*, 1968). The *finO* products are largely interchangeable between IncF plasmids, whereas at least five noninterchangeable alleles of *finP* have been identified; the product of this gene may be an RNA molecule (Willetts and Maule, 1985; Finlay *et al.*, 1986). In contrast, the conjugation systems of IncN, IncP and IncW plasmids are expressed constitutively. Fertility inhibition can also be caused by a coexisting conjugative plasmid of quite different type: for example, the IncI plasmid R62 reduces F transfer by about 100-fold (Meynell, 1973; Gaffney *et al.*, 1983) and IncN plasmids reduce transfer of IncP plasmids by 10^4-fold (Brown, 1981; Winans and Walker, 1985b). The transfer frequency of a new plasmid should therefore be measured using cells containing only this plasmid, and if it is low, this might be due to the presence of *fin* genes, rather than to an inefficient conjugation system.

A. Recognition of Fi⁺(F) plasmids

F itself has an IS*3* element inserted into its *finO* gene (Cheah and Skurray, 1987), and therefore expresses its conjugation system constitutively. However, the *finO* product can be provided *in trans* by other IncF plasmids, reducing F transfer by 100- to 1000-fold and making the cells resistant to F-specific phages. Such plasmids are denoted Fi⁺(F) and this usually (but not always) indicates that they are themselves IncF plasmids compatible with F. The classification Fi⁻(F) has essentially no meaning since it includes all other plasmids.

To test for the Fi⁺(F) phenotype, the plasmid is transferred as described in Section III.E into cells carrying a convenient F prime such as F*lac*. The sensitivity of the resultant strain to F-specific phages (Section II.C) or the transfer frequency of the F prime (Section III.C) is then determined.

The Fi⁺ phenotype of one conjugative plasmid towards a second can be recognized in similar experiments.

B. Preparation of transiently derepressed cultures

In a manner analogous to zygotic induction of a lambdoid prophage or of the *lacZ* gene, cells that have just received a repressed conjugative plasmid lack the *fin* gene products, and after $\frac{1}{2}$ to 1 h delay (for expression of the transfer genes), will retransfer the plasmid at the derepressed frequency. If there is an excess of recipient cells in the culture, repeated retransfer ultimately results in the infectious spread of the plasmid through the population to give a transiently derepressed or high frequency of transfer (HFT) culture (Watanabe, 1963). Such cultures can used to identify pilus-specific phages by titre increase experiments (Section II.E). Repression is ultimately reestablished after about 6 h for IncF plasmids (Willetts, 1974).

Add 10^4 cells from an overnight broth culture of the donor strain and 4×10^5 cells from a similar culture of the plasmid-free recipient strain to 1 ml of fresh broth. Incubate overnight. Next day, dilute 1:20 with fresh broth and grow to exponential phase, to give the HFT culture.

C. Isolation of transfer-derepressed mutants

If transiently derepressed cultures can be prepared, it should also be possible to isolate transfer-derepressed mutants. Such mutants are useful for recognition of the pilus-specific phage group by spot tests (Section II.C) and for measurement of the full transfer potential of the plasmid. They are easily selected by carrying out sequential matings beginning with a mutagenized culture of a strain carrying the repressed plasmid.

Mix 0.5 ml of such a culture grown to exponential phase with 0.5 ml of a (say) Sm^r recipient strain and mate for 1 h in broth or on a membrane filter as appropriate (Section III). Select transconjugants in liquid by adding to 10 ml of broth containing the plasmid-selecting antibiotic plus streptomycin, and incubating overnight at 37°C. This also allows re-establishment of repression in transiently derepressed transconjugant cells. Dilute 1 ml of the culture to 10 ml with fresh broth, and grow to exponential phase. Mate as above with a Sm^r-Nal^r recipient strain, and this time plate out dilutions on selective medium to obtain single transconjugant colonies. Patch these, and screen them for mutants transferring at high frequency by replica-plate mating with an appropriate (Sp^r) recipient strain as described in Section III.E.

VII. Conjugation in Gram-positive organisms

The majority of conjugative plasmids have been found in Gram-negative bacteria, and previous sections have been concerned solely with these. However, conjugative plasmids have been identified in several Gram positive organisms, most notably in various species of *Streptococcus* (Clewell, 1981) and more recently in *Staphylococcus* (reviewed by Lyon and Skurray, 1987). Conjugation in the actinomycetes has also been described, and in a *Streptomyces coelicolor* strain has been shown to be due to a conjugative plasmid (Hopwood and Merrick, 1977). In addition, conjugative plasmids have been found in the anaerobic organisms *Clostridium perfringens* (Rood *et al.*, 1978) and *Bacteroides fragilis* (Welch *et al.*, 1979; Privitera *et al.*, 1979). It seems likely that conjugation systems in other genera will be discovered as investigations broaden.

Conjugative streptococcal plasmids differ from enterobacterial plasmids in that mating pair formation does not involve a pilus. In fact there are two basic types of such streptococcal plasmids: those that can transfer efficiently in broth matings, and those that transfer only in long term membrane filter matings (Clewell, 1981). The mating pair formation systems of the latter plasmids have not been identified, although these plasmids are of some interest because their host range includes *Staphylococcus aureus* as well as a variety of *Streptococcus* species. Plasmids transferring efficiently in broth do so because potential recipient cells excrete sex pheromones, heat stable polypeptides that induce the donor cells to become adherent. The donor cells synthesize a new protein which apparently covers the entire cell surface, and causes them to aggregate both with each other and with recipient cells to form mating aggregates (Ehrenfeld *et al.*, 1986). Since cell-free filtrates of recipient strains induce self-aggregation of donor cells, sex pheromones are sometimes referred to as clumping inducing agents (CIAs). Several pheromones with different plasmid-specificities have been identified, and recipient strains can produce several of these, although their levels or efficiencies in inducing mating by different plasmids vary from strain to strain. Synthesis or function of the corresponding pheromone (but not of any others) is inhibited in cells carrying a conjugative plasmid; this is due to the synthesis of a structurally related inhibitory polypeptide (Clewell *et al.*, 1987b). Conjugation in *Streptococcus* has been reviewed by Clewell *et al.* (1984; 1987a).

Streptococcus also harbours many nonconjugative plasmids, and as in Gram-negative bacteria, these can be mobilized by a coexisting conjugative plasmid; as yet, the mechanisms have not been investigated. One phenomenon peculiar to *Streptococcus* is that of conjugative transposons (Clewell, 1986). These are antibiotic-resistance determinants integrated into the bacterial chromosome, but apparently capable of accurately excising from it, transferring by conjugation to a recipient strain, and there reintegrating into the chromosome.

A. Broth matings in *S. faecalis*

Given the mechanism of mating pair formation described above, there are two ways to carry out matings between a plasmid-carrying donor strain and a recipient strain producing the appropriate CIA: by simply incubating the cultures together for an extended period or by pre-inducing the donor cells

with the CIA and mating for a much shorter period. The protocols given below
were taken from Dunny and Clewell (1975) and Dunny *et al.* (1979).

For the first, mix 0.05 ml of an overnight broth culture of the donor strain
with 0.5 ml of a similar culture of the recipient, and 4.5 ml fresh broth (Oxoid
No. 2). Agitate gently at 37°C for 2 h. Vortex to break up clumps, and plate
out dilutions on selective plates. A recipient strain producing the appropriate
CIA must be chosen, and can be made resistant to streptomycin, rifamycin or
fusidic acid to provide the necessary contraselective marker.

For the second, a cell-free CIA preparation must first be made. Grow CIA-
producing recipient cells to late exponential phase (5×10^8 cells per millilitre),
pellet the cells by centrifugation, and filter the supernatant through a Millipore
filter (0.22-μm pore size). Store the CIA filtrate at 4°C, preferably after
autoclaving. Grow donor and recipient cultures separately to late exponential
phase, and dilute the former 1:10 into 2 ml of a 1:1 mixture of CIA filtrate
and fresh broth. Dilute the recipient culture similarly, but into broth. Incubate
with shaking at 37°C for 45 min. Mix 0.2 ml of the donor culture with 1.8 ml of
the recipient culture, and shake at 37°C for a further 10 or 20 min. Vortex and
plate out as above.

B. Membrane filter matings in *S. faecalis*

A method similar to that for enterobacterial plasmids (Section III.A; Franke
et al., 1978) can be used for conjugative streptococcal plasmids unable to
transfer in broth matings.

VIII. Genetic analysis of conjugation

A full description of the genetic analysis of conjugation is beyond the scope of
this chapter, and the interested reader is referred to other publications
describing the conjugation systems of IncF (Fig. 6, Willetts and Wilkins, 1984;
Ippen-Ihler and Minkley, 1986; Willetts and Skurray, 1987; Silverman, 1987),
IncN (Winans and Walker, 1985a, b), IncP (Barth *et al.*, 1978; Stokes *et al.*,
1981), *Pseudomonas* IncP-10 (Carrigan and Krishnapillai, 1979; Moore and
Krishnapillai, 1982) and *Streptococcus* pAD1 (Ehrenfeld and Clewell, 1987)
plasmids. However, such analysis is essentially similar to the analysis of any
other complex biological function. Some of the important specialized techniques
required are briefly described below.

A. Isolation of *tra* point mutants

Tra⁻ point mutants are best isolated by screening colonies derived from a
mutagenized cell population for transfer ability. It is necessary to test about

72

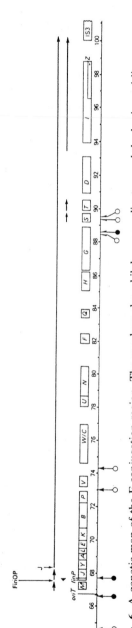

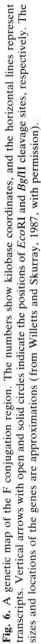

Fig. 6. A genetic map of the F conjugation region. The numbers show kilobase coordinates, and the horizontal lines represent transcripts. Vertical arrows with open and solid circles indicate the positions of *Eco*RI and *Bgl*II cleavage sites, respectively. The sizes and locations of the genes are approximations (from Willetts and Skurray, 1987, with permission).

10^5 colonies, given a mutation frequency of 10^{-3}, to obtain the hundred or so necessary mutants. Immediately after treatment with mutagens such as ethyl methane sulphonate or nitrosoguanidine, the culture is diluted with broth and divided into many small cultures for overnight incubation. Micro titre plates are convenient for this. Ultimately, only one mutant is chosen from each culture, to avoid siblings. Alternatively, isolated plasmid DNA or transducing phages can be treated directly with hydroxylamine; an advantage of this technique is that the plasmid transformants or transductants then obtained are not siblings (Carrigan et al., 1978; Stokes et al., 1981).

Overnight cultures are diluted and plated on nutrient plates to obtain 100–150 colonies per plate. Five to ten plates are prepared per culture. These colonies can then be directly replica plate mated with an appropriate recipient strain as described in Section III.E. Putative Tra⁻ clones are identified by comparing the master and replica plates, patched on a further nutrient plate and retested by replica plate mating before purification. It is worth carrying out these experiments in nonsuppressing hosts at 42°C, so that screening for suppressible and temperature-sensitive mutations can be carried out later.

It is possible to select pilus-specific phage-resistant mutants directly (Ohtsubo et al., 1970; Carrigan et al., 1978), but since phage-sensitive Tra⁻ mutants would be missed, it is preferable to isolate mutants by screening, and then test them for phage sensitivity.

Adaptations of the above techniques can be used to obtain transposon-induced tra mutations (Barth et al., 1978; Winans and Walker, 1985a).

Ultimately, the Tra⁻ plasmid mutants should be tested in a different host strain to ensure that the transfer deficiency is not due to a chromosomal mutation similar, for example, to those that lead to a reduction in F transfer (Beutin et al., 1981; Silverman, 1985).

B. Isolation of polar *tra* point mutants

These mutants are necessary for the identification of *tra* operons. They can be obtained by using standard techniques to insert Mu (Faelen and Toussaint, 1976), lambda (Dempsey and Willetts, 1976) or a transposon (Chapters 10 and 11) into the transfer genes creating *tra* mutations identified as above. Mu is probably the best insertion element, because of its lack of site specificity (compared to lambda) and because of the reliance that can be placed on the mutations being strongly polar (compared to transposons, where some insertions are non-polar or only weakly polar).

C. Isolation of *tra* deletion mutants

Deletion mutants can be constructed by first inserting Mu*cts* or λ*cts* into or nearby the transfer region and then selecting deletion mutants from the resultant lysogens as temperature-resistant survivors after growth at 42°C (Dempsey and Willetts, 1976).

Alternatively, deletions can be made by linearizing the plasmid molecule at a convenient (unique) restriction endonuclease site, and either transforming cells directly with the linear DNA (Thompson and Achtman, 1979), or first treating this with the progressive exonuclease BAL-31 (Lau and Gray, 1979) followed by blunt-end ligation (Chapter 9). Transfer regions often have relatively few cleavage sites for restriction endonucleases, but it may be necessary to delete sites elsewhere on the plasmid, or to clone the transfer region, before this technique can be used.

D. Cloning the transfer region

This plays a central role in facilitating the genetic analysis of conjugation, since such clones allow construction of stable heterozygotes for complementation tests, as well as aiding the genetic and physical mapping and sequencing of *tra* genes. The precise details of the vector and cloning enzyme depends upon the array of restriction endonuclease sites; if necessary, these can be supplemented by transposon insertion (Kleckner *et al.*, 1977). Cleavage sites should be chosen that allow the transfer region to be cloned separately from the incompatibility region, either as a whole or as a collection of segments.

If plasmid derivatives with λ*cts* inserted into or near to the transfer region are available, λ*tra* transducing phages can be obtained by induction. These provide a second method for cloning not dependent upon *in vitro* techniques (Willetts and McIntire, 1978).

E. Complementation analysis of *tra* mutants

If the plasmid transfers at a high frequency (>0.1) in short term broth matings and suppressible mutants are available, it is feasible to utilize a technique for complementation in transient heterozygous cells, where the transfer proficiency of the cell carrying two incompatible *tra* mutants is tested immediately after its construction (Achtman *et al.*, 1972). Alternatively, a second set of *tra* mutants of a compatible *in vitro* recombinant plasmid carrying all or part of the transfer region may be constructed, either by mutagenesis or homozygote formation, to allow strains carrying two plasmids with *tra* mutations to be tested for complementation.

Complementation is carried out, firstly, between point mutants to identify the *tra* genes, secondly, between representative point mutants and deletion mutants to map the order of the *tra* genes (and the *oriT* site) and thirdly, between the representative point mutants and polar insertion mutants to recognize any operons.

Acknowledgements

I am grateful to Lucien Caro for Fig. 3, to David Bradley for Fig. 4 and to Don Clewell and Alan Jacob for advice concerning streptococcal plasmid conjugation.

References

Achtman, M., Willetts, N. and Clark, A. J. (1972). *J. Bacteriol.* **110**, 831–842.
Alfaro, G. and Willetts, N. S. (1972). *Genet. Res.* **20**, 279–289.
Bachmann, B. J. and Low, K. B. (1980). *Microbiol. Rev.* **44**, 1–56.
Bachmann, B. J. (1983). *Microbiol. Rev.* **47**, 180–230.
Barbour, S. D. (1967). *J. Mol. Biol.* **28**, 373–376.
Barth, P. T., Grinter, N. J. and Bradley, D. E. (1978). *J. Bacteriol.* **133**, 43–52.
Beutin, L., Manning, P. A., Achtman, M. and Willetts, N. S. (1981). *J. Bacteriol.* **145**, 840–844.
Bradley, D. E. (1980). *Plasmid* **4**, 155–169.
Bradley, D. E. (1984). *J. Gen. Microbiol.* **130**, 1489–1502.
Bradley, D. E. (1981). *In* "Molecular Biology, Pathogenicity and Ecology of Bacterial Plasmids" (S. Levy, R. C. Clowes, and E. Koenig, Eds), pp. 217–226. Plenum, New York.
Bradley, D. E. (1983). *J. Gen. Microbiol.* **129**, 2545–2556.
Bradley, D. E., Taylor, D. E. and Cohen, D. R. (1980). *J. Bacteriol.* **143**, 1466–1470.
Bradley, D. E., Coetzee, J. N., Bothma, T. and Hedges, R. W. (1981). *J. Gen. Microbiol.* **126**, 389–396 *et seq.*
Bradley, D. E., Sirgel, F. A., Coetzee, J. M., Hedges, R. W. and Coetzee, W. F. (1982). *J. Gen. Microbiol.* **128**, 2485–2498.
Broda, P. (1979). *In* "Plasmids", Chapter 5. Freeman, San Francisco, California.
Broda, P. and Meacock, P. (1971). *Mol. Gen. Genet.* **113**, 166–173.
Brown, A. M. C. (1981). Ph.D. thesis, University of Edinburgh.
Carrigan, J. M. and Krishnapillai, V. (1979). *J. Bacteriol.* **140**, 809–816.
Carrigan, J. M., Helman, Z. M. and Krishnapillai, V. (1978). *J. Bacteriol.* **135**, 911–919.
Chan, P. T., Ohmori, H., Tomizawa, J. and Lebowitz, J. (1985). *J. Biol. Chem.* **260**, 8925–8935.
Cheah, K. C. and Skurray, R. (1987). *J. Gen. Microbiol.* in press.
Clewell, D. B. (1981). *Microbiol. Rev.* **45**, 409–436.
Clewell, D. B. (1986). *Ann. Rev. Microbiol.* **40**, 635–659.
Clewell, D. B., An, F. Y., Mori, M., Ike, Y. and Suzuki, A. (1987b). *Plasmid* **17**, 65–68.
Clewell, D. B., Ehrenfeld, E. E., An, F., Kessler, R. E., Wirth, R., Mori, M., Kitada, C., Fujino, M., Ike, Y. and Suzuki, A. (1987a). *In* "Streptococcal Genetics" in press. American Society for Microbiology, Washington, D.C.
Clewell, D. B., White, B. A., Ike, Y. and An, F. Y. (1984). *In* "Microbial Development" (R. Losick and L. Shapiro, Eds). pp. 133–149. Cold Spring Harbor, N.Y.
Coetzee, J. N., Bradley, D. E., Hedges, R. W., Tweehuizen, M. and Du Toit, L. (1987). *J. Gen. Microbiol.* **133**, 953–960.
Coetzee, J. N., Bradley, D. E., Hedges, R. W., Hughes, V. M., McConnell, M. M., Du Toit, L. and Tweehuysen, M. (1986). *J. Gen. Microbiol.* **132**, 2907–2917.

Coupland, G. M., Brown, A. M. C. and Willetts, N. S. (1987). *Mol. Gen. Genet.* **208**, 219–225.

Cuozzo, M. and Silverman, P. M. (1986). *J. Biol. Chem.* **261**, 5175–5179.

Davidson, N., Deonier, R. C., Hu, S. and Ohtsubo, E. (1975). *Microbiology – 1974*, 56–65.

Dempsey, W. B. and Willetts, N. S. (1976). *J. Bacteriol.* **126**, 166–176.

Denarié, J., Rosenberg, C., Bergeron, B., Boucher, C., Michel, M. and de Bertalmio, M. (1977). *In* "DNA Insertion Elements, Plasmids, and Episomes" (A. Bukhari, J. Shapiro and S. Adhya, Eds), pp. 507–520. Cold Spring Harbor, New York.

Derbyshire, K. M. and Willetts, N. S. (1987). *Mol. Gen. Genet.* **206**, 154–160.

Derbyshire, K. M., Hatfull, G. and Willetts, N. S. (1987). *Mol. Gen. Genet.* **206**, 161–168.

Dunny, G. and Clewell, D. B. (1975). *J. Bacteriol.* **124**, 784–790.

Dunny, G., Craig, R., Carron, R. and Clewell, D. B. (1979). *Plasmid* **2**, 454–465.

Ehrenfeld, E. E. and Clewell, D. B. (1987). *J. Bacteriol.* **169**, 3473–3481.

Ehrenfeld, E. E., Kessler, R. E. and Clewell, D. B. (1986). *J. Bacteriol.* **168**, 6–12.

Faelen, M. and Toussaint, A. (1976). *J. Mol. Biol.* **104**, 525–539.

Finlay, B. B., Frost, L. S., Paranchych, W. and Willetts, N. (1986). *J. Bacteriol.* **167**, 754–757.

Finnegan, D. J. and Willetts, N. S. (1973). *Mol. Gen. Genet.* **127**, 307–316.

Franke, A. E., Dunny, G. H., Brown, B. L., An, F., Oliver, D. R., Damle, S. P. and Clewell, D. B. (1978). *Microbiology – 1978*, 45–47.

Frost, L. S., Paranchych, W. and Willetts, N. S. (1984). *J. Bacteriol.* **160**, 395–401.

Frost, L. S., Finlay, B. B., Opgenorth, A., Paranchych, W. and Lee, J. S. (1985). *J. Bacteriol.* **164**, 1238–1247.

Gaffney, D., Skurray, R. and Willetts, N. (1983). *J. Mol. Biol.* **168**, 103–122.

Guiney, D. G. and Yakobson, E. (1983). *Proc. Natl. Acad. Sci. U.S.A.* **80**, 3595–3598.

Haas, D. and Holloway, B. W. (1978). *Mol. Gen. Genet.* **158**, 229–237.

Harayama, S., Tsuda, M. and Iino, T. (1980). *Mol. Gen. Genet.* **180**, 47–56.

Hayes, W. (1953). *J. Gen. Microbiol.* **8**, 72–88.

Holloway, B. W. (1979). *Plasmid* **2**, 1–19

Holloway, B. W. and Low, K. B. (1987). *In* "*E. coli* and *S. typhimurium*: Cellular and Molecular Biology" (F. C. Neidhardt, Ed.) pp. 1145–1153. American Society for Microbiology, Washington, D.C.

Holloway, B. W., Krishnapillai, V. and Morgan, A. F. (1979). *Microbiol. Rev.* **43**, 73–102.

Hopwood, D. A. and Merrick, M. J. (1977). *Bacteriol. Rev.* **41**, 595–635.

Ippen-Ihler, K. A. and Minkley, E. G. (1986). *Ann. Rev. Genet.* **20**, 593–624.

Johnson, D. A. and Willetts, N. S. (1980). *J. Bacteriol.* **143**, 1171–1178.

Kleckner, N., Roth, J. and Botstein, D. (1977). *J. Mol. Biol.* **116**, 125–159.

Lacey, R. W. (1975). *Bacteriol. Rev.* **39**, 1–32.

Lau, P. P. and Gray, H. B. (1979). *Nucleic Acids Res.* **6**, 331–357.

Low, B. and Wood, T. H. (1965). *Genet. Res.* **6**, 300–303.

Low, K. B. (1973). *Bacteriol. Rev.* **36**, 587–607.

Low, K. B. (1987). *In* "*E. coli* and *S. typhimurium*: Cellular and Molecular Biology" (F. C. Neidhardt, Ed.) pp. 1134–1137. American Society for Microbiology, Washington, D.C.

Lyon, B. R. and Skurray, R. (1987). *Microbiol. Rev.* **51**, 88–134.

Meynell, E. (1973). *J. Bacteriol.* **113**, 502–503.

Meynell, E. and Datta, N. (1966). *Genet. Res.* **7**, 134–140.

Meynell, E., Meynell, G. G. and Datta, N. (1968). *Bacteriol. Rev.* **32**, 55–83.

Monk, M. and Clowes, R. C. (1964). *J. Gen. Microbiol.* **36**, 365–384.

Moore, R. and Krishnapillai, V. (1982). *J. Bacteriol.* **149**, 276–293.

Morgan, T. (1982). *J. Bacteriol.* **149**, 654–661.

Ohtsubo, E., Nishimura, Y. and Hirota, Y. (1970). *Genetics* **64**, 173–188.

Pearce, L. E. and Meynell, E. (1968). *J. Gen. Microbiol.* **50**, 173–175.

Privitera, G., Sebald, M. and Fayolle, F. (1979). *Nature (London)* **278**, 657–659.

Rood, J. I., Scott, V. N. and Duncan, C. L. (1978). *Plasmid* **1**, 563–570.

Silverman, P. M. (1985). *Bioessays* **2**, 254–259.

Silverman, P. M. (1987). *In* "Bacterial Outer Membranes as Model Systems" (M. Inouye, Ed.), Wiley, New York.

Simon, R. (1984). *Mol. Gen. Genet.* **196**, 413–420.

Stokes, H. W., Moore, R. J. and Krishnapillai, V. (1981). *Plasmid* **5**, 202–212.

Thompson, R. and Achtman, M. (1979). *Mol. Gen. Genet.* **169**, 49–57.

Thompson, R., Taylor, L., Kelly, K., Everett, R. and Willetts, N. (1984). *EMBO J.* **3**, 1175–1180.

Toussaint, A., Faelen, M. and Resibois, A. (1981). *Gene* **14**, 115–119.

Tresguerres, E. F., Nandasa, H. G. and Pritchard, R. H. (1975). *J. Bacteriol.* **121**, 554–561.

Van Gijsegem, F. and Toussaint, A. (1982). *Plasmid* **7**, 30–44.

Watanabe, T. (1963). *J. Bacteriol.* **85**, 788–794.

Welch, R. A., Jones, K. R. and Macrina. F. L. (1979) *Plasmid* **2**, 261–268.

Willetts, N. S. (1974). *Mol. Gen. Genet.* **129**, 123–130.

Willetts, N. S. (1977). *J. Mol. Biol.* **112**, 141–148.

Willetts, N. S. (1972). *J. Bacteriol.* **112**, 773–778.

Willetts, N. S. and Johnson, D. (1981). *Mol. Gen. Genet.* **182**, 520–522.

Willetts, N. S. and Maule, J. (1974). *Genet. Res.* **24**, 81–89.

Willetts, N. S. and Maule, J. (1985). *Genet. Res., Camb.,* **47**, 1–11.

Willetts, N. S. and McIntire, S. (1978). *J. Mol. Biol.* **126**, 525–549.

Willetts, N. S. and Skurray, R. (1987). *In* "*E. coli* and *S. typhimurium*: Cellular and Molecular Biology" (F. C. Neidhardt, Ed.) pp. 1110–1133. American Society for Microbiology, Washington, D. C.

Willetts, N. S. and Wilkins, B. (1984). *Microbiol. Rev.* **48**, 24–41.

Willetts, N. S., Crowther, C. and Holloway, B. W. (1981). *Plasmid* **6**, 30–52.

Winans, S. C. and Walker, G. C. (1985a). *J. Bacteriol.* **161**, 402–410.

Winans, S. C. and Walker, G. C. (1985b). *J. Bacteriol.* **161**, 425–427.

Yakobson, E. A. and Guiney, D. G. (1984). *J. Bacteriol.* **160**, 451–453.

4

Bacterial Transformation with Plasmid DNA

J. R. SAUNDERS* AND VENETIA A. SAUNDERS†

*Department of Genetics and Microbiology, University of Liverpool, Liverpool, UK.
†School of Natural Sciences, Liverpool Polytechnic, UK.

I. General introduction

Transformation involves the uptake of exogenous DNA and its subsequent incorporation into the genome of the cell, which thereby acquires an altered genotype. The ability to introduce purified DNA into bacteria has been of vital importance in the development of microbial genetics. Much of our understanding of the mechanisms of transformation has come from studies on the uptake and incorporation of chromosomal DNA by *Bacillus subtilis*, *Haemophilus influenzae* and *Streptococcus pneumoniae* (Goodgal, 1982; Smith *et al.*, 1981; Stewart and Carlson, 1986; Venema, 1979). Interest in plasmids as genetic vectors has led to the development of a variety of methods for introducing plasmid DNA into different bacterial species. This chapter describes techniques available for transforming bacteria with plasmid DNA. Methods for introducing purified plasmid DNA into intact bacteria, bacterial protoplasts and spheroplasts are considered. We have avoided the term transfection (Benzinger, 1978) when referring to the uptake of plasmid DNA, preferring to reserve this term for instances involving uptake and infection by bacteriophage DNA.

Bacterial transformation systems can be divided into two categories, natural (physiological) and artificial (Low and Porter, 1978; Stewart and Carlson, 1986) (Tables I and II). In naturally transformable species the cells become competent (able to take up DNA into a DNase resistant form) at particular stages in the growth cycle or after shifts in nutritional status of cultures. In some cases, competence is dependent on diffusible competence factors but in others depends on the presence of components of the cell envelope such as membrane proteins, autolysins or pili. However, many bacterial species do not become naturally competent under normal culture conditions and must be rendered permeable to DNA artificially. This may involve treating whole cells with high concentrations of divalent cations, Tris or polyethylene glycol (PEG). Alternatively, all or part of the cell envelope exterior to the cytoplasmic membrane may be removed to form protoplasts or spheroplasts, which can then be induced to take up DNA by treatment with divalent cations or polyethylene glycol.

In general, bacteria that can be transformed by chromosomal DNA can also be transformed by plasmid DNA. There are, however, notable differences in the efficiency with which plasmid and chromosomal markers will transform competent cells, due largely to differential processing of the various topological forms of DNA. In naturally competent cells, a single plasmid molecule generally requires interaction with another replicon for successful establishment of the transforming plasmid. (Plasmid transformation refers herein to the utilization of plasmids as transforming DNA.) However, it should be appreciated that transformation in different bacterial species may involve various mechanisms with different requirements for the physiological and genetic state of the recipient and for the topological nature of the input DNA.

Transformation can be divided broadly into three stages: binding of DNA to the outside of the cell, transport of DNA across the cell envelope and establishment of the transforming DNA either as a replicon itself, or by recombination with a resident replicon. Transformation systems generally can be saturated by high concentrations of DNA. Hence, a characteristic dose response curve is observed when a fixed number of competent cells (or protoplasts) is exposed to increasing amounts of transforming DNA (Fig. 1). At non-saturating DNA concentrations, the slope of the dose response curve provides information about the number of DNA molecules required to produce a single transformant. Typically, for single or closely linked genetic markers, this slope = 1 and for two markers it is 2 (Trautner and Spatz, 1973; Weston et al., 1979). If the slope = 1 with a homogeneous population of plasmid DNA molecules, it may be concluded that a single molecule is sufficient to produce a transformant. If, on the other hand, the dose response is >1, then a number of plasmid molecules must react cooperatively to produce a single transformant (Saunders and Guild, 1981a; Weston *et al.*, 1981).

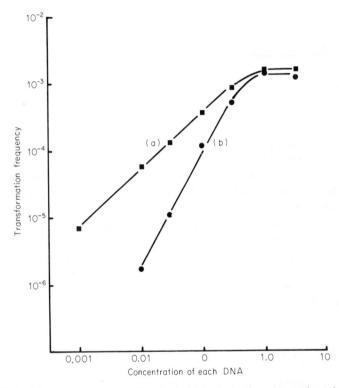

Fig. 1. Typical dose response curves for plasmid transformation. Approximately 5×10^8 *Escherichia coli* cells were transformed with various amounts of plasmid DNA according to the protocol described in Section IV.C. (a) Transformation by DNA of plasmid NTP1 (9.0 kb) encoding Ap[r]. The curve has a slope of 1 below saturation. (b) Simultaneous transformation by DNA of the compatible plasmids NTP1 and NTP11 (9.4 kb) which encodes Km[r]. Selection was for both plasmids (ampicillin + kanamycin) and the curve has a slope of 2 below saturation.

The frequency of plasmid transformation is expressed as transformants per viable cell. With saturating DNA concentrations, this is a measure of the total number of transformable bacteria in a population (Humphreys *et al.*, 1979).

Transformation efficiency is generally expressed as transformants per μg of DNA (Saunders *et al.*, 1987). In this case, measurements must be made at subsaturating concentrations of DNA, otherwise the yield of transformants per μg will be artificially depressed.

II. Mechanisms of transformation

A. Physiologically competent bacteria

1. Gram-positive systems

Streptococcus sanguis, S. pneumoniae, Streptococcus mutans and *B. subtilis* become competent under natural physiological conditions. In *Streptococcus*, competence is a transitory state developing in exponential cultures and is regulated by an extracellular protein of 10 000 daltons termed competence factor (CF). Competence is induced when an effective concentration of CF accumulates in the growth medium. There is a critical cell density for the development of competence which is dependent on medium composition, strain and other factors (McCarty, 1980). Alterations in the initial pH of the culture medium (in the range 6.8 to 8.0) markedly affects the population density for competence development. In media of lower initial pH a single wave of competence occurs, whilst in more alkaline media a series of competence cycles is observed, but with a higher proportion of competent cells in the initial cycle (Chen and Morrison, 1987). CF is involved in governing critical density for competence and its accumulation appears to be dependent on pH. Addition of exogenous CF can provoke competence in otherwise noncompetent strepto-cocci (Chen and Morrison, 1987; Yother *et al.*, 1986).

Development of competence in *Streptococcus* is accompanied by a shift in the pattern of protein synthesis resulting in the appearance of a number of new proteins (competence induced proteins (CIP)) and new properties, such as an efficient DNA processing system for uptake and recombination (Vijayakumar and Morrison, 1986). The half-life of the intracellular CIPs is short and their selective removal from the cells suggests the involvement of a specific protease. (Specific proteolysis has been reported for *Bacillus* spp. (Setlow, 1976, 1979).) Mutants defective in the control of competence (Com⁻), isolated by insertion-duplication mutagenesis (Morrison *et al.*, 1984), have been used to identify and clone the wild-type pneumococcal control region (*com*), a locus of 4.2 to 5.2 kb (Chandler and Morrison, 1987).

Competence of *B. subtilis* for transformation by plasmid or chromosomal DNA develops at the same growth phase in batch culture (Contente and

Dubnau, 1979a). Competence usually appears slowly during the transition from exponential to stationary phase. A competent culture of *B. subtilis* differentiates into two cell types that can be resolved on Renograffin density gradients: the lighter 'competent' fraction comprises 10 to 20% of the culture and consists of small, 'uninucleate' cells (a proportion of which are competent) whilst the heavier fraction contains larger 'multinucleate' cells of low competence (Singh and Pitale, 1967). Competence is associated with the appearance of new proteins and a reduction in the rate of RNA and DNA synthesis. Various competence-defective (Com$^-$) mutants have been described and competence-specific proteins identified (see, for example, Fani *et al.*, 1984; Finn *et al.*, 1985; Hahn *et al.*, 1987; Smith *et al* ., 1983a, 1985). Com$^-$ mutants of *B. subtilis*, obtained by transposon mutagenesis using Tn917*lacZ*, have been grouped into at least seven classes (Hahn *et al.*, 1987). These mutants are poorly transformed by both plasmid and chromosomal DNA. All are defective in DNA uptake and some fail to bind significant amounts of DNA. Expression of *com* genes, as judged by β-galactosidase activity (from the promoter-less *lacZ* element on Tn917*lacZ*) of the mutants, appeared to be dependent on nutritional signals, growth phase and genetic background (Albano *et al.*, 1987). Competence develops late in growth and is maximal in salts-glucose minimal media. Cells grown in complex media, such as LB broth, fail to exhibit competence, even upon addition of glucose. Most of the Com$^-$ mutants exhibited cell-type specific expression of β-galactosidase, which was 10-fold higher in those cells comprising the Renograffin light fraction than in those in the heavy fraction. In *Bacillus* spp., development of competence is apparently dependent on the *spoOH* (probably the structural gene for the sigma factor σ^{30}) and *spoOA* (involved in control of sporulation and in synthesis of products elaborated in stationary phase) gene products. Furthermore, the *com* genes are expressed in a temporal sequence preceding maximal competence of the culture, and control is probably transcriptional (Albano *et al.*, 1987). In addition, DNA-damage inducible (*din*) genes are activated during competence induction, implying some overlap between control of competence and the SOS system (Love *et al.*, 1985).

There is no DNA specificity for transformation in *Bacillus* and *Streptococcus* spp: these bacteria will take up DNA from their own species (homospecific) as well as that from unrelated species (heterospecific) (Smith *et al.*, 1981). Uptake of DNA involves the activity of membrane-bound nucleases that degrade one strand of double-stranded DNA with concomitant entry of the complementary strand. Models for the uptake of DNA explain the role of surface nucleases in transformation of *Bacillus* and streptococci (Lacks, 1977, 1979; Smith *et al.*, 1981, 1985).

A 75 000 dalton membrane-bound protein complex is involved in the binding and entry of donor DNA in *B. subtilis* (Smith *et al.*, 1983b, 1984, 1985). The complex contains two polypeptides, a and b, in approximately equal amounts.

Effective DNA binding involves both polypeptides in the complex, while nuclease activity is restricted to polypeptide b. *In vitro*, polypeptide b produces single and double-stranded breaks in plasmid DNA in the presence of Mg^{2+}, and total DNA degradation in the presence of Mn^{2+} ions. Polypeptide a affects b by reducing its nuclease activity (Smith *et al.*, 1984). A mode for DNA binding and entry (Smith *et al.*, 1985) invokes nicking of one strand of the double-stranded DNA at or close to the sites of binding. Nicking occurs in the presence of divalent cations and involves polypeptide b of the complex. The initial single-stranded nick is followed by a second break, opposite or nearly opposite the first. Acid soluble oligonucleotides are generated by the action of polypeptide b on one of the strands, while the complementary strand enters the cell.

DNA binding and uptake appear to be similar, especially in the initial stages, in *S. pneumoniae*. At least two proteins are involved in transformation of this species: a membrane-located polypeptide has been implicated in DNA binding and the major endonuclease of *S. pneumoniae* is required for DNA entry (Lacks, 1977; Rosenthal and Lacks, 1980). Single strand nicks are introduced on binding of donor DNA. The major endonuclease then acts opposite the initial nicks to introduce breaks, which then probably initiate DNA entry. The *B. subtilis* nuclease resembles the major endonuclease of *S. pneumoniae* in several respects; however, their specificities differ in that the *B. subtilis* nuclease is specific for double-stranded DNA, whereas the *S. pneumoniae* entry nuclease acts on single- and double-stranded DNA. Thus, a segment of single-stranded DNA entering *B. subtilis* will not be susceptible to polypeptide b nuclease activity, whilst the entering single-stranded DNA in streptococci may need protection against the entry nuclease.

Monomeric covalently closed circular (CCC), open circular (OC) or linear plasmid DNA are not capable of transforming naturally competent *B. subtilis* (Canosi *et al.*, 1978). In contrast, plasmid oligomers do transform with high specific activity (Mottes *et al.*, 1979). A single oligomeric plasmid DNA molecule is sufficient to produce a transformant (Contente and Dubnau, 1979a; de Vos *et al.*, 1981). Monomeric circular and linear plasmid DNA can, however, transform *B. subtilis* protoplasts (de Vos and Venema, 1981; Mottes *et al.*, 1979). Genetic markers on monomeric plasmid DNA species can also be rescued by a *recE*-dependent process in competent *B. subtilis* strains that already carry an homologous plasmid (Contente and Dubnau, 1979; Docherty *et al.*, 1981; Gryczan *et al.*, 1980; Weinrauch and Dubnau, 1983). Integration of donor markers appears to proceed by a breakage-reunion mechanism (Weinrauch and Dubnau, 1987). The frequency of marker rescue is dependent on the size of homologous sequences on the rescuing plasmid (Michel *et al.*, 1983). Monomeric CCC forms of plasmid DNA carrying inserts of *B. subtilis* chromosomal DNA are also active in transformation (Canosi *et al.*, 1981;

Duncan *et al.*, 1978; Iglesias and Trautner, 1983) as are monomers containing $\phi105$ DNA when transforming $\phi105$ lysogens (Bensi *et al.*, 1981). Homology between the chromosome and plasmid-borne chromosomal fragments can lead to the integration of such plasmids into the recipient chromosome (Young, 1980). Transformation by monomers of plasmid pBC1, which carries a fragment of the *B. subtilis* chromosome, was found by Canosi *et al.* (1981) to be *recE*-dependent. However, oligomers of the plasmid showed progressively less dependence on *recE* with increasing size. These findings suggest that plasmid DNA is processed extensively during and immediately after uptake. Accordingly, only single-stranded plasmid DNA is found in *B. subtilis* immediately after exposure to monomers, but fully and partially double-stranded DNA can also be found if oligomers are used (de Vos *et al.*, 1981). Reannealing of partially homologous single strands taken up from surface-bound oligomers probably accounts for plasmid transformation in *B. subtilis* (Canosi *et al.*, 1981; de Vos *et al.*, 1981).

In *S. sanguis* (Macrina *et al.*, 1981) and *S. pneumoniae* (Barany and Tomasz, 1980; Saunders and Guild, 1980, 1981a), CCC plasmid oligomers transform efficiently with single-hit kinetics. However, in contrast to *Bacillus*, CCC monomers are active in transformation of both *S. pneumoniae* (Saunders and Guild, 1981a) and *S. sanguis* (Macrina *et al.*, 1981) but with two-hit kinetics. Corresponding results have been obtained with open circular and linear forms of oligomers and monomers, but efficiencies are at least 35-fold lower than with CCC DNA (Saunders and Guild, 1981b). Plasmid molecules linearized at a unique site are inactive in transformation of *S. pneumoniae*, but mixtures of molecules linearized at different sites are active (Saunders and Guild, 1981b). Thus it seems likely that two monomeric plasmid molecules damaged during uptake can cooperate, presumably by reannealing of overlapping single strands, to produce a functional replicon in streptococci. Single-stranded circular plasmid DNA transforms pneumococci at 1% of the frequency obtained with double-stranded DNA (Barany and Boeke, 1983). Homology-dependent facilitation of plasmid transfer has been observed in *S. pneumoniae* (Lopez *et al.*, 1982; Stassi *et al.*, 1981). CCC monomer transfer was increased by more than 10-fold and some linear forms gave rise to intact molecules when the transforming plasmid DNA contained sequences ($>1\,\text{kb}$) homologous to the host chromosome. Entering plasmid fragments apparently interact with the chromosome to establish a plasmid replicon (Lopez *et al.*, 1982).

Plasmid transformation in streptococci shows similar low frequency ($10^{-3}-10^{-4}$ transformants per viable cell at saturation) to that in intact *B. subtilis* (Barany and Tomasz, 1980; Contente and Dubnau, 1979a). However, about 10–50 times more plasmid DNA is required to saturate competent *B. subtilis* than streptococci (Barany and Tomasz, 1980; Macrina *et al.*, 1981). Also, linearization of plasmid DNA causes a greater reduction in transformation frequency in *B. subtilis* (Contente and Dubnau, 1979b; Ehrlich, 1977) than in *S. pneumoniae* (Barany and Tomasz, 1980; Saunders and Guild, 1981b).

2. Gram-negative bacteria

Various Gram-negative bacteria are naturally transformable by plasmid DNA (Table I). In *H. influenzae* and *Haemophilus parainfluenzae*, competence is stimulated by transferring an exponentially growing culture to a medium that does not support growth. Once attained, competence is a stable state that can be maintained for some time. However, shifting competent cells to a rich medium effects rapid loss of transformability. No extracellular competence factor has been detected for these organisms (Smith *et al.*, 1981). Frequencies of transformation of *H. parainfluenzae* (Gromkova and Goodgal, 1977) and *H. influenzae* (Notani *et al.*, 1981) with plasmid DNA are generally lower (by several orders of magnitude) than those obtained with chromosomal DNA from the same organism. Piliated *H. parainfluenzae* cells are transformed by chromosomal or plasmid markers at frequencies that are 20 and 100 times greater, respectively, than non-piliated cells. Furthermore, plasmid (CCC or linearized), but not chromosomal transformation, can be further stimulated 50-fold by the addition of 10 mM $CaCl_2$ and/or 20 mM $MgCl_2$ (Gromkova and Goodgal, 1977; 1979). Such Ca^{2+}-stimulated transformation occurs with heterologous as well as homologous DNA. Ca^{2+} stimulation of *H. parainfluenzae* transformation differs from that observed in *E. coli* by occurring only in cells that are already competent for transformation with chromosomal DNA (Gromkova and Goodgal, 1981).

In *Neisseria gonorrhoeae*, cells are highly competent for transformation at all stages of growth (Sparling, 1977; Dougherty *et al.*, 1979). However, a practical problem is encountered since only piliated cells are competent and piliation is an unstable characteristic. Piliated cell populations can, however, be readily identified by microscopic examination of colonial morphology (Swanson, 1978). Although pili are associated with transformability of gonococci there is no evidence that they play a direct role in DNA uptake.

In most transformation systems, competent bacteria will take up heterospecific as well as homospecific DNA. However, *H. influenzae* (Scocca *et al.*, 1974), *Neisseria meningitidis* (Jyssum *et al.*, 1971) and *N. gonorrhoeae* (Dougherty *et al.*, 1979) can discriminate against heterospecific DNA and normally take up only DNA of the same or closely related species. DNA uptake by competent cells of *H. influenzae* requires the presence of a specific base sequence (uptake site) on double-stranded DNA. The uptake site necessary for transformation in *H. influenzae* has the sequence 5'-AAGTGCGGTCA-3' (Danner *et al.*, 1980; Pifer and Smith, 1985). The genomes of *H. influenzae*

TABLE I
Some plasmid transformation systems involving natural competence

Organism	References
Acinetobacter calcoaceticus	Cruze *et al.* (1979)
Agmenellum quadruplicatum PR6	Buzby *et al.* (1983)
(*Synechococcus* PCC7002)*	
Anacystis nidulans R2	Chauvat *et al.* (1983)
(*Synechococcus* PCC7942 R2)*	Gendel *et al.* (1983)
	Golden and Sherman (1984)
Azotobacter vinelandii	Doran *et al.* (1987)
	Glick *et al.* (1985)
Bacillus subtilis	Albano *et al.* (1987)
	Canosi *et al.* (1978)
	Ehrlich (1977)
	Gryczan *et al.* (1978)
Haemophilus influenzae	Albritton *et al.* (1981)
	Pifer (1986)
	Stuy and Walter (1986)
Haemophilus parainfluenzae	Gromkova and Goodgal (1981)
Neisseria gonorrhoeae	Biswas *et al.* (1986)
	Sox *et al.* (1979)
Pseudomonas stutzeri	Carlson *et al.* (1984)
Rhodobacter sphaeroides	Tucker and Pemberton (1980)
(*Rhodopseudomonas sphaeroides*)*	
Streptococcus mutans	Kuramitsu and Long (1982)
	Murchison *et al.* (1986)
Streptococcus pneumoniae	Barany and Tomasz (1980)
	Chen and Morrison (1987)
	Saunders and Guild (1980;
	1981a, b)
Streptococcus sanguis	LeBlanc and Hassell (1976)
	Behnke (1981)
	Macrina *et al.* (1981)

* Names in parentheses indicate former designations.

and *H. parainfluenzae* contain about 600 copies of this sequence, amounting to about one site per 4 kb of DNA (Goodgal, 1982; Sisco and Smith, 1979). This sequence also occurs in heterologous DNA, but only approximately once per 300 kb, thus substantially reducing the chances of successful transformation. The sequence is recognized by membranous structures on the cell surface called transformasomes, of which there is a limited number (10–12) on each competent *Haemophilus* cell (Smith *et al.*, 1981; Barany and Kahn, 1985). The recognition sequence may not be the sole determining factor for DNA uptake. There is limited uptake of pBR322 and ϕX174 DNA even though these molecules do not contain the 11 bp sequence (Goodgal, 1982). However, *H. influenzae* (Chung and Goodgal, 1979) or *H. parainfluenzae* chromosomal DNA (Sisco and Smith, 1979) cloned on pBR322 in *E. coli* is taken up efficiently. Uptake specificity can be circumvented by transforming *Haemophilus* cells that have been made artificially competent by $CaCl_2$ treatment (Elwell *et al.*, 1977; Gromkova and Goodgal, 1981).

Specificity for uptake in *Neisseria* spp. does not seem to be so stringent as in *Haemophilus* (Dougherty *et al.*, 1979) and a different uptake sequence is required (Burnstein *et al.*, 1988; Graves *et al.*, 1982). Norlander *et al.* (1979) have shown that the 39 kb gonococcal conjugative plasmid may confer low levels of transformability on otherwise nontransformable non-piliated cells, although the mechanism involved is unclear. β-lactamase plasmids isolated from *N. gonorrhoeae* transform piliated (P^+) but not non-piliated (P^-) cells of *N. gonorrhoeae*. However, transformation frequencies are about 1000-fold lower than those achieved with most chromosomal markers (Biswas *et al.*, 1985). Frequencies of transformation by native β-lactamase plasmids are low (approximately 1–5 transformants per μg of plasmid DNA) in *N. gonorrhoeae*, while recombinants formed naturally between these plasmids and the 4.2 kb (2.6-megadalton) cryptic plasmid of *N. gonorrhoeae* transform at 10^3- to 10^4-fold higher frequencies (Sparling *et al.*, 1980). This finding suggests that the 4.2 kb plasmid in these hybrids provides a gonococcal uptake sequence (analogous to that of *Haemophilus*) and/or sequence homology with the resident 4.2 kb plasmid in the recipient, thereby allowing recombinational rescue of transformation-damaged DNA molecules. (It should be noted that this cryptic plasmid is present in >95% of gonococcal strains.) Fragments of gonococcal plasmids that are preferentially taken up by *N. gonorrhoeae* contain the 10 bp sequence 5'-GATGCTCTGT-3' (Burnstein *et al.*, 1988). However this sequence is absent from a restriction fragment of phage M13 RF DNA that is also efficiently taken up. Furthermore, efficiency of uptake of derivatives of pBR322 containing the 10-mer gonococcal 'uptake' sequence is no greater than that of pBR322 itself. This suggests that a single short sequence is not sufficient, by itself, to direct DNA uptake and that other, as yet undefined, structural features of DNA are required in addition or as an alternative.

Transformation frequencies with *Haemophilus* plasmids are dramatically enhanced in *H. influenzae* Rd recipients that already carry homologous plasmid sequences either extrachromosomally or in a chromosomally integrated form (Balganesh and Setlow, 1985; Stuy, 1980; Stuy and Walter, 1986). These results reflect a requirement, as in *Neisseria*, for the recombinational rescue of plasmid molecules that become damaged during or just after uptake. Non-homologous plasmids can establish as independent replicons in naturally (or artificially) competent *H. influenzae*. Such plasmid establishment occurs at lower frequency than marker rescue. Plasmids that do become established appear to circumvent the degradation mechanism that normally accompanies translocation of donor DNA, and enter the cytoplasm as intact duplexes (Pifer, 1986).

In gonococci, about 25% of transformants obtained with β-lactamase plasmids are found to contain deleted plasmids (Sox *et al.*, 1979). The formation of deleted or otherwise damaged plasmids is a complicating factor when transforming these bacteria with natural or recombinant plasmids. Damage occurs because circular plasmid molecules are linearized by an apparently non-specific endonuclease (which does not seem to recognize plasmid DNA that is already linear) during or immediately after uptake (Biswas *et al.*, 1986). Incoming plasmid molecules are then subjected to exonucleolytic processing which may stimulate recombinational rescue to indigenous homologous plasmids. Individual plasmid molecules that have escaped processing may also recircularize perfectly on themselves *in vivo* to form wild-type plasmid replicons. Alternatively, processing of linear molecules may expose regions internal to the termini that contain short homologous sequences that can be matched in a recombinational recyclization reaction (Saunders *et al.*, 1986).

In addition to the well-characterized transformation systems of *Haemophilus* and *Neisseria*, natural plasmid transformation has been demonstrated in *Pseudomonas stutzeri*, where it is dependent upon a *rec* function (Carlson *et al.*, 1984; Ingraham and Carlson, 1985). Simple homology between chromosomal fragments cloned in the plasmid and the recipient genome was sufficient to facilitate transformation. Transformation frequency increased in proportion to the size of the cloned fragment over the range 0.2 to 17.4 kb (Ingraham and Carlson, 1985). *Azotobacter vinelandii* OP, induced to competence by growth in iron- and molybdenum-limited medium, can be transformed by DNA of the IncQ plasmid pKT210 (Doran *et al.*, 1987). CCC and OC plasmid DNA transformed *A. vinelandii* less efficiently than chromosomal DNA, while linearization of plasmid DNA and the presence of a resident DNA sequence homologous to the transforming DNA both stimulated transformation. This suggests the operation of recombinational rescue.

Certain photosynthetic prokaryotes can be transformed by circular DNA (Porter, 1986; Saunders, 1984). Among cyanobacteria, plasmid transformation has been reported in *Anacystis nidulans* R2 (*Synechococcus* PCC7942), *Agmenellum quadruplicatum* PR-6 (*Synechococcus* PCC7002) and *Synechocystis* PCC6803. Bifunctional shuttle vectors (generally based on an *E. coli* replicon and a cyanobacterial plasmid, for example pUH24 for *A. nidulans* R2 or pAQ1 for *A. quadruplicatum*)) have been used to transform *A. nidulans* (Gendel *et al.*, 1983; Kuhlemeier *et al.*, 1981) and *A. quadruplicatum* (Buzby *et al.*, 1983). Recombination with attendant deletion has been observed between transforming hybrid plasmids (such as pUC303, comprising pACYC184 and the cyanobacterial plasmid pUH24) and a resident plasmid (pUH24) in *A.*

nidulans R2, where the incoming and resident replicons carry common sequences (Kuhlemier *et al.*, 1983). Deletion derivatives are not generated upon transformation if the recipient is cured of the resident plasmid pUH24. However, transformation frequencies are about 30-fold lower in cured recipients (Chauvat *et al.*, 1983). The resident plasmid may thus serve to stabilize or rescue the incoming plasmid. Dose-response curves exhibit single-hit kinetics for recipients with and without pUH24 suggesting that there is no cooperativity between incoming plasmid molecules in this system. In *Synechocystis* PCC6803, stable integrative transformation by heterologous plasmid DNA can be enhanced by pretreating the cells with low doses of U.V. light (Dzelzkalns and Bogorad, 1986). The system does not require biphasic cloning vectors, and is independent of homologous recombination. A segment of an introduced plasmid appears to integrate randomly into the recipient genome. Such a system may prove useful for the isolation of mutants where the insertion lesion is tagged by a selectable marker.

The photosynthetic bacterium *Rhodobacter sphaeroides* (until recently *Rhodopseudomonas sphaeroides*) can be transformed with DNA of Rϕ6P, a temperate bacteriophage with a circular genome which encodes β-lactamase production (Tucker and Pemberton, 1980). In this case, competence is dependent on concomitant infection of recipients by the helper phage Rϕ9.

B. Artificial competence in intact cells

1. Calcium-induced competence

Mandel and Higa (1970) first showed that high concentrations of $CaCl_2$ permitted the transfection of intact *E. coli* cells by phage λ DNA. Subsequently, $CaCl_2$ treatment has been used for transfection of *E. coli* by other phages (Benzinger, 1978) and for transformation by chromosomal DNA (Cosloy and Oishi, 1973; Wackernagel, 1973). Cohen *et al.* (1972) first used such treatment to render *E. coli* transformable with plasmid DNA. Cells treated with $CaCl_2$ are exposed to the transforming DNA and subjected to cold shock. A heat pulse (42°C for 2 min) is then supplied. Bacteria are chilled and plated on selective medium, either directly, or following a period of expression at 37°C. Subsequently, this method has been modified and applied to other bacteria, most of which are Gram-negative (Table II). However, the Gram-positive *S. aureus* has been transformed by plasmid DNA in the presence of 100 mM $CaCl_2$ and a helper bacteriophage (Rudin *et al.*, 1974).

The minimum length of time required for full expression of transformed markers varies with strain, plasmid and marker. In particular, resistance to aminoglycoside antibiotics requires at least 1 h for complete expression (Cohen *et al.*, 1972). Variables that affect the transformation procedure, such as optimum times of exposure to Ca^{2+} and DNA and the duration of heat pulse, have been examined in different strains of *E. coli* (Bergmans *et al.*, 1981; Brown *et al.*, 1979; Cohen *et al.*, 1972; Hanahan, 1983; Humphreys *et al.*, 1979; Norgaard *et al.*, 1978).

Transformation of *E. coli* is dependent on the presence of divalent cations. The $CaCl_2$-heat shock procedure causes large (10–50%) losses in cell viability. Increasing the concentration of $CaCl_2$ generally increases the transformation frequency, but, at high concentrations, gains due to increased numbers of transformants are outweighed by losses in viability (Humphreys *et al.*, 1979; Weston *et al.*, 1981). Preferences for different divalent cations and their optimal concentrations vary between strains and species, presumably due to differences in the precise architecture of the cell envelope (Saunders *et al.*, 1984). Higher transformation frequencies have been obtained by using combinations of $CaCl_2$ (30 mM) and $MgCl_2$ (26 mM) in *E. coli* (Bergmans *et al.*, 1980). Prewashing bacteria with $MgCl_2$ before $CaCl_2$ treatment results in improved transformation frequencies in *Salmonella typhimurium* (Lederberg and Cohen, 1974) and *P. aeruginosa* (Sano and Kageyama, 1977; Sinclair and Morgan, 1978). $MgCl_2$ (150 mM) is considerably superior to $CaCl_2$ in promoting transformation of *P. aeruginosa* with RP1-DNA (Mercer and Loutit, 1979). Pretreatment of *Flavobacterium* sp. with protease, α-amylase and other enzymes permits transformation with plasmid DNA in the presence of 50 mM $CaCl_2$ presumably by removing extraneous surface components (Negoro *et al.*, 1980). Kushner (1978) obtained increased transformation frequencies with ColE1-derived plasmids by transforming *E. coli* that had been washed with RbCl prior to $CaCl_2$ treatment. Bagdasarian and Timmis (1981) also included 10 mM RbCl in the prewash and suspension fluids to obtain efficient transformation of *P. aeruginosa*. The addition of 10% (v/v) glycerol to 150 mM $CaCl_2$ wash solution, together with a 45°C heat pulse was found to give efficient transformation of *Pseudomonas syringae* (Gross and Vidaver, 1981).

A highly efficient method for transforming *E. coli* that involves incubating cells with Ca^{2+}, Mn^{2+}, Rb^+ (or K^+), dimethyl sulphoxide, dithiothreitol and hexamine cobalt (III) was devised by Hanahan (1983). This method works very well for most laboratory strains of *E. coli*, for example DH1, DH5 and HB101. The Hanahan procedure gives superior transformation efficiencies (up to 10^9 transformants per μg pBR322 DNA) to those achieved with simpler modifications of the original Cohen *et al.* (1972) method (10^7 or less transformants per μg pBR322 DNA). The proportion of DNA (added to transformation mixtures) that is actually taken up is greater with the more complex Hanahan treatment than with the simpler methods. However, the transformation frequency is not significantly different at saturating DNA concentrations using

TABLE II

Selected plasmid transformation systems involving artificial competence

Organism	Treatment	References
Aerobacter aerogenes	tsPG⁻ Spheroplasts	Suzuki and Szalay (1979)
Agrobacterium tumefaciens	Freeze-thaw	Holsters *et al.* (1978)
Alcaligenes eutrophus	Ca^{2+}, Mg^{2+}	J. R. Saunders
		unpublished observations
Amycolatopsis orientalis	Protoplasts/PEG	Matsushima *et al.* (1987)
(*Nocardia orientalis*)*		
Azotobacter vinelandii	Ca^{2+}	David *et al.* (1981)
Bacillus brevis	Tris, PEG	Takahashi *et al.* (1983)
Bacillus megaterium	Protoplasts, PEG	Brown and Carlton (1980)
		Vorobjeva *et al.* (1980)
Bacillus stearothermophilus	Protoplasts, PEG	Imanaka *et al.* (1982)
Bacillus subtilis	Protoplasts, PEG	Chang and Cohen (1979)
Bacillus thuringiensis	Tris, PEG, Sucrose	Heierson *et al.* (1987)
	Protoplasts, PEG	Fischer *et al.* (1984)
		Martin *et al.* (1981)
Bacteroides fragilis	PEG, Mg^{2+}	Smith (1985a)
Brevibacterium		
lactofermentum	Protoplasts, PEG	Santamaria *et al.* (1985)
		Smith *et al.* (1986)
Brevibacterium flavum	Protoplasts, PEG	Katsumata *et al.* (1984)
Citrobacter intermedius	Ca^{2+}	Prieto *et al.* (1979)
Clostridium acetobutylicum	Protoplast, PEG	Lin and Blaschek (1984)
Clostridium		
thermohydrosulfuricum	PEG	Soutschek-Bauer *et al.* (1985)
Corynebacterium glutamicum	Protoplasts, PEG	Katsumata *et al.* (1984)
		Ozaki *et al.* (1984)
Corynebacterium herculis	Protoplasts, PEG	Katsumata *et al.* (1984)
Corynebacterium lilium	Protoplasts, PEG	Flickering *et al.* (1985)
		Smith *et al.* (1986)
Erwinia carotovora	Ca^{2+}	Hinton *et al.* (1985)
Erwinia amylovora	Ca^{2+}, Rb^+, DMSO	Bauer and Beer (1983)
Erwinia herbicola	Ca^{2+}	Lacy and Sparks (1979)
Escherichia coli	Ca^{2+}	Brown *et al.* (1979)
		Cohen *et al.* (1972)
		Dagert and Ehrlich (1979)
	Ca^{2+}, Rb^+, DMSO, DTT,	Hanahan (1983)
	hexamine cobalt	
	Rb^+, Ca^{2+}	Kushner (1978)
	Freeze-thaw	Dityatkin and Ilyashenko (1979)
	Ca^{++}, freeze-thaw	Merrick *et al.* (1987)
	PEG, Mg^{2+}	Klebe *et al.* (1983)
	tsPG⁻ Spheroplasts	Suzuki and Szalay (1979)
Flavobacterium sp. K172	Protease/amylase, Ca^{2+}	Negoro *et al.* (1980)
Haemophilus influenzae	Ca^{2+}	Elwell *et al.* (1977)
		Stuy, 1979
Haemophilus parainfluenzae	tsPG⁻ Spheroplasts	Suzuki and Szalay (1979)
Klebsiella pneumoniae	Ca^{2+}, freeze-thaw	Merrick *et al.* (1987)
	tsPG⁻ Spheroplasts	Suzuki and Szalay (1979)

TABLE II—*continued*

Organism	Treatment	References
Lactobacillus acidophilus	Protoplasts, PEG	Morelli *et al.* (1987)
Lactobacillus reuteri	Protoplasts, PEG	Morelli *et al.* (1987)
Microbacterium ammoniaphilum	Protoplasts, PEG	Katsumata *et al.* (1984)
Micrococcus luteus	tsPG⁻ Spheroplasts	Suzuki and Szalay (1979)
Mycobacterium smegmatis	Spheroplasts, PEG	Jacobs *et al.* (1987)
Paracoccus denitrificans	Ca^{2+}, Mg^{2+}	Spence and Barr (1981)
Pseudomonas aeruginosa	Mg^{2+}, Ca^{2+}	Sano and Kageyama (1977)
		Sinclair and Morgan (1978)
	Mg^{2+}	Mercer and Loutit (1979)
	Rb^+, Ca^{2+}	Bagdasarian and Timmis (1981)
Pseudomonas phaseolicola	Ca^{2+}	Gantotti *et al.* (1979)
Pseudomonas putida	Ca^{2+}	Chakrabarty *et al.* (1975)
Pseudomonas savastanoi	Ca^{2+},	Comai and Kasuge (1980)
	Mg^{2+}	Comai and Kasuge (1980)
Pseudomonas syringae	$CaCl_2$, glycerol	Gross and Vidaver (1981)
Proteus mirabilis	Ca^{2+}, freeze-thaw	Merrick *et al.* (1987)
Proteus vulgaris	Ca^{2+},	Gnedoi *et al.* (1977)
Providencia stuartii	tsPG⁻ Spheroplasts	Suzuki and Szalay (1979)
Rhodobacter sphaeroides	Tris, Ca^{2+}, PEG	Fornari and Kaplan (1982)
Rhizobium meliloti	Ca^{2+}	Kiss and Kalman (1982)
	Freeze-thaw	Selvaraj and Iyer (1981)
Saccharopolyspora erythraeus (*Streptomyces erythraeus*)*	Protoplasts, PEG	Yamamoto *et al.* (1986)
Salmonella typhimurium	Mg^{2+}, Ca^{2+}	Lederberg and Cohen (1974)
	Ca^{2+}, freeze-thaw	Merrick *et al.* (1987)
	tsPG⁻ Spheroplasts	Suzuki and Szalay (1979)
Serratia marcescens	Ca^{2+}	Reid *et al.* (1982)
Staphylococcus aureus	Ca^{2+}, helper phage	Rudin *et al.* (1974)
Staphylococcus carnosus	Protoplasts, PEG	Thudt *et al.* (1985)
Streptococcus faecalis	Protoplasts, PEG	Smith (1985c) Wirth *et al.* (1986)
Streptococcus lactis	Protoplasts, PEG	Kondo and McKay (1982; 1984) Simon *et al.* (1986)
	PEG	Sanders and Nicholson (1987)
Streptomyces ambofaciens	Protoplasts, PEG	Matsushima and Baltz (1985)
Streptomyces coelicolor	Protoplasts, PEG	Bibb *et al.* (1978)
Streptomyces fradiae	Protoplasts, PEG	Matsushima and Baltz (1985)
Streptomyces lividans	Protoplasts, PEG	Thompson *et al.* (1982)
Streptomyces parvulus	Protoplasts, PEG	Bibb *et al.* (1978)
Streptomyces wadayamensis	Protoplasts, PEG	Acebal *et al.* (1986)
Streptomyces peucetius	Protoplasts, PEG	Lampel and Strohl (1986)
Thermomonospora fusca	Protoplasts, PEG	Pidcock *et al.* (1985)
Xanthomonas campestris	Ca^{2+}, Rb^+, PEG	Murooka *et al.* (1987)

Ca^{2+}, Mg^{2+}, Rb^+, calcium, magnesium, rubidium ions; PEG, polyethylene glycol; DMSO, dimethyl sulphoxide; DTT, dithiothreitol; Tris, Tris (hydroxymethyl) aminomethane; tsPG⁻, temperature sensitive peptidoglycan mutant.
* Names in parentheses indicate previous designations.

either of the two methods (England, Humphreys and Saunders, unpublished observations). This implies that the Hanahan method improves the permeability of the cell surface rather than increasing the number of potentially transformable cells. The amount of plasmid DNA required to saturate cells is less for this method than for the other procedures. The Hanahan procedure is therefore to be preferred where efficient use of input DNA is required, as for example, in transformation by ligation mixtures to form gene libraries in plasmid vectors. On the other hand, the conventional methods for preparing competent *E. coli* are certainly simpler to perform and less time consuming than the Hanahan technique.

Marginal improvements in transformation frequencies can be made by ensuring that the pH of the divalent cation solutions employed is within the optimum range (pH 6.0–7.5 for *E. coli* C600), buffering where necessary with 10 mM 3-(*N*-Morpholino) propane-sulphonic acid (MOPS), pH 6.5 (Bagdasarian and Timmis, 1981; Humphreys *et al.*, 1979) or 12 mM *N*-2-hydroxyethylpiperazine-*N'*-ethanesulphonic acid (HEPES) pH 6.0 (Bergmans *et al.*, 1981). Addition of glucose (0.5% w/v) to competent cells in 75 mM $CaCl_2$ improves frequencies two to four-fold in *E. coli* (Humphreys *et al.*, 1979). Addition of 10–20 mM Mg^{2+} to the growth medium improves transformation efficiency without affecting growth rate (Hanahan, 1983).

Kushner (1978) found that *E. coli* strain SK1590 was more transformable than other commonly used strains, such as C600. By contrast, Dagert and Ehrlich (1979) found only marginal differences in frequency between these strains. However, it is difficult to compare the efficiency of transformation for individual strains of *E. coli* unless all have been harvested under growth conditions permitting maximal competence (Brown *et al.*, 1979). The growth temperature may affect transformation efficiencies markedly. For example, *P. aeruginosa* grown at 15 to 30°C is transformed more efficiently than if grown at higher temperatures (Berry and Kropinski, 1986). Somewhat higher transformation frequencies are obtained when *E. coli* C600 is grown at 30°C rather than 25°C or 37°C (Saunders *et al.*, 1987). The growth phase of recipient cultures also has a dramatic effect on transformation frequencies of $CaCl_2$-treated *E. coli*. The transformation frequency varies over a 100-fold range during batch growth at 37°C (Brown *et al.*, 1979; Hanahan, 1983). Transformation frequencies (1–7% of cells transformed with pBR322 DNA giving 10^6–10^7 transformants per μg) rise to a maximum in early exponential phase ($OD_{660} = 0.15$ to 0.2) and decline thereafter for *E. coli* treated by the conventional procedure (Brown *et al.*, 1979; Humphreys *et al.*, 1981; Fig. 2). The larger the dilution of inoculum used, the greater the magnitude and duration of maximal competence obtained (Saunders *et al.*, 1987). There is no evidence for either a diffusible competence factor or an inhibitor of transformation in *E. coli* (Brown *et al.*, 1979). However, maximum competence is associated with a time in a batch culture when the mean cell volume of the culture is at its greatest (Saunders *et al.*, 1987) and the cells are most susceptible to killing by $CaCl_2$ treatment (Brown *et al.*, 1979). Repeated two-fold dilution of cultures at the time of maximal transformability leads to maintenance of both transformability and mean cell volume (Brown *et al.*, 1979). In general, maximal growth

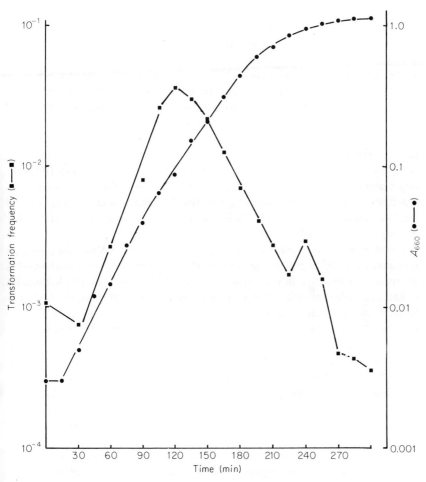

Fig. 2. 1 ml of an overnight static culture of *E. coli* C600 grown at 37°C in nutrient broth was inoculated into 500 ml of prewarmed nutrient broth and incubation continued with vigorous aeration at 37°C. At the times indicated, samples (4–80 ml as appropriate) were removed, harvested by centrifugation, washed in 10 mM $CaCl_2$ and concentrated to a constant cell density of about 10^9 bacteria per millilitre in 75 mM $CaCl_2$. 5×10^8 cells were then mixed with saturating amounts (3 μg) of pBR322 DNA and transformed as described in the text ●, A_{660}; ■, transformants per viable cell.

rates in batch culture provide maximum numbers of competent cells (Hanahan, 1983; Saunders *et al.*, 1987). *E. coli* exhibits maximum transformability (equivalent to the highest achieved in batch culture) in carbon-, nitrogen- or phosphate-limited chemostat cultures growing at maximum dilution rate (Jones *et al.*, 1981). There is a direct relationship between modal cell volume (which is generally maximal at higher growth rates) and maximum transformability although the reasons for this are unclear (Humphreys *et al.*, 1981; Saunders *et al.*, 1987). It is obvious however, that larger cells have increased surface area for binding DNA and may have stressed cell envelopes.

Dagert and Ehrlich (1979) have shown that prolonged incubation in $CaCl_2$ at 0°C improves the competence of *E. coli* by as much as ten-fold but their results have been difficult to reproduce for all strains. Cells incubated overnight at 0°C in Ca^{2+} adsorb ten times more radiolabelled plasmid DNA in a given time than those incubated at 0°C for only 20 min. (Sabelnikov and Domaradsky, 1981). It is possible to maintain competent $CaCl_2$-treated *E. coli* (and other transformable species) by rapidly freezing the cells in $CaCl_2$ containing 15% (v/v) glycerol and storing them at −70°C (Morrison, 1977). Transformability and viability are retained for a year or more and it is therefore possible to maintain a constant population of competent bacteria that can be thawed and used as desired. For very high efficiencies there is, however, no substitute for making competent cells freshly as required.

Covalently closed circular and open circular plasmid DNA species are taken up equally well by *E. coli* (Cohen *et al.*, 1972), but variations in transformation ability occur between different preparations of the same plasmid DNA. Double-stranded circular DNA is taken up by Ca^{2+} treated *E. coli* without conversion to the single-stranded form (Strike *et al.*, 1979). There appears to be no specificity for homologous DNA uptake (Brown *et al.*, 1981; Hanahan, 1983). Furthermore, monomeric plasmid DNA molecules are active in transformation with single-hit kinetics (Mottes *et al.*, 1979). DNA is taken up in the form in which it is presented, whether as single-stranded circular, CCC, OC or double-stranded linear (Cohen *et al.*, 1972; Hanahan, 1983; Strike *et al.*, 1979). Transformation of wild-type *E. coli* with linearized plasmid DNA occurs with frequencies that are about 100-fold lower than with CCC or OC DNA (Bergmans *et al.*, 1980; Cohen *et al.*, 1972; Conley and Saunders, 1984). Much of the damage to linear DNA is attributable to the *recBC* product (exonuclease V) which normally severely inhibits transformation of *E. coli* with chromosomal DNA (Cosloy and Oishi, 1973; Hoekstra *et al.*, 1980; Wackernagel, 1973). However, it is possible to transform *recB*, *recC*, *sbcB* (or *sbcA*) strains of *E. coli* with linear chromosomal fragments at high frequency due to the absence of exonuclease V and the presence of a functional *recF* pathway (Cosloy and Oishi, 1973), but note that ColEl and its derivatives are stable in *recBC*, *sbcA* strains but unstable in *recBC*, *sbcB* background (Vapnek *et al.*, 1976).

At best, using a small plasmid such as pBR322, only about 5% of bacteria in an *E. coli* population can be transformed (Humphreys *et al.*, 1981; Kushner, 1978). Transformation frequencies with larger plasmids (>15 kb) are dramatically reduced (Hanahan, 1983; Humphreys *et al.*, 1979), due to damage incurred during uptake and establishment. Dose-response experiments suggest that a single (monomeric) plasmid DNA molecule is sufficient to produce one transformant (Hanahan, 1983; Mottes *et al.*, 1979; Weston *et al.*, 1979, 1981). However, considerably more plasmid DNA molecules bind tightly to the outside of the competent cells than enter, as judged by resistance of transforming DNA molecules to DNase and by the number of transformants obtained (Sabelnikov and Domaradsky, 1981; Weston *et al.*, 1981). Even under optimum conditions, only about 10^{-2}–10^{-4} of the plasmid DNA molecules in a transformation mixture will give rise to transformants (Hanahan, 1983). Experiments using pairs of plasmid DNA species to transform *E. coli* have suggested that competent cells may take up more than one plasmid molecule. Using the conventional procedure, the capacity of the competent subfraction to be transformed by more than one plasmid molecule is limited; >80% of competent cells can take up and establish only one molecule, <20% two and <1% three or more (Weston *et al.*, 1979, 1981). These results presumably reflect the low efficiency with which externally bound plasmid DNA molecules are transported across the cell envelope to the cytoplasm, where they become established. It is significant, therefore, that cells treated by the Hanahan (1983) method apparently have a greater capacity for being multiply transformed. The possibility of multiple transformation events, especially at saturating or just subsaturating DNA concentrations, should be appreciated, particularly when constructing gene libraries where any individual cell may become transformed by two or more different plasmid recombinants.

Divalent cations may be required for a number of processes that act either alone or in concert to mediate transformation. Such ions are necessary to promote DNA binding (presumably by neutralization of negative charges) to cell membranes as a prelude to DNA transport (Weston *et al.*, 1981). Ca^{2+} also causes dramatic rearrangements of the lipopolysaccharide and protein of the outer membrane of Gram-negative bacteria (van Alphen *et al.*, 1978), which might allow the binding and/or inward transport of DNA molecules. In this respect it is interesting that lipopolysaccharide-defective (rough) mutants of *S. typhimurium* (MacLachlan and Sanderson, 1985) and *P. aeruginosa* (Berry and Kropinski, 1986) exhibit differences in efficiency of transformation using the $CaCl_2$-technique compared with smooth, wild-type strains. In *S. typhimurium*, *galE* or *rfaF* mutants give 8 to 630-fold higher efficiencies of transformation than smooth strains. However, other rough mutations, e.g. *rfaC* and *rfaE*, confer decreased transformability with respect to the wild-type (MacLachlan and Sanderson, 1985). Treatment of isolated outer membranes with trypsin dramatically reduces Ca^{2+}-promoted binding of plasmid DNA and proteases reduce transformation frequencies in intact Ca^{2+}-treated *E. coli* (Saunders *et al.*, 1984; Weston *et al.*, 1981). This suggests that binding of DNA to protein exposed by Ca^{2+} may be a prerequisite for DNA transport to the cytoplasm. The combination of low temperature and Ca^{2+} also causes freezing of membrane lipids (Sabelnikov and Domaradsky, 1979). Subsequent heating of cells during the heat pulse would cause unfreezing of the lipids and possible ingestion of DNA molecules that might have bound to the outside of the inner membrane. Ca^{2+}-DNA complexes may also be transported down a Ca^{2+} gradient extending across the inner membrane to the interior of the cell (Grinius, 1980). Santos and Kaback (1981) concluded that the proton motive force and, in particular, the membrane potential, plays a critical role in Ca^{2+}-mediated transformation of *E. coli* by plasmid DNA. Conversely, Sabelnikov and Domaradsky (1979, 1981) proposed that the proton motive force does not play any significant role in uptake of the plasmid DNA. Interestingly, in *B. subtilis* the proton motive force and especially the pH component apparently provides a driving force for DNA uptake (van Nieuwenhoven *et al.*, 1982). Plasmid DNA responsible for the production of transformants does not enter *E. coli* until the heat pulse (Weston *et al.*, 1981). Interestingly, a heat pulse carried out in the absence of DNA renders cells transformable by chromosomal DNA, suggesting that induction of competence is separable from DNA uptake *per se* (Bergmans *et al.*, 1981).

2. *Polyethylene glycol (PEG) treatment of intact cells*

PEG has been used to facilitate the uptake of transforming DNA by intact Gram-positive and Gram-negative bacteria (see also Section IIB.3 and Table II). A procedure that employs PEG has been developed for transformation of intact *E. coli* (Klebe *et al.*, 1983). The efficiency of the process (10^6 to 10^7 transformants per μg DNA) is comparable with that obtained using the standard $CaCl_2$-treatment procedure (Cohen *et al.*, 1972) as modified by Dagert and Ehrlich (1979). However, the PEG method differs from the $CaCl_2$ method in several respects. PEG-induced transformation occurs optimally at 22°C and is severely inhibited at 6°C, whereas the $CaCl_2$ method requires treatment of *E. coli* at 0 to 4°C for 30 min or more. Lower concentrations (below 30 mM) of divalent cations are required for the PEG method and Mg^{2+} is most effective. In addition, the pH optimum for the PEG procedure is markedly more acidic (pH 4.4) than for the $CaCl_2$ method (see Section IIB.1).

PEG-induced transformation of *Bacteroides fragilis* has also been achieved (Smith, 1985a) by using a modification of the method of Klebe *et al.* (1983). Transformation is optimal with a mid-exponential phase culture of *B. fragilis* grown in brain heart infusion broth supplemented with 20 mM $MgCl_2$ at a cell density of 8×10^8 cells ml^{-1}, concentrated 10-fold in the reaction mixture. PEG 1000 and PEG 6000 were found to be almost equivalent in the standard transformation system, but PEG 6000 was superior for the transformation of frozen cells. No transformants were obtained when transformation mixtures were incubated at 0°C and there was a reduction of >60% in the number of transformants when incubation was above 30°C or below 15°C. Transformation efficiency for *B. fragilis* is dependent upon the source of transforming DNA. Using pBFTM10 from *B. fragilis* the efficiency was about 4.2×10^3 transformants μg DNA^{-1} (Smith, 1985a). However, more than 10^5 transformants per μg DNA could be obtained using hybrid *E. coli/B. fragilis* plasmids that are smaller than pBFTM10. When the DNA of such hybrids was purified from *E. coli* rather than from *B. fragilis* efficiencies were 1000-fold lower, presumably due to restriction (Salyers *et al.*, 1987; Smith, 1985b).

3. Tris-induced competence

Exposure of intact cells of the photosynthetic bacterium *R. sphaeroides* to high concentrations of Tris (500 mM) has provided a means of inducing competence (Fornari and Kaplan, 1982). Transformation frequencies obtained with Tris-treated cells were dependent on the accompanying concentrations of $CaCl_2$ and PEG 6000 used in the transformation mixture. The mechanism(s) involved in this nonphysiological DNA uptake in *R. sphaeroides* is complex. Tris is known to disrupt the outer membrane of Gram-negative bacteria such as *E. coli* and *Salmonella* spp. (Irvin *et al.*, 1981; Voss, 1967). However, Tris-induced effects on the outer membrane of *E. coli* are apparently different from the physical changes produced by washing cells in 50 mM $CaCl_2$ (Sabelnikov and Ilyashenko, 1979).

Tris has also been used to induce competence in intact cells of *Bacillus brevis* (Takahashi *et al.*, 1983) and *Bacillus thuringiensis* (Heierson *et al.*, 1987). 50 mM Tris-Cl buffer of alkaline pH (containing 30% (w/v) sucrose in the case of *B. thuringiensis*) was found to be effective in the presence of PEG. Addition of $MgCl_2$ or $CaCl_2$ proved inhibitory for *B. brevis* and reduced the frequency of transformation for *B. thuringiensis*. Unlike *B. subtilis*, where natural competence is induced in late exponential phase, there is a broad optimum between early and late exponential phase for Tris-induced competence in *B. thuringiensis*. However, the highest transformation frequencies were achieved with late exponential phase cultures of low culture density during DNA uptake (Heierson *et al.*, 1987). Growth of *B. thuringiensis* in a minimal medium containing casamino acids gave 35 times more transformants than growth in rich medium, such as brain heart infusion or LB medium. Addition of glucose to minimal medium reduced transformation, possibly by repressing the synthesis of some component(s) required for competence. CCC DNA was more efficient than linear DNA in transforming Tris-treated *B. thuringiensis*, and oligomeric forms gave decreased transformation frequencies compared with CCC monomeric DNA. This contrasts with natural transformation of *B. subtilis* where CCC monomers are inefficient (see Section IIA.1).

Transformation of Tris-sucrose-treated vegetative cells is applicable to a number of *B. thuringiensis* subspecies including *gelechiae, kurstaki, galleriae, thuringiensis* and *israelensis*. The procedure is relatively rapid and can give frequencies of up to 10^{-3} transformants per viable cell or 10^4 transformants per μg DNA (Heierson *et al.*, 1987). This compares with 10 transformants per μg DNA obtained with protoplasts of a number of *B. thuringiensis* subspecies (Fischer *et al.*, 1984).

4. Transformation with liposome-entrapped plasmid DNA

It is possible to entrap biologically active plasmid DNA in liposomes (Dimitriadis, 1979; Fraley *et al.*, 1979). Liposome-entrapped pBR322 DNA transforms Ca^{2+}-treated *E. coli* at frequencies that are about 1% of those achieved with free pBR322 DNA (Fraley *et al.*, 1979). Makins and Holt (1981) demonstrated high frequencies of transformation of *Streptomyces* protoplasts by encasing chromosomal DNA in liposomes. However, such transformation systems have not proved to be generally useful in either *E. coli* or *Streptomyces*. Nevertheless, the use of liposome-entrapped-DNA may be applicable to bacterial species that are normally not transformable due to a lack of competence or to the production of extracellular nucleases. DNA-free liposomes have, however, found application in stimulating transfection and transformation of *Streptomyces* protoplasts (Rodicio and Chater, 1982).

5. Transformation of frozen and thawed bacteria

Dityatkin and Iliyashenko (1979) showed that *E. coli* HB101 could be transformed in the absence of added Ca^{2+} by freezing cells with pMB9 DNA at $-70°C$ or $-196°C$ and then thawing at $42°C$. Optimal yields of transformants were obtained with $1-5 \times 10^9$ cells ml^{-1} and $0.05-0.5$ μg ml^{-1} plasmid DNA in reaction mixtures containing $0.5-1\%$ (w/v) bactopeptone at pH 7.5. The freeze-thaw procedure presumably produces sufficient stress in the cell envelope to allow penetration of DNA. In *E. coli*, transformation frequencies are substantially lower than with $CaCl_2$-heat shock techniques, but freeze-thaw has in the past proved useful in organisms such as *Agrobacterium tumefaciens*

(Holsters *et al.*, 1978) and *Rhizobium meliloti* (Selvaraj and Iyer, 1981) when alternative procedures were not available. A freeze-thaw cycle carried out in the presence of 50 mM $CaCl_2$ proved vastly superior to the Cohen *et al.* (1972) procedure for transforming *Klebsiella pneumoniae* (Merrick *et al.*, 1987). Transformation of *E. coli* and *S. typhimurium* can also be achieved using this freeze-thaw procedure but somewhat less efficiently than with standard $CaCl_2$ competence induction regimes.

6. Transformation of spheroplasts

Spheroplasts of *E. coli* and other Gram-negative bacteria can be readily transfected in the presence of low concentrations of divalent cations (Benzinger, 1978). Unfortunately, spheroplasts generated by lysozyme treatment do not regenerate and cannot be used as recipients for transformation by chromosomal or plasmid DNA. Suzuki and Szalay (1979) isolated mutants with temperature-sensitive defects in peptidoglycan biosynthesis that produce spheroplasts that will regenerate. Spheroplast formation is controlled by shifting the temperature of cultures maintained in 0.25 M sucrose. Such spheroplasts of *E. coli* can be transformed with efficiencies that are similar to those obtained with Ca^{2+} treatment. The method is also applicable to the transformation of *Aerobacter aerogenes*, *H. parainfluenzae*, *K. pneumoniae*, *Micrococcus luteus*, *Providencia stuartii* and *S. typhimurium*.

An efficient PEG-mediated spheroplast transfection method that may be applicable with plasmids has been developed for the archebacterium *Halobacterium halobium* (Cline and Doolittle, 1987). Spheroplasts were obtained by treating cells with Tris, NaCl, KCl, sucrose and EDTA. The efficiency of transfection is 5×10^6 to 2×10^7 transfectants per μg of linear phage DNA.

7. Transformation of protoplasts

The ability of PEG to promote the uptake of plasmid DNA into bacterial protoplasts was first described in *Streptomyces coelicolor* and *Streptomyces parvulus* (Bibb *et al.*, 1978; Hopwood, 1981). Protoplast transformation has been achieved for several *Streptomyces* species, see, for example, Acebal *et al.* (1986), Lampel and Strohl (1986), Matsushima and Baltz (1985), Thompson *et al.* (1982) and for other actinomycete genera, including *Saccharopolyspora* (Yamamoto *et al.*, 1986), *Thermomonospora* (Pidcock *et al.*, 1985) and *Amycolatopsis* (Matsushima *et al.*, 1987). Chang and Cohen (1979) demonstrated PEG-protoplast transformation of *B. subtilis* by a procedure that was later modified to introduce plasmid DNA into a number of Gram-positive bacteria including, *Bacillus megaterium* (Brown and Carlton, 1980; Vorobjeva *et al.*, 1980), *Bacillus stearothermophilus* (Imanaka *et al.*, 1982), *B. thuringiensis* (Fischer *et al.*, 1984; Loprasert *et al.*, 1986; Martin *et al.*, 1981), *Lactobacillus* spp. (Morelli *et al.*, 1987) and streptococci (Smith, 1985c; Wirth *et al.*, 1986).

Protoplasts are normally prepared by enzymic degradation of the cell wall in an osmotically buffered medium. They are then exposed to DNA in the presence of PEG. (The optimum molecular weight and concentration of PEG varies between systems.) Filter sterilizing the PEG (rather than autoclaving) improves protoplast survival in some systems (Wirth *et al.*, 1986). PEG is then removed by washing and protoplasts are incubated to allow expression of transformed markers and plated on a complex medium that favours cell wall regeneration. Different osmotic stabilizers may be required for formation and regeneration of protoplasts from different strains (Hooley and Wellington, 1985; Mirdamadi-Tehrani *et al.*, 1986). The choice of stabilizer may be dependent upon the ability of the organism to metabolize the carbon source concerned. Furthermore, some stabilizers may affect the expression of certain resistance determinants, particularly those for aminoglycosides (see, Saunders *et al.*, 1984). Various factors influence the transformation of *Streptomyces* protoplasts, notably the growth phase of the mycelium at the time of protoplasting, the temperature at which regenerating protoplasts are incubated and the composition and dryness of the regeneration medium. It is necessary to wash protoplasts immediately prior to addition of transforming DNA, and PEG should be added directly after mixing DNA and protoplasts. Presumably such procedures remove or inhibit any nucleases that may have been liberated. RNase, used in various plasmid isolation protocols, reduces the frequency of protoplast regeneration and transformation. Transformants may be detected by pock[1] formation, when plasmid-containing regenerants develop in a plasmid-free background, or by expression of antibiotic resistance or other phenotypic markers. Antibiotic resistance can be detected (after sporulation) by replication of regenerated protoplasts on selective medium or by applying antibiotic-containing overlays to regeneration plates following a period of phenotypic expression (for details see Chater *et al.*, 1982; Hopwood *et al.*, 1985).

Plasmid monomers are taken up efficiently by *Streptomyces* protoplasts, with transformation efficiencies of 10^6 to 10^7 transformants per μg CCC plasmid DNA for *Streptomyces lividans* and *Strep. coelicolor* A3(2). Transformation with open circular or linearized plasmids with cohesive termini is 10–100-fold less efficient than with CCC DNA. No transforming activity is apparently detectable using linearized plasmid molecules with incompatible ends or after alkaline phosphatase treatment.

---

[1] A pock is a circular zone of growth inhibition formed when plasmid-containing individuals grow in a lawn of mycelium of a strain lacking the same (or closely related) plasmid. The inhibition reaction has been called the lethal zygosis reaction (Ltz). Many *Streptomyces* conjugative plasmids give rise to pocks and are hence Ltz+.

B. subtilis protoplasts are transformed at the same efficiency by plasmid monomers as by dimers or higher oligomers (de Vos and Venema, 1981; Mottes *et al.*, 1979). Single-stranded plasmid DNA also transforms protoplasts efficiently (but fails to transform naturally competent cells (Rudolph *et al.*, 1986). Linear monomers, open circular and nonsupercoiled double-stranded DNA molecules (constructed by *in vitro* ligation) all give rise to transformants but at frequencies between one and three orders of magnitude lower than those observed with native plasmid preparations (Chang and Cohen, 1979; Gryczan *et al.*, 1980). Native plasmid DNA enters *B. subtilis* protoplasts in a double-stranded and predominantly CCC form, with each plasmid molecule that enters being sufficient to produce a transformant (de Vos and Venema, 1981). The high efficiency of protoplast transformation in *B. subtilis* (up to 80%) and lack of requirement for oligomeric DNA make this method an alternative (for CCC DNA) to transforming intact competent cells. In the case of *Bacillus brevis* and certain strains of *B. thuringiensis*, low transformation frequencies were obtained using protoplasts. However, high frequencies were obtained when lysozyme-sensitive strains, for example *B. thuringiensis* 0 016, were employed (Loprasert *et al.*, 1986).

The development of PEG-protoplast systems has permitted the efficient transformation of bacteria that cannot be made competent by other means. In addition, protoplast fusions can be used to transfer nonconjugative plasmids between cells of *E. coli* (Vorobjeva and Khmel, 1979) or between *Bacillus* species (Dancer, 1980). Plasmids can also be transferred from bacteria to eucaryotes by fusion of *E. coli* spheroplasts or protoplasts, for example with yeast protoplasts (Broach *et al.*, 1979; Kingsman *et al.*, 1979) or mammalian cells (Schaffner, 1980). It should be noted that the act of protoplasting itself may lead to curing of resident plasmids in *S. aureus* (Novick *et al.*, 1980) and *Streptomyces* species (Hopwood, 1981).

III. Applications of plasmid transformation

A. Gene transfer

Plasmid transformation is a useful method of genetic exchange, especially where phage vectors for transduction or conjugative plasmids for mobilization are not available. Cell surface alterations, which are at their extreme in protoplasting, may also overcome barriers to genetic exchange, such as the absence of phage or sex pilus receptors. A limitation of transformation, especially in intact cell systems, is that frequencies tend to decrease with increasing plasmid size (Hanahan, 1983; Humphreys *et al.*, 1979). Thus, with plasmids of >75 kb, frequencies may be prohibitively low and the transformants isolated are more likely to contain deleted plasmid molecules. In the case of transformation with *in vitro* ligation mixtures, there will be an inevitable bias towards the formation of transformants containing smaller recombinant plasmids. Transformation is, however, an efficient method for transferring nonmobilizable derivatives of small nonconjugative plasmids. Incompatibility hierarchies among ColE1-derived replicons have been determined by transforming cell lines already carrying homologous plasmids (Warren and Sherratt, 1978). In this case transformation has the advantage of permitting the introduction of limited numbers (usually one) of plasmid DNA molecules into bacteria.

B. Identification of plasmid function

Transformation provides a means of reintroducing chemically pure DNA into bacteria and can be used as a direct assay of biological activity of plasmid DNA preparations (Mooibroek and Venema, 1982; Roberts and Strike, 1981). Phenotypes specified by plasmids can be identified by separation of individual DNA species (e.g. by agarose gel electrophoresis) which can then be used to transform standard recipients. This may be of value in analysing wild-type isolates harbouring a number of plasmids with unassigned functions.

C. Selection of cryptic plasmids by cotransformation

Cotransformation (congression) is a valuable technique for the indirect selection of plasmids lacking easily identifiable or selectable phenotypes. Kretschmer *et al.* (1975) transformed Ca^{2+}-treated *E. coli* cells with a mixture of a small amount of DNA of an indicator plasmid pSC201 (which specifies resistance to kanamycin and tetracycline and is temperature-sensitive for replication) and an excess (10^3 to 10^5-fold) of the DNA of a second plasmid that did not have an identifiable marker. Between 50 and 85% of cells selected on the basis of acquisition of pSC201 were found also to have acquired the unmarked plasmid. Cell lines carrying this plasmid alone could then be isolated simply by growing the transformants at 42°C and eliminating pSC201 by segregation. As an alternative to an indicator plasmid, Bergmans *et al.* (1980) used *E. coli* chromosomal DNA as an indicator in conjunction with a genetically marked hyper-transformable strain of *E. coli* as recipient. Selection for a particular chromosomal marker after transformation with chromosomal DNA in the presence of a 10 to 20-fold excess of unmarked plasmid DNA leads to a high frequency (5–10%) of coinheritance of the plasmid. The advantage of this method is that there is no need to remove an indicator plasmid from the cotransformed strain. In principle, both these methods could be adapted to any other organism where suitable indicator plasmid or chromosomal systems are available.

D. Formation of *in vivo* deletions

Deletions and other rearrangements that occur during and just after uptake of plasmids may present a hindrance in genetic manipulations. However, the formation of deletions is also a valuable tool for analysing the genetic structure of plasmids or any cloned inserts they might carry. In some bacteria such as gonococci and *B. subtilis*, deletion occurs spontaneously during or just after transformation. However, deletion of circular transforming DNA does not occur in *E. coli*. To obtain deletants, it is first necessary to linearize the plasmid DNA by random mechanical shearing or by specific cleavage with restriction endonucleases. *In vivo* processing of the linear DNA during circularization in the recipient generates plasmid deletants in a proportion of the transformants. Alternatively, deletants may be formed by truncating plasmid DNA *in vitro* by exonucleolytic digestion of the terminal sequences followed by ligation, prior to transformation (Saunders and Saunders, 1987).

Transformation of bacteria with linearized plasmid DNA molecules occurs fortuitously whenever recombinant ligation mixtures from cloning experiments are used as transforming DNA. Linear molecules will be present even when the ligation conditions are adjusted to favour the formation of desired circular recombinant (vector-insert) combinations. The efficiency of transformation of *E. coli* is reduced by 100 to 1000-fold in wild-type and by a further 10 to 40-fold in various *rec*-backgrounds using linear plasmid DNA compared with the analogous circular form (Conley and Saunders, 1984). The majority of transformants obtained with linear molecules are perfectly recircularized and lose no nucleotides from their exposed termini. However, a proportion, ranging from about 9% in wild-type to 42% in a *recBC, lop11* mutant contain deletions that extend from one or both sides of the linearization site. The frequency of deletion is affected by the nature of the termini of linear plasmid molecules: those with cohesive termini produce many more transformants and a lower proportion of deletants than blunt (flush) -ended molecules, where the few surviving transformants exhibit deletion frequencies in excess of 85% even in wild-type backgrounds (Conley *et al.*, 1986a). Deletion occurs as the result of exonucleolytic erosion of termini followed by recombination at sites where there are short (>3 bp) homology matches between the exposed single-stranded arms of the linearized molecule (Conley *et al.*, 1986b). Most deletant plasmids produced in this way have a molecular weight that is < monomeric and exhibit simple mono-(type Ia) or bidirectional (type Ib) excision of terminal sequences as a result of *in vivo* recyclization. However, a small proportion of deletant plasmids are > monomeric but < dimeric and contain both deletions and duplications of plasmid sequence in either direct (type IIa) or indirect (type IIb) repeat (Conley and Saunders, 1984; Conley *et al.*, 1986a). A third type of deletant (type III) are plasmids that are > dimeric but < trimeric; examples are encountered rarely (Conley *et al.*, 1986a). Class II and III deletants arise by complex recombination events involving two or more plasmid molecules (Conley *et al.*, 1986b). Interestingly, type IIb deletants predominate when exonuclease III-deficient (*xth*) *E. coli* are transformed with linearized plasmid DNA. This provides further evidence for the role of exonucleolytic processing in the recircularization process (Conley *et al.*, 1986a, b).

The formation of deletions and other rearrangements in plasmids due to recombinational recyclization following the transformation of bacteria with linearized molecules accounts for many of the aberrant recombinants that have been reported in gene cloning experiments. Furthermore, *in vivo* recircularization of linear plasmid molecules may be exploited in genetic manipulation. Transformation of *E. coli* with linearized recombinant plasmid DNA has been used deliberately as a means of introducing 'random' deletions from fixed points within the *tra* genes of the F plasmid (Thompson and Achtman, 1979). The generation of directed deletions is also possible if one free terminus contains a specific tract of sequence for which there is an homologous match in the other arm of a linear molecule (Conley *et al.*, 1986b). Recombination *in vivo* between partly homologous gene sequences carried on each arm of a linear plasmid molecule can be exploited to generate hybrid genes that might otherwise be difficult to create by normal *in vitro* cloning techniques (Weber and Weissmann, 1983).

E. Introduction of recombinant DNA into bacteria

Transformation is the main route by which plasmid cloning vectors containing DNA inserted *in vitro* are reintroduced into bacteria. It is therefore a vital technique in both cloning and reversed genetics/allele replacement (Weissmann *et al.*, 1979) procedures. The development of plasmid transformation regimes is thus an important prerequisite for the extension of recombinant DNA technology to previously unexplored bacteria. Transformation systems are, however, limited in most cases by the low efficiency of the process and by the maximum length of plasmid DNA that can be successfully taken up by recipient cells. Plasmid chimeras containing large DNA inserts may be positively discriminated against, thus hindering the formation of representative gene libraries. Such problems can be overcome by using cosmid or phage cloning vectors.

IV. Transformation methods

Representative transformation methods are described in this section. Such methods are applicable, with some modification, to a wide range of bacteria. One of the main difficulties with transformation systems in general is that strains of the same bacterial species may require different procedures. Protocols are provided that have proved effective for commonly used strains of the following organisms:

A. calcoaceticus
B. subtilis
E. coli
N. gonorrhoeae
P. aeruginosa
S. typhimurium
Strep. lividans

A. Media and buffers

DTT

Dithiothreitol	2.25 M
Potassium acetate	40 mM

pH 6.0. Filter sterilize and store at −20°C.

GC supplement

Glucose	40 g
L-Glutamine	0.5 g
Ferric nitrate	50 mg
Thiamine pyrophosphate	2 mg

Distilled water to 1000 ml
pH adjusted to 7.2 after filter sterilization.

L-Broth

Difco Tryptone	10 g
Difco Yeast Extract	5 g
NaCl	0.5 g
1 M NaOH	2 ml

Distilled water to 1000 ml
Adjust pH to 7.0 with 1 M NaOH
Autoclave at 115°C for 20 min.
Add 10 ml of sterile glucose (20% w/v).

Minimal Salts Agar

As for phosphate-buffered minimal salts (below) but containing 120 mM Tris-HCl (pH 7.0) 100 μM phosphate and 1% (w/v) Davis New Zealand Agar.

MOPS-buffered minimal salts

As for phosphate-buffered minimal salts (below) but containing 120 mM MOPS (pH 7.0) and 100 μM phosphate.

Nutrient Agar

| Lab M (or Oxoid) Nutrient Broth No. 2 | 25 g |
| Davis New Zealand Agar | 10 g |

Autoclave at 121°C for 15 min.

When cooled to about 50°C, antibiotics are added where appropriate. Selective concentrations for *E. coli* are ampicillin (20 μg ml^{-1}), chloramphenicol (20 μg ml^{-1}), kanamycin (30 μg ml^{-1}), streptomycin (40 μg ml^{-1}), tetracycline (10 μg ml^{-1}). Plates are dried at 37°C for 1 h before use.

Nutrient Broth

| Lab M (or Oxoid) Nutrient Broth No. 2 | 25 g |
| Distilled water to 1000 ml | |

Autoclave at 121°C for 15 min.

P medium

Sucrose	10.3 g
K_2SO_4	0.025 g
$MgCl_2.6H_2O$	0.203 g
Trace element solution*	0.2 ml
Distilled water	80 ml

Autoclave at 115°C for 20 min.
Then add sequentially, sterile solutions of:

KH_2PO_4 (0.5% w/v)	1 ml
$CaCl_2.2H_2O$ (3.68% w/v)	10 ml
Tris-HCl 10 mM, EDTA 1 mM, NaCl 0.25 M, pH 7.2	10 ml

*Trace element solution

$ZnCl_2$	40 mg
$FeCl_3.6H_2O$	200 mg
$CuCl_2.2H_2O$	10 mg
$MnCl_2.4H_2O$	10 mg
$Na_2B_4O_7.10H_2O$	10 mg
$(NH_4)_6Mo_7O_{24}.4H_2O$	10 mg

Distilled water to 1000 ml
Autoclave at 115°C for 20 min.

Penassay Broth

Lab Lemco	1.5 g
Difco Yeast Extract	1.5 g
Difco Bacto Peptone	5 g
Glucose	1 g
NaCl	3.5 g
K_2HPO_4	3.68 g
KH_2PO_4	1.32 g

Distilled water to 1000 ml
pH adjusted to 7.0 and autoclave at 115°C for 20 min.

Phosphate-buffered minimal salts

KH_2PO_4	6 g
K_2HPO_4	6 g
NH_4Cl	2 g
$MgSO_4.7H_2O$	50 mg
$FeSO_4.H_2O$	5 mg

Distilled water to 1000 ml
Autoclave at 115°C for 30 min.
Then add sterile glucose solution to 0.2% (w/v).

PP Agar

As for PP broth but containing $10 g l^{-1}$ Davis New Zealand Agar.

PP Broth

Difco Protease Peptone No. 3	15 g
NaCl	5 g
K_2HPO_4	4 g
$KH_2PO_4.2H_2O$	1 g
Soluble starch	1 g

Distilled water to 1000 ml
Autoclave at 115°C for 30 min.
Then add 1% (v/v) sterile GC supplement.

Protoplast Maintenance Buffer (PMB)

4 × strength Penassay broth	50 ml
2 × SMM	50 ml

Protoplast Regeneration Agar (PRI)

Difco Technical Casamino acids	15 g
Difco Yeast Extract	15 g
Difco Bacto Agar	34 g

Distilled water to 1000 ml
Autoclave at 115°C for 20 min.

Protoplast Regeneration Sorbitol (PRII)

Sorbitol (Sigma)	182 g
Oxoid Gelatin	40 g

Distilled water to 1000 ml
pH adjusted to 8.0 and autoclave at 115°C for 30 min.

R2YE plates

Sucrose	10.3 g
K_2SO_4	0.025 g
$MgCl_2.6H_2O$	1.012 g
Glucose	1.0 g
Difco Casamino acids	0.01 g
Difco Agar	2.2 g
Distilled water	80 ml

Autoclave at 115°C for 20 min.
Then add sequentially, sterile solutions of:

Trace element solution* (p. 108)	0.2 ml
KH_2PO_4(0.5% w/v)	1.0 ml
$CaCl_2.2H_2O$ (3.68% w/v)	8.02 ml
L-Proline (20% w/v)	1.5 ml
Tris-HCl 10 mM, EDTA 1 mM, NaCl 0.25 M, pH 7.2	10 ml
1 M NaOH	0.5 ml
Yeast Extract (10% w/v)	5 ml

Regeneration plates

Molten PRI	150 ml
Molten PRII	250 ml
5 × SS	100 ml
Glucose (10% w/v)	10 ml

Where appropriate, selective concentrations of either chloramphenicol (5 μg ml^{-1}) or neomycin (7.5 μg ml^{-1}). Plates should be dried at 37°C for 1 h before use.

2 × SMM

Sucrose	342.3 g
Maleic acid	4.64 g
$MgCl_2.6H_2O$	8.13 g
Oxoid gelatin	10 g

Distilled water to 1000 ml
pH adjusted to 6.5 and autoclave at 115°C for 30 min.

SOB medium

Bacto-tryptone	2 g
Yeast extract	0.5 g
NaCl	0.58 g
KCl	0.19 g

Distilled water to 990 ml
Autoclave at 115°C for 30 min.
Add 10 ml filter-sterilized 2 M stock solution of Mg^{2+} ions (1M $MgCl_2.6H_2O$ [203.3 g l^{-1}] plus 1 M $MgSO_4.7H_2O$ [246.47 g l^{-1}]) to cooled medium. Final pH is 6.8.

SOC medium

SOB medium containing 20 mM glucose. Add 10 ml filter-sterilized 2 M glucose solution (306.32 g l^{-1}).

Spizizen salts (1 × SS)

K_2HPO_4	14 g
KH_2PO_4	6 g
$Na_3C_6H_5O_7.2H_2O$	1 g
$(NH_4)_2SO_4$	2 g
$MgSO_4.7H_2O$	0.2 g

Distilled water to 1000 ml
Autoclave at 115°C for 20 min.

Streptomyces growth medium (YEME)

Difco Yeast Extract	3 g
Difco Peptone	5 g
Oxoid Malt Extract	3 g
Glucose	10 g

Distilled water to 1000 ml
Autoclave at 115°C for 20 min.

TFB

2-N-morpholinoethanesulphonic acid (to pH 6.2 with KOH)	0.812 g
RbCl	5 g
$MnCl_2$	3.7 g
$CaCl_2$	0.611 g
Hexamine cobalt (III) chloride	0.33 g

Add 416 ml distilled water and sterilize through a 0.22 μm filter. Final pH should be 6.2. This solution keeps for up to 1 year at 4°C.

B. Isolation of transforming DNA

Plasmid DNA isolated by any of the methods outlined in Chapter 5 can be used for transformation. For routine preparation of large amounts of plasmid DNA of high specific activity for transformation, it is often safer to use a method employing purification by ethidium bromide-CsCl gradient centrifugation. Plasmid DNA preparations should be dialyzed exhaustively against distilled water or buffers (pH7.0–8.0) containing 1 mM EDTA, and preserved at −70°C. Repeated freezing and thawing of plasmid DNA preparations should be avoided as this may lead to nicking of CCC DNA molecules; accordingly, plasmid DNA should be stored in conveniently sized aliquots.

C. Transformation protocols

1. Transformation of Acinetobacter calcoaceticus

(a) Transformation test (Juni, 1972). Strains that are naturally competent must first be identified. Such strains should be stable auxotrophs. A crude chromosomal DNA preparation is first made from an *A. calcoaceticus* prototroph by suspending several colonies in 0.5 ml sterile 0.5% (w/v) sodium dodecylsulphate and then heating at 60°C for 1 h. A small amount of a bacterial suspension of an *A. calcoaceticus* auxotroph is then applied to the surface of a nutrient agar plate and a loopful of the crude DNA is mixed in and spread to an area 5 to 10 mm in diameter. After overnight incubation, the cells are streaked out on a succinate minimal agar plate to check for the formation of prototrophs.

(b) Transformation per se (Cruze et al., 1979). 5 ml of an overnight nutrient broth culture is used to inoculate a 500 ml flask containing 150 ml fresh L-broth plus 0.5 ml 50% (w/v) glucose and incubation continued with aeration at 37°C until an OD_{600} of 0.9 has been reached. 1 ml of the resulting culture is then added to 20 µl (about 1 µg) plasmid DNA. Incubation is then carried out statically at room temperature for 2 h and the mixture plated out on selective medium.

2. Plasmid transformation of competent Bacillus subtilis cells (Saunders et al., 1984)

(a) Preparation of competent cells. A single colony of the strain to be transformed is used to inoculate a 250 ml conical flask containing 100 ml of double-strength L-broth. After incubation at 37°C overnight with gentle shaking, the culture should have an OD_{575} of 1.5–1.7. A 5 ml portion of this culture is used

to inoculate a further flask containing 100 ml of prewarmed 1 × SS supplemented with 0.5% (w/v) glucose and the amino acids and vitamins (each at 50 μg ml^{-1}) required by the strain. (If competent cells of a strain already harbouring a plasmid are to be prepared, the medium should also contain selective concentrations of the relevant antibiotics.) The culture is then shaken vigorously at 37°C. Maximal competence occurs 2.5 h after the cessation of exponential growth. The precise time of maximal competence for individual strains should be determined by removing 0.2 ml samples of the culture and transforming with plasmid DNA as described below.

Cells are harvested at the time of maximal competence by centrifugation at 4°C and concentrated 10-fold by resuspension in 10 ml of culture supernatant containing 10% (v/v) glycerol. Cell suspensions are then either used directly or divided into 0.5 ml aliquots and shell frozen with ethanol-dry ice. Frozen competent cells kept at −60 to −80°C remain viable for at least nine months.

(b) Transformation. 0.2 ml of fresh or frozen cells thawed slowly on ice and then washed to remove glycerol is mixed with plasmid DNA (1–5 μg in 5–50 μl) and incubated at 37°C for 30 min with gentle shaking. 0.8 ml of double-strength L-broth is added and incubation continued at 37°C for 90 min. Transformants are obtained by plating dilutions of the transformation mixture on appropriate selective media.

3. Standard procedure for transforming Escherichia coli *cells by calcium shock treatment (modified from Humphreys* et al., *1979)*

(a) Preparation of fresh competent cells. 100 ml of prewarmed nutrient broth in a 250 ml conical flask are inoculated with 1 ml of a culture of *E. coli* grown stationary for 18 h at 37°C in nutrient broth. Incubation is continued at 37°C with aeration (rotary shaker at 160 rev/min) until OD$_{660}$ = 0.15–0.2. The culture is then chilled on ice and the cells harvested by centrifugation (12 000 × g for 5 min at 4°C). The cell pellet is resuspended in 25 ml of ice-cold 10 mM CaCl$_2$, centrifuged, and finally resuspended in 5 ml of ice-cold 75 mM CaCl$_2$ (optionally containing 0.5% (w/v) glucose, 10 mM MOPS, pH 6.5). This produces a suspension of competent cells (2.5–5.0 × 10^9 ml^{-1}) which can be kept at 0 to 4°C for up to 48 h without significant loss of viability.

(b) Transformation. 0.2 ml of competent cells is mixed in a sterile glass or polypropylene tube with plasmid DNA (0.01–10 μg) and made to 0.5 ml with ice-cold 75 mM CaCl$_2$ (plus 0.5% (w/v) glucose, 10 mM MOPS at pH 6.5 if required). (Note that this system is saturated with 0.5–1 μg DNA per 5 × 10^8– 1 × 10^9 viable cells.) The transformation mixture is kept on ice for 45 min and

then transferred to a 42°C water bath or dry block for 1.5 to 10 min. 0.5 ml of nutrient broth is then added and incubation continued at 37°C for 2 h with shaking to allow expression. Appropriate dilutions of the transformation mixture are plated on nutrient agar plates containing antibiotics where necessary.

4. Efficient transformation of E. coli by the Hanahan method (modified from Hanahan, 1983)

A single colony of the strain to be transformed is picked from the surface of a nutrient agar plate and the cells resuspended by vortexing in 1 ml SOB. A loopful of the suspension is used to inoculate 10 ml fresh SOB medium and the culture is incubated at 37°C with shaking at 200 rev/min. for about 3 h until the OD_{550} reaches 0.45 to 0.55. The culture is then decanted into a sterile Oak Ridge 50 ml centrifuge tube and placed on ice for 10–15 min. The cells are pelleted by centrifugation at $1000 \times g$ for 12 min at 4°C. The pellet is resuspended in 3 ml TFB and kept on ice for 10–15 min. The cells are repelleted in the same way and resuspended in 1 ml TFB. 35 μl fresh glass distilled dimethylsulphoxide (DMSO) are added, mixed and then left on ice for 10 min. 35 μl DTT are added and the mixture left on ice for a further 10 min. A second 35 μl of DMSO are added, mixed and incubation continued at 4°C for 5 min. 210 μl of the resulting cell suspension are then transferred to a chilled microcentrifuge tube. DNA, in a volume of 10 μl, is then added. Plasmid DNA (10 μg to 100 ng) can be added in H_2O. Where the DNA is present in a ligation mixture it is better to dilute 1:10 with distilled H_2O to about 10–100 ng, since ligase and components of ligation buffers inhibit transformation. The cell-DNA mixture is then kept on ice for 30–60 min before being subjected to a heat shock at 42°C for 90 sec (in a water bath) or 120 sec (in a dry block) after which it is placed on ice for 1–2 min. 800 μl SOC are then added and the whole mixture incubated at 37°C with shaking at 225 rev/min for 1–2 h. Dilutions are then plated out on nutrient agar containing appropriate antibiotics to select for the plasmid concerned.

5. Transformation of E. coli directly on the surface of agarose gels (modified from Niesel and Hertzog, 1983)

This procedure, whilst inefficient in terms of DNA molecules taken up, is useful for the direct transformation of specific plasmid molecules that have been separated on agarose gels. The electrophoresis tank is sterilized by immersion in H_2O_2 (7% v/v) and both agarose and electrophoresis buffers are sterilized by autoclaving. Duplicates of each DNA sample are separated electrophoretically. One set of samples is stained and photographed, whilst the other is immersed in sterile 100 mM $CaCl_2$ at 4°C overnight. The $CaCl_2$-treated gel is then placed in a sterile dish and blotted with 10 sheets of sterile Whatman No. 1 filter paper for about 10 min. This reduces the gel to about 50% of its original thickness. The gel is then chilled to 4°C for 30 min. Approximately 10^7 competent *E. coli*, prepared by either of the methods described above, are added in a volume of 50–100 μl to one end of the gel. The bacterial suspension is then dispensed across the surface of the gel in one direction only by the single spreading action of a sterile glass spreader. (If multiple strokes of the spreader are used, potential transformants will be distributed randomly instead of remaining above specific plasmid DNA bands.) The gel is held at 4°C overnight, and then carefully transferred, inoculated side up, on to the surface of a square agar plate. The plate comprises a basal agar layer containing an appropriate antibiotic that has subsequently been overlaid with a 1 mm layer of 0.5% (w/v) antibiotic-free agar, all contained in a square plastic petri dish. After incubation at 37°C for 18–24 h, it is usually possible to visualize specific transformant colonies growing on the agarose surface immediately above the plasmid DNA bands.

6. Transformation of S. typhimurium *(Lederberg and Cohen, 1974)*

(a) Preparation of competent cells. 50 ml of prewarmed L-broth in a 250 ml conical flask are inoculated with 0.5 ml of a culture of *S. typhimurium* grown statically at 37°C overnight. Incubation is continued at 37°C with shaking until $OD_{600} = 0.6$. The culture is then chilled and the cells harvested by centrifugation (12 000 × g for 5 min at 4°C). The pellet is resuspended and washed by centrifugation in 50 ml of ice-cold 100 mM $MgCl_2$ and then resuspended in 25 ml of 100 mM $CaCl_2$. The suspension is held on ice for 20 min and then harvested by centrifugation. The cells are finally resuspended in 2.5 ml of 100 mM $CaCl_2$.

(b) Transformation. 0.1 ml of plasmid DNA (0.01–10 μg) in distilled water or TEN buffer [0.02 M Tris (hydroxymethyl)aminomethane, 1 mM ethylenediaminetetraacetic acid, 0.02 M NaCl, pH 8.0] is mixed with 0.2 ml of competent cells. The mixture is incubated on ice for 30 min and then subjected to a 42°C heat pulse of 2 min to induce uptake of DNA. After chilling on ice, the mixture is diluted 10-fold with prewarmed L-broth and incubated at 37°C with shaking for 2 h to allow expression of transforming DNA. Dilutions of the culture are then plated on appropriate media to select for transformants and to estimate viable cell counts.

7. Selection of cryptic plasmids by cotransformation

(a) Using a competing plasmid (based on the method of Kretschmer et al., 1975). 1 μg of the nonselected (cryptic) plasmid is mixed with 0.5 ng of pSC201

indicator DNA (or DNA of any suitable plasmid that is temperature-sensitive for replication or which can be removed from transformants as required). 0.2 ml competent cells of *E. coli* K12, prepared by the conventional or Hanahan method outlined above, is then added and transformation carried out as described above, except that expression of the indicator plasmid (e.g. pSC201) must be at 32°C. At the end of the expression period selection is made for a genetic marker on the indicator plasmid. Transformants can then be examined directly, e.g. by agarose gel electrophoresis of plasmid preparations (Chapter 5), for the coinheritance of the unmarked plasmid (normally 50–80% of transformants carrying pSC201 also inherit the unmarked plasmid). The indicator plasmid can be removed from appropriate transformant clones by subculturing into liquid media and incubating at 42°C for sufficient time to allow segregation (at least ten generations). Loss of the indicator and retention of the cryptic plasmid is confirmed by agarose gel electrophoresis of plasmid DNA preparations.

(b) Using competing chromosomal DNA (Bergmans et al., 1980). E. *coli* strain AM1268, a Rec$^+$Sbc$^+$ derivative of the transformable strain AM1095 (*recB21, recC22, sbc15, leu*), is grown in phosphate-buffered minimal salts medium containing leucine with shaking at 37°C until early exponential phase. Cells are chilled, harvested by centrifugation, washed once with 10 mM NaCl and resuspended in 0.05 × original culture volume of 20 mM HEPES buffer pH 6.0. 0.3 ml of the cell suspension is then mixed with chromosomal DNA from a prototroph (3.2 μg ml^{-1}) and cryptic plasmid DNA (50 μg ml^{-1}) and with CaCl$_2$ and MgCl$_2$ to final concentrations of 30 mM and 26 mM, respectively, in a final volume of 0.5 ml. The mixture is incubated at 0°C for 10 min, heated to 42°C for 6 min, chilled and incubated at 0°C for a further 30 min. It is then diluted 1:10 in supplemented minimal salts medium buffered with 120 mM MOPS pH 7.0 and incubated at 37°C with aeration for 90 min to allow expression of DNA taken up by the cells. Leu$^+$ transformants are selected on minimal agar lacking leucine. Leu$^+$ transformants are screened for the presence of plasmid DNA. Approximately 5–10% are cotransformed with the plasmid.

8. Transformation of P. aeruginosa *(essentially the modification of Bagdasarian and Timmis (1981) of Kushner's (1978) method for E. coli)*

(a) Preparation of competent cells. Cultures of *P. aeruginosa* are grown with shaking at 30°C in L-broth to early exponential phase (approximately 2×10^8 cells ml^{-1}), chilled and harvested by centrifugation at 0°C. Cells are washed by resuspension in an equal volume of cold 10 mM MOPS pH 7.0, 10 mM RbCl, 100 mM MgCl$_2$ and collected by centrifugation. The cell pellet is then resuspended in an equal volume of cold 100 mM MOPS pH 6.5, 10 mM RbCl,

$100\,mM$ $CaCl_2$ and kept at $0°C$ for 30 min. The cells are then harvested by centrifugation and resuspended in $1:10$ volume of the $CaCl_2$ buffer at $0°C$.

(b) Transformation. 0.2 ml portions of this suspension are mixed with plasmid DNA ($0.2–1.0\,\mu g$), incubated at $0°C$ for 45 min and then at $42°C$ for 1 min. The transformation mixture is diluted with 3 ml of L-broth and incubated at $30°C$ with aeration for 90 min to allow expression of plasmid genes before plating on selective media.

9. Plasmid transformation of N. gonorrhoeae *(based on the method of Sox* et al., *1979)*

(a) Preparation of competent cells. Piliated colonial forms of *N. gonorrhoeae* are grown at $37°C$ for 18 h, in the presence of $5–10\%$ (v/v) CO_2, on the surface of PP agar plates. Colonies growing on the surface of the plates are resuspended to a density of between 5×10^7 and 1×10^8 colony forming units (cfu) in PP broth containing $10\,mM$ $MgCl_2$. (Note that cell densities of $>10^8\,cfu.\ ml^{-1}$ lead to reduced frequencies because of competition with DNA from lysing cells.)

(b) Transformation. $0.1–5\,\mu g$ of plasmid DNA are added to 1 ml of competent cell suspension and incubation continued at $37°C$ for 30 min. Transformation is then terminated by adding DNase I ($25\,\mu g\ ml^{-1}$). The suspension is diluted $1:10$ with PP broth and incubation continued with shaking at $37°C$ in $5–10\%$ (v/v) CO_2 for 3–6 h. Dilutions are then spread on PP agar plates containing the selective agent (normally benzyl penicillin at $0.2\,\mu g\ ml^{-1}$ for β-lactamase plasmids). Alternatively, the transformed culture can be incorporated in soft agar overlayers. Transformant colonies normally appear within 48 h on incubation at $37°C$ in $5–10\%$ (v/v) CO_2.

10. Transformation of B. subtilis *protoplasts (based on Chang and Cohen, 1979)* 1979)

(a) Preparation of protoplasts. Cultures of *B. subtilis* strains are grown at $37°C$ in 50 ml of Penassay Broth to $OD_{575} = 0.7–1.0$. The cells are then harvested by centrifugation and resuspended in 4 ml of PMB. 1 ml of filter-sterilized lysozyme solution ($10\,mg\ ml^{-1}$ in PMB) is added. After gentle shaking at $37°C$ for 120 min, essentially 100% of cells should have been converted to protoplasts (as monitored by phase-contrast microscopy). 5 ml of PMB are added and the protoplasts harvested by centrifugation at approximately $2500 \times g$ for 10 min at room temperature. The protoplasts are resuspended in fresh PMB and the washing procedure repeated twice in order to remove residual lysozyme. (Note that biological inactivation of DNA by lysozyme has been reported by Wilson

and Bott, 1970.) The washed protoplasts are finally resuspended in 2 ml of PMB.

(b) Transformation. Plasmid DNA (5 μg) in 5–50 μl of 10 mM Tris-HCl 0.1 mM EDTA, pH 7.5 is added to 0.5 ml of protoplast suspension. PEG 1500 (1.5 ml of 40% w/v in 1 × SMM) is then added with thorough mixing. PEG 1500 is less viscous than PEG 6000 and allows easier mixing without reducing transformation frequencies. After 2 min, 5 ml of PMB are added and mixed thoroughly. The protoplasts are harvested as described above, resuspended in 1 ml of fresh PMB and incubated at 37°C with gentle shaking for 90 min. 0.1 ml dilutions in PMB are then plated on regeneration media to assay for transformants. Dilutions may also be plated on ordinary nutrient agar to determine the number of osmotically stable nonprotoplasted cells in the preparations. On sorbitol-containing regeneration plates, colonies appear after overnight incubation at 37°C. Typically, from each protoplast suspension 5×10^7 colonies can be regenerated, of which less than 0.001% grow on nutrient agar. When plasmids such as pUB110 or pDB64 (Gryczan *et al.*, 1980) are used in this procedure, between 0.1 and 10% of regenerated protoplasts give rise, after direct selection, to neomycin- ($7.5 \, \mu g \, ml^{-1}$) or chloramphenicol- ($5 \, \mu g \, ml^{-1}$) resistant transformants. (Note that disposable plastic pipettes should be used when handling protoplasts to prevent problems arising from traces of detergent in washed glass pipettes.)

11. Transformation of Streptomyces *protoplasts (method of Chater et al., 1982; see also Hopwood et al., 1985)*

The protocols are for *Strep. lividans* but with modification they should be applicable to other *Streptomyces* species.

(a) Preparation of protoplasts. About 0.1 ml *Streptomyces* spore suspension is added to 25 ml of YEME containing 34% (w/v) sucrose, 0.005 M MgCl$_2$ and 0.5% (w/v) glycine (note that different concentrations of glycine are required for different strains) in a 250 ml baffled flask and incubated with shaking at 30°C for 36–40 h. The cells are washed twice by resuspension in 10.3% (w/v) sucrose and centrifugation. The mycelium is then resuspended in 4 ml of lysozyme solution ($1 \, mg \, ml^{-1}$) in P medium, with CaCl$_2$ and MgCl$_2$ concentrations reduced to 0.0025 M, and incubated at 30°C for 15–60 min. The solution is then mixed by pipetting three times in a 5 ml pipette and incubation continued for a further 15 min. 5 ml of P medium are added and the solution again mixed by pipetting. The solution is filtered through cotton wool and the protoplasts sedimented gently (7 min at 800 × g) at room temperature. The protoplasts are resuspended in 4 ml of P medium and recentrifuged. This is

repeated and the protoplasts are finally suspended in the drop of P medium remaining after decanting the supernatant. (Tap the side of the tube repeatedly until the protoplasts are dispersed.) For storage, add 2 ml of P medium to the suspended pellet. Protoplasts can be used for several days after preparation if stored at 4°C and for many months if stored at −70°C (slow freezing and rapid thawing are best).

Note that the growth stage is important for obtaining high transformation frequencies. Late exponential phase mycelium is best for *Strep. lividans* but the optimum growth time should be determined for each species.

(b) Transformation. The protoplasts (4×10^9, viability normally 1%) are resuspended in the drop of P medium left after decanting the supernatant. Where stored protoplasts are used, these should be washed by centrifugation immediately prior to transformation to remove substances that have leaked out during storage. DNA in $< 20 \mu l$ of buffer (10 mM Tris-HCl, 1 mM EDTA, pH 8.0) is added. Immediately, 0.5 ml of PEG 1000 solution [2.5 g of PEG dissolved in 7.5 ml of 2.5% (w/v) sucrose, 0.0014 M K_2SO_4, 0.1 M $CaCl_2$, 0.05 M Tris-maleic acid pH 8.0 plus trace elements] is added and the mixture pipetted once. After 1 min, 5 ml of P medium are added and the protoplasts pelleted by gentle centrifugation. The pellet is resuspended in 1 ml of P medium. 0.1 ml samples are spread on to R2YE plates (plates should be dried to 85% of their fresh weight) and incubated at 30°C for one to three days. Transformants are detected either by the formation of pocks by plasmid-containing regenerants in a plasmid-free background (Bibb *et al.*, 1977) or by screening for drug resistance markers by overlaying plates with soft agar containing the appropriate drug, after a suitable time of growth to allow phenotypic expression of drug resistance.

V. Ground rules for transforming previously untransformed species

Widespread interest in the use of molecular genetic techniques for analysing potentially useful genes in different microorganisms creates a need to introduce DNA into bacteria that have previously not been transformed. Therefore, we have proposed a series of guidelines for the development of novel transformation systems and the improvement of existing ones (Fig. 3). It is intended as an outline guide and the protocols and original references should be consulted for details. Table III lists major factors that should be considered when devising or improving transformation protocols. One of the most important of these is the strain of the particular species concerned. Failure to achieve success with one strain does not mean that a species is untransformable and, provided it is feasible, a series of separate isolates may have to be tested. The source of transforming DNA can also have a dramatic effect in determining the success

120

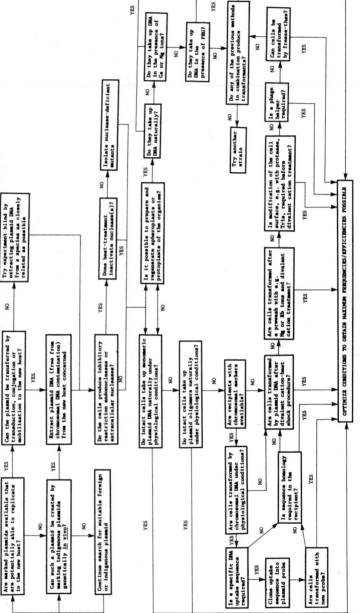

Fig. 3. Flow chart for devising a plasmid transformation protocol.

TABLE III
Factors affecting plasmid transformation systems

---

1. The nature of plasmid DNA used as probe
    Small size (ideally <15 kb)
    Stable replication in new host (wide host range plasmids useful)
    Plasmid DNA preferably already modified for new host
    Uncontaminated with chromosomal DNA or RNA
    Topological state (e.g. CCC, OC, linear, monomer, oligomer)
2. The state of recipient cells
    Growth phase
    Nutritional status
    Aeration
    Extracellular nucleases (remove by mutation if present)
    Restriction — modification status
    Homologous replicons in recipient for rescue
    Presence of incompatible replicon
3. Preparation procedures
    Concentration of divalent and other ions, pH
    Concentration, molecular weight and pH of PEG
    Prewashing (e.g. with RbCl, MgCl$_2$, Tris, etc.)
    Presence of basic proteins (e.g. spermidines)
4. Exposure to DNA
    Concentration relative to saturation
    Duration of exposure to DNA
5. Regeneration and recovery of transformants
    Washing to remove competence inducers
    Choice of stabilisers and regeneration media
    Expressivity of marker gene
    Duration of expression period
    Concentration of selective agents

---

of transformation. Restrictionless mutants are obviously of great use where the transforming DNA comes from a heterologous species. Most of the strains of *E. coli* commonly used for cloning are therefore defective in the K restriction and/or modification systems. DNA modified by site-specific methylases may be cleaved by methylation specific endonucleases and induce the SOS response in transformants, producing drastic inhibition of growth (Heitman and Model, 1987). This phenomenon is a common cause of failure when attempting to clone genes from certain bacteria (notably *Neisseria* and *Haemophilus* spp., which methylate their DNA extensively) in *E. coli*. This problem may be overcome in the case of cytosine-methylated DNA by using *rglB⁻* (*mcrB⁻*) mutants (Raleigh and Wilson, 1986) or by using *mrr⁻* mutants (Heitman and Model, 1987) for adenine-methylated DNA. Such mutants lack restriction endonucleases that recognize and cleave incoming DNA bearing the appropriate methylation pattern.

References

Acebal, C., Rubio, V. and Marquez, G. (1986). *FEMS Microbiol. Lett.* **35**, 79–82.
Albano, M., Hahn, J. and Dubnau, D. (1987). *J. Bacteriol.* **169**, 3110–3117.
Albritton, W. L., Bendler, J. W. and Setlow, J. K. (1981). *J. Bacteriol.* **145**, 1099–1101.
Bagdasarian, M. and Timmis, K. N. (1981). *Curr. Top. Microbiol. Immunol.* **96**, 47–67.
Balganash, M. and Setlow, J. K. (1985). *J. Bacteriol.* **161**, 141–146.
Barany, F. and Boeke, J. D. (1983). *J. Bacteriol.* **153**, 200–210.
Barany, F. and Kahn, M. C. (1985). *J. Bacteriol.* **161**, 72–79.
Barany, F. and Tomasz, A. (1980). *J. Bacteriol.* **144**, 698–709.
Bauer, D. W. and Beer, S. V. (1983). *Phytopathology* **73**, 1342.
Behnke, D. (1981). *Mol. Gen. Genet.* **182**, 490–497.
Bensi, G., Iglesias, A., Canosi, U. and Trautner, T. A. (1981). *Mol. Gen. Genet.* **184**, 400–404.
Benzinger, R. (1978). *Microbiol. Rev.* **42**, 194–236.
Bergmans, H. E. N., Kooijman, D. H. and Hoekstra, W. P. M. (1980). *FEMS Microbiol. Lett.* **9**, 211–214.
Bergmans, H. E. N., Van Die, I. M. and Hoekstra, W. P. M. (1981). *J. Bacteriol.* **146**, 564–570.
Berry, D. and Kropinski, A. M. (1986). *Can. J. Microbiol.* **32**, 436–438.
Bibb, M. J., Ward, J. M. and Hopwood, D. A. (1977). *Mol. Gen. Genet.* **154**, 155–156.
Bibb, M. J., Ward, J. M. and Hopwood, D. A. (1978). *Nature (London)* **274**, 398–400.
Biswas, G. D., Burnstein, K. and Sparling, P. F. (1985). *In* "The Pathogenic *Neisseriae*" (Proceedings 4th International Symposium Asilomar California 1984) (G. K. Schoolnik, Ed.) pp. 204–208. American Society for Microbiology, Washington, D.C.
Biswas, G. D., Burnstein, K. and Sparling, P. F. (1986). *J. Bacteriol.* **168**, 756–761.
Broach, J. R., Stathem, J. and Hicks, J. B. (1979). *Gene* **8**, 121–123.
Brown, B. J. and Carlton, B. C. (1980). *J. Bacteriol.* **142**, 508–512.
Brown, M. G. M., Saunders, J. R. and Humphreys, G. O. (1981). *FEMS Microbiol. Lett.* **11**, 97–100.
Brown, M. G. M., Weston, A., Saunders, J. R. and Humphreys, G. O. (1979). *FEMS Microbiol. Lett.* **5**, 219–222.
Burnstein, K. L., Dyer, D. W. and Sparling, P. F. (1988). *J. Gen. Microbiol.* (in press.)
Buzby, J. S., Porter, R. D. and Stevens, S. E. Jr. (1983). *J. Bacteriol.* **154**, 1446–1450.
Canosi, U., Iglesias, A. and Trautner, T. A. (1981). *Mol. Gen. Genet.* **181**, 434–440.
Canosi, U., Morelli, G. and Trautner, T. A. (1978). *Mol. Gen. Genet.* **166**, 259–268.
Carlson, C. A., Steenbergen, S. M., Ingraham, J. L. (1984). *Arch. Microbiol.* **140**, 134–138.
Chakrabarty, A. M., Mylroie, J. R., Friello, D. A. and Vacca, J. G. (1975). *Proc. Natl. Acad. Sci. U.S.A.* **72**, 3647–3651.
Chandler, M. S. and Morrison, D. A. (1987). *J. Bacteriol.* **169**, 2005–2011.
Chang, S. and Cohen, S. N. (1979). *Mol. Gen. Genet.* **168**, 111–115.
Chater, K. F., Hopwood, D. A., Kieser, T. and Thompson, C. J. (1982). *Curr. Top. Microbiol. Immunol.* **96**, 69–95.
Chauvat, F., Astier, C., Vedel, F. and Joset-Espardellier, F. (1983). *Mol. Gen. Genet.* **191**, 39–45.
Chen, J-D. and Morrison, D. A. (1987). *J. Gen. Microbiol.* **133**, 1959–1967.
Chung, B. and Goodgal, S. H. (1979). *Biochem. Biophys. Res. Commun.* **88**, 208–214.
Cline, S. W. and Doolittle, W. F. (1987). *J. Bacteriol.* **169**, 1341–1344.

Cohen, S. N., Chang, A. C. Y. and Hsu, L. (1972). *Proc. Natl. Acad. Sci. U.S.A.* **69**, 2110–2114.

Comai, L. and Kosuge, T. (1980). *J. Bacteriol.* **143**, 950–957.

Conley, E. C. and Saunders, J. R. (1984). *Mol. Gen. Genet.* **194**, 211–218.

Conley, E. C., Saunders, V. A. and Saunders, J. R. (1986a). *Nucl. Acids Res.* **14**, 8905–8917.

Conley, E. C., Saunders, V. A., Jackson, V. and Saunders, J. R. (1986b). *Nucl. Acids Res.* **14**, 8919–8932.

Contente, S. and Dubnau, D. (1979a). *Mol. Gen. Genet.* **167**, 251–258.

Contente, S. and Dubnau, D. (1979b). *Plasmid* **2**, 555–571.

Cosloy, S. D. and Oishi, M. (1973). *Proc. Natl. Acad. Sci. U.S.A.* **70**, 84–87.

Cruze, J. A., Singer, J. T. and Finnerty, W. R. (1979). *Curr. Microbiol.* **3**, 129–132.

Dagert, M. and Ehrlich, S. D. (1979). *Gene* **6**, 23–28.

Dancer, B. N. (1980). *J. Gen. Microbiol.* **121**, 263–266.

Danner, D. B., Deich, R. A., Sisco, K. L. and Smith, H. O. (1980). *Gene* **11**, 311–318.

David, M., Tranchet, M. and Denairie, J. (1981). *J. Bacteriol.* **146**, 1154–1157.

de Vos, W. and Venema, G. (1981). *Mol. Gen. Genet.* **182**, 39–43.

de Vos, W. M., Venema, G., Canosi, H. and Trautner, T. A. (1981). *Mol. Gen. Genet.* **181**, 424–433.

Dimitriadis, G. J. (1979). *Nucleic Acids Res.* **6**, 2697–2708.

Dityatkin, S. Y. A. and Ilyashenko, B. N. (1979). *Genetika* **15**, 220–225.

Docherty, A. J. P., Grandi, G., Grandi, R., Gryczan, T. J., Shivakumar, A. G. and Dubnau, D. (1981). *J. Bacteriol.* **145**, 129–137.

Doran, J. L., Bingle, W. H., Roy, K. L., Hiratsuka, K. and Page, W. J. (1987). *J. Gen. Microbiol.* **133**, 2059–2072.

Dougherty, T. J., Asmus, A. and Tomasz, A. (1979). *Biochem. Biophys. Res. Commun.* **86**, 97–104.

Duncan, C. H., Wilson, G. A. and Young, F. E. (1978). *Proc. Natl. Acad. Sci. U.S.A.* **75**, 3664–3668.

Dzelzkalns, V. A. and Bogorad, L. (1986). *J. Bacteriol.* **165**, 964–971.

Ehrlich, S. D. (1977). *Proc. Natl. Acad. Sci. U.S.A.* **74**, 1680–1682.

Elwell, L. P., Saunders, J. R., Richmond, M. H. and Falkow, S. (1977). *J. Bacteriol.* **131**, 356–362.

Fani, R., Mastromei, G., Polsinelli, M. and Venema, G. (1984). *J. Bacteriol.* **157**, 152–157.

Finn, C. W. Jr. and Landman, O. E. (1985). *Mol. Gen. Genet.* **198**, 329–335.

Fischer, H-M., Luthy, P. and Schweitzer, S. (1984). *Arch. Microbiol.* **139**, 213–217.

Flickering, J. L., Smith, M. D. and Lineberger, D. W. (1985). *Plasmid* **13**, 221–222.

Fornari, C. S. and Kaplan, S. (1982). *J. Bacteriol.* **152**, 89–97.

Fraley, R. T., Fornari, C. S. and Kaplan, S. (1979). *Proc. Natl. Acad. Sci. U.S.A.* **76**, 3348–3352.

Gantotti, B. V., Patil, S. S. and Mandel, M. (1979). *Mol. Gen. Genet.* **174**, 101–103.

Gendel, S., Straus, N., Pulleyblank, D. and Williams, J. (1983). *J. Bacteriol.* **156**, 148–154.

Glick, B. R., Brooks, H. E. and Pasternak, J. J. (1985). *J. Bacteriol.* **162**, 276–279.

Gnedoi, S. N., Babushkina, L. M., Frolov, V. N. and Levashev, U. S. (1977). *ZH. Mikrobiol. Epidemiol. Immunobiol.* **1**, 99–104.

Golden, S. S. and Sherman, L. A. (1984). *J. Bacteriol.* **158**, 36–42.

Goodgal, S. H. (1982). *Annu. Rev. Genet.* **16**, 169–192.

Graves, J. F., Biswas, G. D. and Sparling, P. F. (1982). *J. Bacteriol.* **152**, 1071–1077.

124   J. R. SAUNDERS AND V. A. SAUNDERS

Grinius, L. (1980). *FEBS Microbiol. Lett.* **113**, 1–10.

Gromkova, R. and Goodgal, S. H. (1977). *In* "Modern Trends in Bacterial Transformation" (A. Portoles, R. Lopez and M. Espinosa, Eds), pp. 299–306, North-Holland Publ., Amsterdam.

Gromkova, R. and Goodgal, S. (1979). *Biochem. Biophys. Res. Commun.* **88**, 1428–1434.

Gromkova, R. and Goodgal, S. (1981). *J. Bacteriol.* **146**, 79–84.

Gross, D. C. and Vidaver, A. K. (1981). *Can. J. Microbiol.* **27**, 759–765.

Gryczan, T. J., Contente, S. and Dubnau, D. (1978). *J. Bacteriol.* **134**, 318–329.

Gryczan, T., Contente, S. and Dubnau, D. (1980). *Mol. Gen. Genet.* **177**, 459–467.

Hanahan, D. (1983). *J. Mol. Biol.* **166**, 557–580.

Hahn, J., Albano, M. and Dubnau, D. (1987). *J. Bacteriol.* **169**, 3104–3109.

Heierson, A., Landen, R. and Lovgren, A., Dalhammar, G. and Boman, H. G. (1987). *J. Bacteriol.* **169**, 1147–1152.

Heitman, I. and Model, P. (1987). *J. Bacteriol.* **169**, 3243–3250.

Hinton, J. C. D., Perombelon, M. C. M. and Salmond, G. P. C. (1985). *J. Bacteriol.* **161**, 786–788.

Hoekstra, W. P. M., Bergmans, J. E. N. and Zuidweg, E. M. (1980). *J. Bacteriol.* **143**, 1031–1032.

Holsters, M., De Waele, D., Depicker, A., Messens, E., van Montagu, M. and Schell, J. (1978). *Mol. Gen. Genet.* **163**, 181–187.

Hooley, P. and Wellington, E. M. H. (1985). *Lett. Appl. Microbiol.* **1**, 77–80.

Hopwood, D. A. (1981). *Annu. Rev. Microbiol.* **35**, 237–272.

Hopwood, D. A., Bibb, M. J., Chater, K. F., Kieser, T., Bruton, C. J., Kieser, H. M., Lydiate, D. J., Smith, C. P. and Ward, J. M. (1985). "Genetic Manipulation of *Streptomyces*". John Innes Foundation, Norwich.

Humphreys, G. O., Saunders, J. R., Brown, M. G. M. and Royle, P. M. (1981). *In* "Transformation 1980" (M. Polsinelli and G. Mazza, Eds), pp. 307–312, Cotswold Press, Oxford.

Humphreys, G. O., Weston, A., Brown, M. G. M. and Saunders, J. R. (1979). *In* "Transformation 1978" (S. W. Glover and L. O. Butler, Eds), pp. 254–279, Cotswold Press, Oxford.

Iglesias, A. and Trautner, T. A. (1983), *Mol. Gen. Genet.* **189**, 73–76.

Imanaka, T., Fujii, M., Aramori, I. and Aiba, S. (1982). *J. Bacteriol.* **149**, 824–830.

Ingraham, J. L. and Carlson, C. (1985). *In* "The Molecular Biology of Bacterial Growth" (M. Schaechter, F. C. Neidhardt, J. L. Ingraham and N. D. Kjeldgaard, Eds) pp. 360–366, Boston, Jones & Bartlett.

Irvin, R. T., MacAlister, T. J. and Costerton, J. W. (1981). *J. Bacteriol.* **145**, 1397–1403.

Jacobs Jr., W. R., Tuckman, M. and Bloom, B. R. (1987). *Nature (London)* **327**, 532–534.

Jones, I., Primrose, S. B., Robinson, A. and Elwood, D. C. (1981). *J. Bacteriol.* **146**, 841–846.

Juni, E. (1972). *J. Bacteriol.* **112**, 917–931.

Jyssum, K., Jyssum, S. and Gundersen, W. B. (1971). *Acta Pathol. Microbiol. Scand. Sect. B* **79**, 563–571.

Katsumata, R., Ozaki, A., Oka, T. and Furuya, A. (1984). *J. Bacteriol.* **159**, 306–311.

Kingsman, A., Clarke, L., Mortimer, R. K. and Carbon, J. (1979). *Gene* **7**, 141–152.

Kiss, G. B. and Kalman, Z. (1982). *J. Bacteriol.* **150**, 465–470.

Klebe, R. J., Harriss, J. V., Sharp, Z. D. and Douglas, M. G. (1983). *Gene* **25**, 333–341.

Kondo, J. K. and McKay, L. L. (1982). *Appl. Environ. Microbiol.* **43**, 1213–1215.

Kondo, J. K. and McKay, L. L. (1984). *Appl. Environ. Microbiol.* **48**, 252–259.

Kretschmer, P. J., Chang, A. C. Y. and Cohen, S. N. (1975). *J. Bacteriol.* **124**, 225–231.

Kuhlemeier, C. J., Borrias, W. E., van den Hondel, C. A. M. J. J. and van Arkel, G. A. (1981). *Mol. Gen. Genet.* **184**, 249–254.

Kuhlemeier, C. J., Thomas, A. A. M., van der Ende, A., van Leen, R. W., Borrias W. E., van den Hondel, C. A. M. J. J. and van Arkel, G. A. (1983). *Plasmid* **10**, 156–163.

Kuramitsu, H. K. and Long, C. M. (1982). *Infect. Immun.* **36**, 435–436.

Kushner, S. R. (1978). *In* "Genetic Engineering" (H. B. Boyer and S. Nicosa, Eds), pp. 17–23, Elsevier, Amsterdam.

Lacks, S. A. (1977). *In* "Receptors and Recognition, Series B" (J. L. Reissig, Ed.). Vol. 3, Microbiol. Interactions, pp. 177–232, Chapman & Hall, London.

Lacks, S. (1979). *J. Bacteriol.* **138**, 404–409.

Lacy, G. H. and Sparks, R. B. (1979). *Phytopathology* **69**, 1293–1297.

Lampel, J. S. and Strohl, W. R. (1986). *Appl. Environ. Microbiol.* **51**, 126–131.

Le Blanc, D. J. and Hassell, F. P. (1976). *J. Bacteriol.* **128**, 347–355.

Lederberg, E. M. and Cohen, S. N. (1974). *J. Bacteriol.* **119**, 1072–1074.

Lin, Y. L. and Blaschek, H. P. (1984). *Appl. Environ. Microbiol.* **48**, 737–742.

Lopez, P., Espinosa, M., Stassi, D. L. and Lacks, S. A. (1982). *J. Bacteriol.* **150**, 692–701.

Loprasert, S., Pantuwatana, S. and Bhumiratana, A. (1986). *J. Invert. Pathol.* **48**, 325–334.

Love, P. E., Lyle, M. J. and Yasbin, R. E. (1985). *Proc. Natl. Acad. Sci. U.S.A.* **82**, 6201–6205.

Lovett, P. S., Duvall, E. J. and Keggins, K. M. (1976). *J. Bacteriol.* **127**, 817–828.

Low, K. B. and Porter, D. D. (1978). *Annu. Rev. Genet.* **12**, 249–287.

MacLachlan, P. R. and Sanderson, K. E. (1985). *J. Bacteriol.* **161**, 442–445.

Macrina, F. L., Jones, K. R. and Welch, R. A. (1981). *J. Bacteriol.* **146**, 826–830.

Makins, J. F. and Holt, G. (1981). *Nature (London)* **293**, 671–672.

Mandel, M. and Higa, A. (1970). *J. Mol. Biol.* **53**, 157–162.

Martin, P. A. W., Lohr, J. R. and Dean, D. H. (1981). *J. Bacteriol.* **145**, 980–983.

Matsushima, P. and Baltz, R. H. (1985). *J. Bacteriol.* **163**, 180–185.

Matsushima, P., McHenney, M. A. and Baltz, R. H. (1987). *J. Bacteriol.* **169**, 2298–2300.

McCarty, M. (1980). *Ann. Rev. Genet.* **14**, 1–15.

Mercer, A. A. and Loutit, J. S. (1979). *J. Bacteriol.* **140**, 37–42.

Merrick, M. J., Gibbins, J. and Postgate, J. R. (1987). *J. Gen. Microbiol.* **133**, 2053–2057.

Michel, B., Niaudet, B. and Ehrlich, S. D. (1983). *Plasmid* **10**, 1–10.

Miramadi-Tehrani, J., Mitchell, J. I., Williams, S. T. and Ritchie, D. A. (1986). *Lett. Appl. Microbiol.* **3**, 27–30.

Mooibroek, H. and Venema, G. (1982). *Mol. Gen. Genet.* **185**, 165–168.

Morelli, L., Cocconcelli, P. S., Bottazzi, V., Damiani, G., Ferretti, L. and Sgaramella, V. (1987). *Plasmid* **17**, 73–75.

Morrison, D. A. (1977). *J. Bacteriol.* **132**, 349–351.

Morrison, D. A., Trombe, M-C., Hayden, M. K., Waszak, G. A. and Chen, J-D. (1984). *J. Bacteriol.* **159**, 870–876.

Mottes, M., Grandi, G., Sgaramella, V., Canosi, U., Morelli, G. and Trautner, T. A. (1979). *Mol. Gen. Genet.* **174**, 281–286.

Murchison, H. H., Barrett, J. F., Cardineau, G. A. and Curtiss III, R. (1986). *Infect. Immun.* **54**, 273–282.

Murooka, Y., Iwamoto, H., Hamamoto, A. and Yamaguchi, T. (1987). *J. Bacteriol.* **169**, 4406–4409.

Negoro, S., Shinagawa, H., Hagata, A., Kinoshita, S., Hatozaki, T. and Okada, H. (1980). *J. Bacteriol.* **143**, 245–248.

Niesel, D. W. and Hertzog, N. K. (1983). *Focus (BRL Technical Bulletin)* **5**, 6–7.

Norgaard, M. V., Keem, K. and Monahan, J. J. (1978). *Gene* **3**, 279–292.

Norlander, L., Davies, J. and Ormark, S. (1979). *J. Bacteriol.* **138**, 756–761.

Notani, N. K., Setlow, J. K., McCarthy, D. and Clayton, N-L. (1981). *J. Bacteriol.* **148**, 812–816.

Novick, R. P., Sanchez-Ruras, C., Gruso, A. and Edelman, R. (1980). *Plasmid* **3**, 348–358.

Ozaki, A., Katsumata, R., Oka, T. and Furuya, A. (1984). *Mol. Gen. Genet.* **196**, 175–178.

Pidcock, K. A., Montenecourt, B. S. and Sands, J. A. (1985). *Appl. Environ. Microbiol.* **50**, 693–695.

Pifer, M. L. (1986). *J. Bacteriol.* **168**, 683–687.

Pifer, M. L. and Smith, H. O. (1985). *Proc. Natl. Acad. Sci. U.S.A.* **101**, 72–79.

Porter, R. D. (1986). *CRC Crit. Rev. Microbiol.* **13**, 111–132.

Prieto, M. J., Jofre, J., Tomas, J. and Pares, R. (1979). *In* "Transformation 1978" (S. W. Glover and L. O. Butler, Eds), pp. 339–348. Cotswold Press, Oxford.

Raleigh, E. A. and Wilson, G. (1986). *Proc. Natl. Acad. Sci. U.S.A.* **83**, 9070–9074.

Reid, J. D., Stoufer, S. D. and Ogrydziak, D. M. (1982). *Gene* **17**, 107–112.

Roberts, R. J. and Strike, P. (1981). *Plasmid* **5**, 213–220.

Rodicio, M. R. and Chater, K. F. (1982). *J. Bacteriol.* **151**, 1078–1085.

Rosenthal, A. L. and Lacks, S. A. (1980). *J. Mol. Biol.* **141**, 133–147.

Rudin, L., Sjostrom, J. E., Lindberg, M. and Philipson, L. (1974). *J. Bacteriol.* **118**, 155–164.

Rudolph, C. F., Schmidt, B. J. and Saunders, C. W. (1986). *J. Bacteriol.* **165**, 1015–1018.

Sabelnikov, A. G. and Domaradsky, I. V. (1979). *Mol. Gen. Genet.* **172**, 313–318.

Sabelnikov, A G. and Domaradsky, I. V. (1981). *J. Bacteriol.* **146**, 435–443.

Sabelnikov, A. G. and Ilyashenko, B. N. (1979). *Bull. Exp. Biol. Med.* **88**, 65–68.

Salyers, A. A., Shoemaker, N. B. and Guthrie, E. P. (1987). *CRC Crit. Rev. Microbiol.* **14**, 49–71.

Sanders, M. E. and Nicholson, M. A. (1987). *Appl. Environ. Microbiol.* **53**, 1730–1736.

Sano, Y. and Kageyama, M. (1977). *J. Gen. Appl. Microbiol.* **23**, 183–186.

Santamaría, R. I., Gil, J. A. and Martin, J. F. (1985). *J. Bacteriol.* **162**, 463–467.

Santos, E. and Kaback, R. H. (1981). *Biochem. Biophys. Res. Commun.* **99**, 1153–1160.

Saunders, C. W. and Guild, W. R. (1980). *Mol. Gen. Genet.* **180**, 573–578.

Saunders, C. W. and Guild, W. R. (1981a). *Mol. Gen. Genet.* **181**, 57–62.

Saunders, C. W. and Guild, W. R. (1981b). *J. Bacteriol.* **146**, 517–526.

Saunders, J. R., Docherty, A. and Humphreys, G. O. (1984). *Meth. Microbiol.* 1st edn. **17**, 61–75.

Saunders, J. R., Hart, C. A. and Saunders, V. A. (1986). *J. Antimicrob. Chemother.* **18**, 57–66.

Saunders, J. R., Ball, A., Brown, M. G. M., Humphreys, G. O. and Edwards, C. (1987). *FEMS Microbiol. Lett.* **42**, 125–128.

Saunders, V. A. (1984). *In* "Aspects of Microbial Metabolism and Ecology" (Special Publication of Soc. Gen. Microbiol. 11) pp. 241–276 (G. A. Codd, Ed.) Academic Press, London.

Saunders, V. A. and Saunders, J. R. (1987). "Microbial Genetics Applied to Biotechnology. Principles and Techniques of Gene Transfer and Manipulation". pp. 188–192. Croom Helm, Beckenham, Kent.

Schaffner, W. (1980). *Proc. Natl. Acad. Sci. U.S.A.* **77**, 2163–2167.

Scocca, J. J., Polland, R. L. and Zoon, K. C. (1974). *J. Bacteriol.* **118**, 369–373.

Selvaraj, G. and Iyer, V. N. (1981). *Gene*, **15**, 279–283.

Setlow, P. (1976). *J. Biol. Chem.* **51**, 7853–7862.

Setlow, P. (1979). *In* "Limited proteolysis in microorganisms" (G. N. Cohen and H. Holzer, Eds) pp. 109–113. U.S. Department of Health Education and Welfare, Bethesda, Md.

Simon, D., Rouault, A. and Chopin, M-C. (1986). *Appl. Environ. Microbiol.* **52**, 394–395.

Sinclair, M. I. and Morgan, A. F. (1978). *Aust. J. Biol. Sci.* **31**, 679–688.

Singh, R. N. and Pitale, M. P. (1967). *Nature (London)* **213**, 1262–1263.

Sisco, K. L. and Smith, H. O. (1979). *Proc. Natl. Acad. Sci. U.S.A.* **76**, 972–976.

Smith, C. J. (1985a). *J. Bacteriol.* **164**, 466–469.

Smith, C. J. (1985b). *J. Bacteriol.* **164**, 294–301.

Smith, M. D. (1985). *J. Bacteriol.* **162**, 92–97.

Smith, H. O., Danner, D. B. and Deich, R. A. (1981). *Annu. Rev. Biochem.* **50**, 41–68.

Smith, H. O., de Vos, W. and Bron, S. (1983a). *J. Bacteriol.* **153**, 12–20.

Smith, H. O., Wiersma, K., Bron, S. and Venema, G. (1983b). *J. Bacteriol.* **156**, 101–108.

Smith, H. O., Wiersma, K., Venema, G. and Bron, S. (1984) *J. Bacteriol.* **157**, 733–738.

Smith, H., Wiersma, K., Venema, G. and Bron, S. (1985). *J. Bacteriol.* **164**, 201–206.

Smith, M. D., Flickinger, J. L., Lineberger, D. W. and Schmidt, B. (1986). *Appl. Environ. Microbiol.* **51**, 634–649.

Soutschek-Bauer, E., Hartl, L. and Saudenbauer, W. L. (1985). *Biotechnol. Lett.* **7**, 705–710.

Sox, T. E., Mohammed, W. and Sparling, P. F. (1979). *J. Bacteriol.* **138**, 510–518.

Sparling, P. F. (1977). *In* "The Gonococcus" (R. Roberts, Ed.), pp. 111–135, Wiley, New York.

Sparling, P. F., Biswas, G., Graves, J. and Blackman, E. (1980). *In* "Genetics and Immunobiology of Pathogenic *Neisseria*" (S. Normark and D. Danielsson, Eds), pp. 123–125, University of Umea.

Spencer, D. W. and Barr, G. C. (1981). *FEMS Microbiol. Lett.* **12**, 159–161.

Stassi, D. L., Lopez, P., Espinosa, M. and Lacks, S. A. (1981). *Proc. Natl. Acad. Sci. U.S.A.* **78**, 7028–7032.

Stewart, G. J. and Carlson, C. A. (1986). *Annu. Rev. Microbiol.* **40**, 211–235.

Strike, P., Humphreys, G. O. and Roberts, R. J. (1979). *J. Bacteriol.* **138**, 1033–1035.

Stuy, J. H. (1979). *J. Bacteriol.* **139**, 520–529.

Stuy, J. H. (1980). *J. Bacteriol.* **142**, 925–930.

Stuy, J. H. and Walter, R. B. (1986). *Mol. Gen. Genet.* **203**, 288–295.
Suzuki, M. and Szalay, A. A. (1979). *In* "Methods in Enzymology" (R. Wu, Ed.).
    Vol. 68, Recombinant DNA, pp. 331–342. Academic Press, New York and London.
Swanson, J. (1978). *Infect. Immun.* **19**, 320–331.
Takahashi, W., Yamagata, H., Yamaguchi, K., Tskukagoshi, N. and Udaka, S. (1983).
    J. Bacteriol. **156**, 1130–1134.
Thompson, C. J., Ward, J. M. and Hopwood, D. A. (1982). *J. Bacteriol.* **151**, 668–677.
Thompson, R. and Achtman, M. (1979). *Mol. Gen. Genet.* **16**, 49–57.
Thudt, K., Schleifer, K. H. and Gotz, F. (1985). *Gene,* **37**, 163–169.
Trautner, T. A. and Spatz, H. C. (1973). *Curr. Top. Microbiol. Immunol,* **62**, 61–88.
Tucker, W. T. and Pemberton, J. M. (1980). *J. Bacteriol.* **143**, 43–49.
van Alphen, L., Verkleij, A., Leunissen-Bijvelt, J. and Lugtenberg, B. (1978). *J.
    Bacteriol.* **134**, 1089–1098.
van Nieuwenhoven, M. H., Hellingwerf, K. J., Venema, G. and Konings, W. N. (1982).
    J. Bacteriol. **151**, 771–776.
Vapnek, D., Alton, N. K., Bassett, C. L. and Kushner, S. R. (1976). *Proc. Natl. Acad.
    Sci. U.S.A.* **73**, 3492–3496.
Venema, G. (1979). *Adv. Microb. Physiol.* **19**, 245–331.
Vijayakumar, M. N. and Morrison, D. A. (1986). *J. Bacteriol.* **165**, 689–695.
Vorobjeva, I. P. and Khmel, I. A. (1979). *In* "Advances in Protoplast Research" (L.
    Ferenczy and G. L. Farkas, Eds). pp. 37–41, Pergamon, Oxford.
Vorobjeva, I. P., Khmel, I. A. and Alfoldi, L. (1980). *FEMS Microbiol. Lett.* **7**, 261–
    263.
Voss, J. G. (1967). *J. Gen. Microbiol.* **48**, 391–400.
Wackernagel, W. (1973). *Biochem. Biophys. Res. Commun.* **51**, 306–311.
Warren, G. and Sherratt, D. (1978). *Mol. Gen. Genet.* **161**, 39–47.
Weber, H. and Weissmann, C. (1983). *Nucl. Acids. Res.* **11**, 5661–5669.
Weinrauch, Y. and Dubnau, D. (1983). *J. Bacteriol.* **154**, 1077–1087.
Weinrauch, Y. and Dubnau, D. (1987). *J. Bacteriol.* **169**, 1205–1211.
Weissmann, C., Nagata, S., Tanaguchi, T., Weber, H. and Meyer, F. (1979). *In* "Genetic
    Engineering, Principles and Methods". Vol. 1 (J. K. Setlow and A. Hollaender, Eds)
    pp. 133–150. Plenum Press, New York.
Weston, A., Humphreys, G. O., Brown, M. G. M. and Saunders, J. R. (1979). *Mol.
    Gen. Genet.* **172**, 113–119.
Weston, A., Brown, M. G. M., Perkins, H. R., Saunders, J. R. and Humphreys, G.
    O. (1981). *J. Bacteriol.* **145**, 780–787.
Wilson, G. A. and Bott, K. F. (1970). *Biochim. Biophys. Acta.* **199**, 464–475.
Wirth, R., An, F. Y. and Clewell, D. B. (1986). *J. Bacteriol.* **165**, 831–836.
Yamamoto, H., Maures, K. H. and Hutchinson, C. R. (1986). *J. Antibiot.* **39**, 1304–
    1313.
Yother, J., McDaniel, L. S. and Briles, D. E. (1986). *J. Bacteriol.* **168**, 1463–1465.
Young, F. E. (1980). *J. Gen. Microbiol.* **119**, 1–15.

5

Preparation and Electrophoresis of Plasmid DNA

J. GRINSTED AND P. M. BENNETT

Department of Microbiology and Unit of Molecular Genetics, University of Bristol, Medical School, Bristol, UK

I. Introduction

The main problem to overcome in the preparation of plasmid DNA is its separation from chromosomal DNA, since the plasmid usually comprises no more than about 5% of the total DNA. The physical characteristics that permit separation are the relatively small size of plasmids, their covalently closed circular structure, and the fact that they are not bound to other cellular components in a lysate. [It should be noted that chromosomal DNA is very large (in *Escherichia coli*, for example, about 4700 kb), so that lysis of the bacteria and manipulation of the lysate inevitably results in random shearing of the chromosomal molecules. In general, the size of the resulting fragments will depend on the vigour with which the lysate has been handled.] In this chapter, we start by describing the properties of covalently closed circular

METHODS IN MICROBIOLOGY
VOLUME 21   ISBN 0-12-521521-5

(CCC) DNA that permit its separation from other forms of DNA. Methods that are successfully used in this laboratory for the isolation of plasmid DNA are then described. Other chapters in this volume refer to related techniques and these will be referred to as appropriate. There are a number of books on the market that contain advice on the sort of techniques that are described here; we especially recommend the manual entitled 'Molecular Cloning' from Cold Spring Harbor (Maniatis et al., 1982).

II. Nature of CCC DNA

Plasmids consist of double-stranded DNA and are generally isolated as circular molecules in which each of the single strands is a covalently closed circle. Since the single strands are also wound round each other to give the double helix, they are inextricably linked together, and the strands cannot escape from each other even if the forces that normally keep them together in the double helix are broken. Such links have been called topological bonds and require that both of the single strands be covalently closed, since a single break in just one strand will introduce a point of free rotation that will allow the now discontinuous strand to wind off the other and escape. This topological binding is the basis of methods of plasmid isolation that use high pH (see below). At pH 12.5 the forces that hold the double helix together are disrupted (i.e. the DNA is denatured). When the solution is neutralized, the DNA can renature, provided that complementary strands can find each other again. With linear DNA, or circular DNA which is not covalently closed in both strands, the complementary strands would have completely separated from each other, but with CCC DNA this is not the case. Thus, CCC DNA will preferentially renature and can be separated from other forms of DNA, which will still be single-stranded.

The topological linking of the two single strands in CCC DNA is described by the linking number (α), which is the net number of times one of the strands crosses the other. The linking number of a CCC DNA molecule is a constant: it cannot be changed unless there is a break in one of the strands so that this strand can rotate with respect to the other. The linking number is related to the double helix in the following way:

$$\alpha = \alpha_0 + \Delta\alpha \qquad\qquad (5\text{-}1)$$

where α_0 is approximately numerically equal to the number of turns of the double helix in a linear molecule of identical size, and $\Delta\alpha$ is the difference between α and α_0. With almost all naturally occurring CCC DNA, $\alpha < \alpha_0$. Thus, $\Delta\alpha$ is negative and the molecules are supercoiled in a right-handed sense with $\Delta\alpha$ approximately equal to the number of supercoils. (Recently, positively supercoiled circular DNA has been discovered in the archaebacterium, *Sul-*

folobus, which grows at 80°C (Nadal *et al.*, 1986).) Supercoiled molecules are strained with respect to the open form. Thus, supercoils can only be maintained if the two single strands of the double helix are constrained from rotating with respect to each other, as in CCC DNA. Wang (1980) and Bauer *et al.* (1980) give straightforward descriptions of the properties of CCC DNA.

α_0 is determined by the dimensions of the double helix (see above) and can be changed by the addition of ethidium bromide (EB); this drug intercalates between the base pairs of the double helix, lengthening and unwinding it (Radloff *et al.*, 1967). Thus, in the presence of EB, α_0 of CCC DNA is decreased, with the consequence that $\Delta\alpha$ increases (since α is a constant). With natural CCC DNA (where $\alpha < \alpha_0$ in the absence of EB, see above), the consequence of adding increasing concentrations of EB will be that $\Delta\alpha$, which starts at some negative value, will become zero and then, as the concentration of EB continues to increase, will become positive. To put it another way, right-handed supercoils will be removed and will eventually be replaced by left-handed supercoils. Since supercoiled molecules are in a state of strain (see above), unwinding of the supercoils is energetically favoured and the establishment of supercoils is disfavoured. Thus, when $\alpha < \alpha_0$ (at low concentrations of EB) EB will preferentially bind to CCC DNA (compared to other forms of DNA); but when $\alpha > \alpha_0$ (at high concentrations of EB) the reverse is the case since binding of the drug to CCC DNA will now result in the generation of supercoils. Thus, at high concentrations of EB, CCC DNA binds less of the drug than do other forms of DNA. This is the basis of CsCl/EB gradients (see below): binding of EB to DNA reduces its density and since, at high drug concentrations CCC DNA binds less of the drug than other forms of DNA, its density will be reduced less. This density difference is exploited with the CsCl density gradient.

III. Rapid methods for analysis of plasmid DNA

There are numerous methods by which plasmids can be separated from chromosomal DNA without the need for CsCl/EB gradients. It should be noted that such rapid preparations will invariably be contaminated to a greater or lesser extent with chromosomal DNA. For most purposes, however, this does not matter and such preparations will usually be adequate for analysis with restriction enzymes, for transformation and also in many cases for cloning. The great advantage of these methods of preparation is their speed: a single worker could reasonably expect to make hundreds of preparations in a day. It should be noted, however, that these methods are designed assuming the use of a microfuge (a small bench-top centrifuge capable of 10000 r/min), and their rapidity and convenience requires this piece of equipment. (There are now many firms

producing such centrifuges; we find the MSE MicroCentaur to be very satisfactory: it is inexpensive, and quiet in operation, which is an important factor
in a machine that is used in working areas.) Scaled-up versions of the methods
for removing the bulk of chromosomal DNA are also used as the first stage in
the preparation of purified plasmid DNA (Section IV).

In general, the smaller the plasmid, the better the yield of plasmid DNA.
There are two reasons for this: firstly, smaller plasmids are less prone to physical
damage during the preparation procedure; secondly, there are usually many
copies of the smaller plasmids present in each cell but few of large plasmids
(see Chapter 1). Indeed, large plasmids can be very difficult to isolate at all.
Even so, there are reports of reliable methods for preparation of very large
plasmids in some circumstances. (For example, plasmids of up to 200 kb can
be isolated from *Rhizobium* (Hirsch *et al.*, 1980).) Giant linear plasmids
have been resolved from the chromosomes of *Streptomyces* by pulsed-field
electrophoresis (Kinashi *et al.*, 1987).

A. Analysis of plasmid content of bacteria

In certain circumstances (e.g. in the preliminary screening of strains) a simple
procedure that just displays the plasmids contained by a strain is required. In
principle the simplest method is that of Eckhardt (1978), in which bacteria are
lysed by treatment with lysozyme followed by sodium dodecyl sulphate (SDS)
in the wells of the agarose gel to be used for analysis. When the gel is
electrophoresed, chromosomal DNA hardly penetrates into the gel, so that
the plasmids are displayed. This technique has the obvious advantage of
convenience, since there is minimal manipulation. However, it is notoriously
unreproducible.

In practice, it is generally simpler, and more dependable, to use one of the
standard rapid methods for plasmid preparation (see below). These involve a
step to remove the chromosomal DNA but they can be very fast: the method
described in detail in B, for instance, gives a plasmid preparation in about
30 min. An advantage of this is that the preparation is then available for
restriction enzyme analysis or other procedures after the display of the plasmids.
The preparations can be made even quicker by omitting a later stage of the
procedure, in particular, precipitation steps (e.g. see Kado and Liu, 1981); this
would mean, however, that the preparations would not be suitable for further
analysis.

B. Rapid methods for the preparation of plasmid DNA

To isolate plasmid DNA, first the cells have to be lysed, and second the bulk of the chromosomal DNA must be removed. There are two main categories of method that achieve this: in the first, lysis is as gentle as possible and the chromosomal DNA can be removed by centrifugation (a 'clearing spin'); in the second, one aims for complete lysis and then removes the chromosomal DNA by selective precipitation.

For laboratory strains of E. coli, lysozyme treatment followed by gentle lysis with Triton X-100, as described in detail in the first edition of this book, gives a lysate from which the chromosomal DNA can be removed by a short spin. However, some bacteria are not lysed by this procedure, whereas others (including some strains of E. coli) are lysed but the chromosomal DNA does not pellet when the lysate is centrifuged. The success of this method, therefore, depends on the actual strain being tested. A similar procedure, substituting lysostaphin for lysozyme, can be used to isolate plasmids from Staphylococcus aureus. A more rapid version of the Triton method is that of Holmes and Quigley (1981), in which the bacteria are resuspended in a solution containing Triton X-100 and lysozyme and this mixture is then incubated briefly at 100°C. Centrifugation then removes the bulk of the protein and chromosomal DNA.

More general methods that are applicable to a wide range of bacterial species are those that involve lysis with SDS followed by denaturation and selective renaturation; SDS should lyse most bacteria and dissociate DNA from proteins etc., and selective renaturation is dependent only on the properties of the DNA, so that the separation of the plasmid should not depend on the particular properties of the bacteria. Such methods (e.g. Birnboim and Doly, 1979; Kado and Liu, 1981; Ish-Horowicz and Burke, 1981) usually use a lysing mixture of SDS dissolved in NaOH, sometimes after treatment with lysozyme. This treatment will lyse many types of bacteria (the prior lysozyme treatment is necessary with some bacilli, for instance), and the high pH is provided at the same time. The chromosomal DNA is then selectively precipitated from the lysate either by neutralization with acetate, which gives coprecipitation of the SDS and chromosomal DNA (Birnboim and Doly, 1979; Ish-Horowicz and Burke, 1981), or by direct extraction with unneutralized phenol (Kado and Liu, 1981; Kieser, 1984), which results in the chromosomal DNA banding at the interface (in this case, the lysate is heated at about 60°C prior to the phenol extraction).

In the first edition of this book, we described a Triton X-100 clearing spin method in detail, this being the method that we used at the time. Since then we have changed our standard procedures and now use a method based on lysis with SDS/NaOH followed by precipitation by addition of acetate. This is described in detail below; it is based on the method described by Ish-Horowicz and Burke (1981). The described procedure is for 10 ml of an overnight culture in nutrient broth in a universal bottle. All manipulations are carried out at room temperature on the bench. Apart from the initial harvesting, 1.5 ml plastic

Eppendorf tubes are used throughout and all centrifugation is carried out in a microfuge.

1. Harvest the culture in a bench-top centrifuge (spin for 10 min) and resuspend the bacterial pellet in 0.2 ml of water. Transfer the suspension to a 1.5 ml plastic Eppendorf tube.
2. Add 0.4 ml of 1 % (w/v) SDS in 0.2 M NaOH. Mix by inversion. Lysis usually occurs immediately, although in some cases the mixture may have to be left for a few minutes. In general, lysis is not complete, so a transparent solution should not be expected.
3. Add 0.3 ml of 3 M potassium acetate, 2 M acetic acid, 1 mM EDTA and mix. A heavy white precipitate will be seen as soon as the solution is added.
4. Spin for 5 min. Remove supernatant with an automatic pipette or by decanting. Discard pellet.
5. Add 0.5 ml of buffered phenol solution (Section V describes its preparation). Vortex, then spin for 1 min (this is a phenol extraction to remove protein). Remove the aqueous (top) phase containing DNA, avoiding collecting the denatured protein at the interface (but don't worry if some of this is sucked up). Discard phenolic (bottom) phase.
6. Add 0.1 ml of 4 M sodium acetate (pH 6) and 0.5 ml of isopropanol. Mix by inversion and then spin for 5 min. Remove supernatant by decanting, or with a pipette, and discard.
7. Add 1 ml of ethanol and spin for 1 min. Discard supernatant.
8. Add 1 ml of ether and spin for 1 min. Discard supernatant. Leave pellet to dry (for a few minutes) and then dissolve in $100 \mu l$ of Tris-HCl (pH 7.5), EDTA (0.1 mM).

The pellet should be readily visible after the isopropanol precipitation; and it tends to come loose from the side of the tube when ethanol and, particularly, ether is added, so care must be taken at this stage. It should also be noted that the precipitated DNA is often distributed up the sides of the tube and will not become visible until it is dried. Such preparations will contain RNA, which will show up on a gel as a very bright fast running band. The RNA can be removed by RNase A: $1 \mu l$ of a boiled solution of RNase A ($1 \, mg \, ml^{-1}$) could be added to the final solution above; this could then be phenol extracted and precipitated (in which case the size of the pellet would be dramatically reduced), or the solution could be stored and used containing RNase A, which does not affect most manipulations. An alternative to RNase treatment is to heat the lysate prior to the precipitation step; since the solution is at high pH, this will degrade the RNA, and the degradation products will be removed at the final isopropanol precipitation.

$10 \mu l$ of such a preparation should be sufficient for analysis with restriction enzymes (Chapter 6). In addition to being susceptible to restriction enzymes, such preparations can be used for transformation and in cloning experiments. As regards the speed of preparation, one person with two microfuges can make 48 preparations in 2–3 h.

The major difference compared with the original method (Ish-Horowicz and

Burke, 1981) is that we resuspend the bacteria in water rather than glucose (50 mM), EDTA (10 mM), Tris-HCl (25 mM, pH 8.0), which was said to be necessary as a buffer. Other differences are in the temperatures used (we use room temperature throughout, the original carries out many of the steps on ice) and the times mixtures are left (we generally add solutions and immediately process the mixtures, while the original leaves various incubation times); finally, it was originally recommended that the NaOH/SDS solution be made up fresh each week; in our experience this is not necessary.

The procedure could be speeded up by using the microfuge to spin down the bacteria (about 15 s is adequate; longer will give a pellet that is hard to resuspend). In this case, 1.1–1.5 ml of culture could be transferred to an Eppendorf tube (or even grown in it if aeration is not a problem); under these circumstances, volumes given above would be halved. Single colonies or patches on plates could also be used; they would be scraped off plates and resuspended in SDS/NaOH in an Eppendorf tube. The procedure would also be shortened by removing the phenol step: we normally include this to remove the last traces of protein, but it is not strictly necessary (Ish-Horowicz and Burke, 1981). For really large scale screening of clones, it is possible to grow cultures and process them in microtitre plates, if there is a suitable centrifuge available (Gibson and Sulston, 1987); a single plate would give 96 preparations, and two plates can be easily processed in a day.

IV. Preparation of pure plasmid DNA

Pure plasmid DNA is prepared by running lysates in CsCl gradients in the presence of high concentrations of EB (Section II). This treatment usually results in a density differential of $0.04–0.05 \, \text{g ml}^{-1}$ (depending on the EB concentration) between the chromosomal DNA and CCC DNA (the plasmid in this case). CsCl gradients are established by centrifugation and, therefore, have to be run for several hours to establish the equilibrium gradient. Angle rotors used to be used for CsCl gradients; the time required to establish equilibrium in these rotors is a minimum of 40 h. However, recently, vertical rotors have been introduced. In these the centrifuge tubes stand vertically in the rotor so that their length is parallel to the axis of rotation; CsCl gradients can be formed in these in as little as 5 h. During spinning, the gradient and any bands of DNA therein are obviously oriented with respect to the axis of rotation (i.e. the gradient is formed across the width of the tube). Surprisingly, perhaps, the gradient reorientates smoothly as the rotor slows down so that at rest, the bands are horizontal.

Prior to CsCl/EB gradient centrifugation, bacteria are treated with one of the methods discussed in Section III to remove the bulk of the chromosomal DNA, otherwise the gradients will be overloaded. In the method that is described below, a scale-up of the method described in III.B (without the phenol extraction) is used for this purpose. The following is for 250 ml of an overnight culture grown in nutrient broth.

1. Harvest the culture and resuspend in 5 ml of water.
2. Add 10 ml 1% (w/v) SDS in 0.2 M NaOH. Mix and leave for a few minutes at room temperature.
3. Add 7.5 ml of 3 M K acetate, 2 M acetic acid, 1 mM EDTA. Mix and leave at room temperature for a few minutes.
4. Spin at 15 000 r/min and 5°C for 5 min. Decant supernatant and discard pellet.
5. Add 0.4 ml of 4 M sodium acetate and 12 ml of isopropanol. Mix and then spin for 5 min at 15 000 r/min and 5°C. Decant and discard supernatant.
6. Fill tube with ethanol and spin briefly. Decant and discard supernatant.
7. Dry pellet and resuspend in 3.5 ml of Tris-HCl (10 mM, pH 7.5), EDTA (0.1 mM). Add 0.2 ml of EB solution (10 mg ml^{-1} in water) and 3.7 g of CsCl. Agitate gently to dissolve CsCl.
8. Check the refractive index of the solution. This should be about 1.386 (density 1.55 g ml^{-1}). We use a simple sugar refractometer and aim for a reading of 34–35%. If the preparation contains a lot of protein, the refractive index at this stage would be about 38% on our instrument.
9. The solution is put into a 5 ml polyallomer centrifuge tube (Dupont-Sorvall Cat. No. 03127). About 4.6 ml is required to fill the tube. If necessary, add a CsCl solution of the same density. Spin in the vertical Sorvall rotor (Catalogue TV865) at 55 000 r/min and 15°C for 5 h, or at 48 000 r/min and 15°C overnight, in a Sorvall OTD65 centrifuge.

The tube is removed from the rotor and illuminated with U.V. light; the tube is transparent to U.V. light and the DNA appears as bright orange fluorescent bands. There is usually also a purple pellicle on top of the solution and running down the inside of the tube (this is protein) and a bright fluorescent strip running down the other side of the tube (RNA). The pellicle can be removed (it sticks very nicely to a disposable plastic spatula); the upper band of chromosomal DNA is then removed with a syringe and discarded; then the lower band of plasmid DNA is removed with a clean syringe. If a second CsCl/EB spin is needed, this plasmid solution can immediately be diluted with a CsCl solution of the correct density (1.55 g ml^{-1}, see above), a little EB is added and the mixture spun again as before.

To extract the plasmid from the CsCl/EB solution the following procedure is used.

1. Dilute the solution three-fold with Tris-HCl (10 mM, pH 7.5) and distribute this solution into 1.5 ml plastic Eppendorf tubes such that each tube contains 0.8 ml (typically, two or three tubes will be required).
2. To each tube add 70 μl of sodium acetate (4 M, pH 6) and 0.5 ml of isopropanol. Mix by inversion. Spin for 5 min in a microfuge at room temperature. Decant supernatant and discard.

3. Add 1 ml of isopropanol solution (40%(v/v) containing 0.2 M sodium acetate, pH 6) to remove last drops of CsCl solution. Spin for 1 min. Decant and discard supernatant.
4. Add 1 ml of ethanol, spin and decant and discard the supernatant.
5. Dissolve and combine the pellets in a total of 0.8 ml of Tris-HCl (10 mM, pH 7.5).
6. Add equal volume of buffered phenol (Section V). Mix thoroughly and spin for 1 min. Remove upper aqueous phase and discard lower phase.
7. Add 70 μl of sodium acetate (4 M, pH 6) and 0.5 ml of isopropanol. Mix and spin for 1 min. Remove and discard supernatant.
8. Add 1 ml of ethanol, spin for 1 min, and remove and discard supernatant.
9. Add 1 ml of ether, spin for 1 min and remove and discard the supernatant.
10. Dry pellet (a few minutes on the bench) and dissolve in about 500 μl of Tris–HCl (10 mM, pH 7.5), EDTA (0.1 mM).

This procedure for the extraction of plasmids from CsCl/EB solutions is very fast (about 15 min from rotor to a clean solution ready for analysis). It should be noted that no special effort is expended to remove the EB; the precipitations and the phenol treatment are quite sufficient.

Sometimes this method results in a red non-fluorescent band (it can be seen in daylight) close to the plasmid band in the CsCl/EB gradient (if it is coincident it quenches the fluorescence of the plasmid in the U.V.). It is presumably a carbohydrate-EB complex (it does not occur if polyethylene glycol is used instead of isopropanol to precipitate the DNA from the cleared lysate). Its position relative to the plasmid is easily manipulated by adjusting the amount of EB in the gradient (i.e. by changing the density of the plasmid), and it is not normally a problem. Some workers have dealt with this by doing an initial CsCl gradient in very low concentrations of EB (so that all the DNA bands together), and then taking this band and running it in a high concentration of EB (Halford and Johnson, 1981).

The yield from this method is variable. Typically, starting from one litre of a culture of a strain containing a high copy number plasmid and ending with a 500 μl preparation. 1–2 μl should be sufficient for visualization on an agarose gel.

V. Electrophoresis of plasmid DNA

Plasmid preparations can be simply checked by electrophoresis through agarose gels. This will give an indication of the amount and size of other DNA and RNA present. With strains containing multiple plasmids (such as many from natural sources), it will also show the number and sizes of the plasmids present (the 'plasmid profile').

A. Agarose gels

Details of agarose gel electrophoresis are discussed in Chapter 6. [For rapid analysis of a preparation, a 'mini-gel' (which can be cast on a microscope slide) can be used.] Usually, the gels are run without ethidium and stained with the dye afterwards (see Chapter 6). In general, a 0.7% or 1% gel would be used, depending on the size of the plasmids to be displayed. (The larger the plasmids, the lower the percentage agarose needed to resolve them.) If necessary, it is possible to go down to 0.3% agarose with horizontal gels, but the agarose does get very sloppy and difficult to handle at the lower concentrations. But, if there is more than one large plasmid present (greater than 60 kb, for example), it may be difficult to resolve them satisfactorily. This can be a problem with natural isolates, which can contain both large and small plasmids that one wants to display on the same gel to give a plasmid profile. Thus, to optimize resolution of a number of large plasmids, electrophoresis should be allowed to continue for as long as possible. (For example, plasmids of between 100 kb and 200 kb from *Rhizobium* can be resolved on a 0.7% gel (Hirsch *et al.*, 1980).) Running the gel for extended periods will result in loss of the smaller plasmids off the end. (It should be noted that the recently introduced techniques that employ pulsed-field electrophoresis to resolve very large DNA molecules (see Chapter 6) is only applicable to linear DNA.)

B. Appearance of contaminants

After electrophoresis, the gel is illuminated with U.V. light so that the DNA can be visualized (see Chapter 6). If the plasmid preparation is clean, all that will be seen is a band corresponding to the CCC DNA and, maybe, another, fainter band which is the open circular form (see below). Contaminating chromosomal DNA usually shows as a relatively tight band (although not as sharp as the plasmid), in a position that will depend on the method of preparation. Single-stranded plasmid DNA runs well ahead of the native plasmid, and denatured CCC DNA runs ahead of the native CCC DNA (the latter two are sometimes seen in preparations from methods that use NaOH—see Kieser (1984)). Sometimes, preparations are contaminated with fragments of random size, so that there is a smear right down the track. (Plasmid preparations sometimes degrade on storage and display such an appearance.) (The paper by Kieser (1984) gives examples of all these sorts of contaminants.) If there is RNA present there will be a diffuse, bright area well in front of the plasmids.

C. Different plasmid forms and size determination

A fresh plasmid preparation usually contains mostly CCC DNA. There is often a small amount of the open circular (OC) form also present (the proportion is likely to be higher with larger plasmids). As the preparation ages, the proportion of the OC form increases, and sometimes even the linear form appears eventually. (See Section II for the relationship between the forms.) The problem this presents is that these forms have different electrophoretic mobilities, so that the same plasmid can appear on a gel as three bands that are widely separated. The electrophoretic mobility of DNA molecules changes with different conditions, such as change of the running buffer, or agarose concentration, or voltage (see Longo and Hartley, 1986), but the changes for the three forms of plasmid are not uniform, and their relative positions on the gel can change dramatically, depending on the gel and the conditions of electrophoresis. For example, the mobilities are usually in the order OC DNA (slowest), linear, CCC DNA (fastest), but under some conditions, the linear and the CCC species run together.

These factors result in a number of potential problems for interpreting the pattern of bands seen in gels. Firstly, care must be taken when the size of plasmids is to be determined. The size of DNA molecules in gels is estimated by comparing the mobility of the unknown with the mobilities of a set of marker molecules of known size run in the same gel (see Chapter 6). The differing mobilities of different forms of DNA requires that the size markers must be the same form as the unknown. So, in the case of plasmids, the markers must be CCC DNA that are supercoiled to a similar extent (see below): the markers should be plasmids that have been isolated from similar strains. There are some marker 'ladders' (mixtures of supercoiled plasmids of various sizes) available commercially. (BRL, for instance, provides a 'supercoiled ladder' with 11 rungs between 2 kb and 16 kb.)

A second difficulty can be in deciding on the identity of bands in the gel, particularly where there is more than one plasmid present. Thus, it may not be clear whether a particular band is the OC form of one plasmid or the CCC form of another, larger, plasmid. Various methods have been published to distinguish OC from CCC molecules, based generally on converting the CCC form to its corresponding OC form. For instance, Hintermann et al. (1981) described a method that involves cutting the track from the gel, illuminating it with U.V. light (to convert CCC forms to OC forms), and then embedding the track in another gel and running the DNA at right angles to the first dimension. In this way, the CCC bands in the first dimension can be distinguished.

Some plasmids in recombination-proficient strains exist as a mixture of oligomers: thus, in addition to the expected monomer, there may be other DNA molecules present on the gel with integral multiples of the size of the monomer. This happens particularly with small plasmids such as those based on ColEl.

The mobility of the DNA in the gel depends on its shape, which is why the compact CCC form travels faster than the OC form. However, a CCC DNA molecule can exist as different topoisomers, which differ only in their linking number (see Section II), and these topoisomers vary in the extent of supercoiling and, thus, compactness. So, different topoisomers of a particular CCC DNA should be separable on an agarose gel. In fact, to get good separation on agarose gels of topoisomers, they have to have not more than about 20 supercoils. CCC DNA isolated from natural sources usually is more highly supercoiled than this, so that, in spite of the fact that plasmids are actually a mixture of topoisomers, they run as a single CCC DNA band. However, some naturally occurring plasmids are not very highly supercoiled and, thus, show their topoisomers on an agarose gel. An example of this is the *Streptomyces* plasmids shown by Kieser (1984). The gels show a ladder of the various topoisomers. If the topoisomers of a more usual plasmid were to be displayed on an agarose gel, the supercoils would have to be reduced. This can be done by adding an intercalating drug (see Section II). Thus, gels containing carefully controlled concentrations of ethidium or chloroquine can be used to analyse the topoisomers of a CCC DNA preparation (Shure *et al.*, 1977).

VI. Miscellaneous observations

To complete this chapter, we comment on various aspects of the methods discussed above.

A. Flexibility

As with any technique, much time can be saved if important variables are distinguished from unimportant ones. Many of the details in the procedures described should not be considered as prescriptions; they are simply descriptions of what was done rather than what had to be done. Alternatively, relatively small changes in the important variables can have dramatic effects. So, if a particular technique does not immediately work with a particular strain, small changes in a published protocol may result in success. The reader is directed to the article by Kieser (1984) for a systematic study of the development of a rapid method and a discussion of the factors that can be important.

B. Precipitation of DNA

DNA is usually precipitated from solutions either with ethanol or with iso-propanol; the former is probably the most popular. It is usually recommended that, after addition of the alcohol, the solution is cooled to −20°C or −70°C prior to spinning (with isopropanol solutions this will mean that freezing will occur). It is our experience that isopropanol precipitation at room temperature, leaving only 1–2 min for the precipitation to occur, works very satisfactorily for most purposes. Besides saving time, this has the advantage that isopropanol precipitation needs a smaller volume of the alcohol than does ethanol precipitation (say 0.6 volumes as opposed to 2), which can be very useful, and that the lower concentration of alcohol and the higher temperature gives a degree of selective precipitation of the DNA (e.g. RNA may stay in solution under these conditions). This method of precipitation is especially useful when purifying the DNA from CsCl/EB solutions (see above). Systematic studies on the requirements for precipitation of DNA with ethanol have recently been published. These come to broadly the same conclusions as above: using either sodium acetate (Zeugin and Hartley, 1985) or ammonium acetate (Crouse and Amorese, 1987) as the salt, low temperatures and extended incubations are not required for ethanol precipitation. It was found, not surprisingly, that low concentrations of DNA (less than about 0.1 μg in a final volume of 1 ml) were not as efficiently recovered, and that the best way of improving recovery if this were necessary, was usually by extending the centrifugation time.

C. Phenol

We do not saturate our phenol with buffer. The solution is made up by taking 500 g of commercial phenol and adding to it 7.5 ml of NaOH (2 M), 130 ml of water, 6 ml of Tris-HCl (1 M, pH 7.5) and leaving the mixture overnight to liquefy. This gives a solution of pH 7.5 and 10 mM Tris. We do not redistil the phenol; in our experience, even darkly coloured phenol works satisfactorily. Following extractions with phenol, many procedures recommend that the solution is extracted with chloroform or ether to extract residual phenol prior to precipitation. This is a waste of time: precipitation of the DNA followed by the normal washes (see above) removes all phenol.

Some workers use a mixture of phenol and chloroform for deproteinization. This could be prepared simply by adding an equal volume of chloroform to the phenol solution above, shaking, leaving to settle, and then discarding the top layer. (Some people also add a small amount of 8-hydroxyquinoline as a preservative.) It is said that this mixture gives better extraction and separation of the phases than does phenol alone. The mixture is used in exactly the same way as described for phenol alone.

It should also be noted that the material that many centrifuge tubes are made of is sensitive to many organic solvents, including phenol. So check before adding; it would be disastrous, for example, if a phenol extraction was done in a polycarbonate centrifuge tube.

D. Chloramphenicol amplification

The relative amount of small multicopy plasmids (compared to chromosomal DNA) can often be increased by chloramphenicol amplification (Maniatis *et al.*, 1981). This involves incubation of the culture, prior to lysis, in the presence of chloramphenicol, under which conditions the chromosome stops replicating while the plasmids continue to replicate. This method can give very high yields of such small plasmids. The method is not applicable to large plasmids. In our experience, amplification is seldom worth the effort, and one often finds that the increased yield per cell is offset by the decreased number of cells (they do not increase in number in the presence of chloramphenicol, of course, and there is also often loss due to lysis).

References

Bauer, W. R., Crick, F. H. C. and White, J. H. (1980). *Sci. Am.* **243**, 100–113.
Birnboim, H. C. and Doly, J. (1979). *Nucleic Acids Res.* **7**, 1513–1523.
Crouse, J. and Amorese, D. (1987). *Focus (BRL Technical Bulletin)* **9**, no. 2, 3–5.
Eckhardt, T. (1978). *Plasmid* **1**, 584–588.
Gibson, T. and Sulston, J. (1987). *Gene Anal. Tech.* **4**, 41–44.
Halford, S. E. and Johnson, N. P. (1981). *Biochem. J.* **199**, 767–777.
Hintermann, G., Fischer, H.-M., Crameri, R. and Hutter, R. (1981). *Plasmid* **5**, 371–373.
Hirsch, P. R., van Montagu, M., Johnston, A. W. B., Brewin, N. J. and Schell, J. (1980). *J. Gen. Microbiol.* **120**, 403–412.
Holmes, D. S. and Quigley, M. (1981). *Anal. Biochem.* **114**, 193–197.
Ish-Horowicz, D. and Burke, J. F. (1981). *Nucleic Acids Res.* **9**, 2989–2998.
Kado, C. I. and Liu, S.-T. (1981). *J. Bact.* **145**, 1365–1373.
Kieser, T. (1984) *Plasmid* **12**, 19–36.
Kinashi, H., Shimaki, M. and Sakai, A. (1987). *Nature (London)* **328**, 454–456.
Longo, M. C. and Hartley, J. L. (1986). *Focus (BRL Technical Bulletin)* **8**, no. 3, 3–4.
Maniatis, T., Fritsch, E. F. and Sambrook, J. (1982). "Molecular Cloning. A Laboratory Manual". Cold Spring Harbor, New York.
Nadal, M., Mirambeau, G., Forterre, P., Reiter, W.-D. and Duguet, M. (1986). *Nature (London)* **321**, 256–258.
Radloff, R., Bauer, W. and Vinograd, J. (1967). *Proc. Natl. Acad. Sci. U.S.A.* **57**, 1514–1521.
Shure, M. P., Pulleybank, D. E. and Vinograd, J. (1977). *Nucleic Acids Res.* **4**, 1183–1205.
Wang, J. C. (1980). *Trends Biochem. Sci.* **5**, 219–221.
Zeugin, J. A. and Hartley, J. L. (1985). *Focus (BRL Technical Bulletin)* **7**, no. 4, 1–2.

6

Analysis of Plasmid DNA with Restriction Endonucleases

J. GRINSTED AND P. M. BENNETT

Department of Microbiology and Unit of Molecular Genetics, University of Bristol, Medical School, Bristol, UK

I. Introduction

Restriction endonucleases are enzymes that recognize specific sequences of DNA (usually 4, 5 or 6 base pairs) and subsequently cut the phosphodiester backbone of the molecule. We are concerned exclusively with Class II restriction endonucleases in this chapter; these require only magnesium ions for activity and cleave both strands of the DNA at specific points either within the recognition sequence or just outside it. The term 'restriction' derives from the discovery of such enzymes: strains that contain them restrict bacteriophage growth by initiating degradation of the bacteriophage DNA, thus rendering it biologically inactive. A necessary corollary for a cell that contains a restriction endonuclease is some system that will ensure that the cell's own DNA is not degraded. This protection is usually provided by modification of the DNA such that the sites recognized by the endonuclease are methylated. These methylated sites are not susceptible to the enzyme. A formal restriction role for most

METHODS IN MICROBIOLOGY
VOLUME 21   ISBN 0-12-521521-5

'restriction endonucleases' has not been demonstrated, and only in a few cases have a corresponding modification methylase been shown.

Since restriction endonucleases cut at specific sequences, there are only a limited number of sites in any particular DNA molecule which are susceptible to a particular enzyme. This specificity of restriction endonucleases is the cornerstone of molecular genetics. In particular, genetic engineering and DNA sequencing would be impossible without these enzymes. And now, even in the context of a simple study of plasmids, determination of fragment patterns generated by restriction endonucleases and of maps of the sites for particular enzymes on the plasmids is almost *de rigueur*.

II. Digestion with restriction endonucleases

A. Enzymes

The selection of the actual restriction endonucleases to be used depends on the DNA that is to be analysed, and on the purpose of the digestion. The procedure taken with a new DNA sequence, such as a newly-isolated plasmid, would usually be to construct a map of the sites recognized by various enzymes. One might start with enzymes that recognize 6 bp sequences (and, thus, cut on average once every 4096 bp (i.e. 4^6)) to find a few points of reference to work from; then other enzymes that might cut more frequently could be used. If simply trying to show if two plasmids are the same or not, it is usually sufficient to digest both with an enzyme or mixture of enzymes that give many fragments, and compare the patterns obtained. Complete lists of known restriction endonucleases are published regularly in, for instance, the journal *Gene* (Kessler and Holtke, 1986). In addition, many of the firms that sell the enzymes provide useful charts of the available enzymes, their properties, and conditions for their use.

A large number of restriction endonucleases is available commercially from numerous firms. In our experience, commercially-available enzymes are usually dependable, and our main criterion in selecting the source is cost. One of the consequences of the rapidly-increasing use of these enzymes and the competition between the various firms supplying them is that the cost is decreasing, in some cases quite dramatically. However, restriction enzymes are still expensive; if this is a problem, they can be simply prepared (Diver and Grinsted, 1984). But, in general, the time taken for such preparations is now not worth the effort.

There is often a choice in the concentration of enzyme in commercial preparations. The point is that people who use the enzyme simply for analytical purposes, generally wish simply to use the enzyme directly from the stock solution without having to go to the trouble of diluting it (see below). So, if one just wants to use the enzyme analytically, buy the low concentration if it is available.

Restriction endonucleases are usually supplied and stored in 50% glycerol solutions, so that they can be kept at $-20°C$ without freezing. Before removing a sample from the stock solution, it is sensible to give it a quick spin (1 s) in a microfuge to ensure that all the solution is at the bottom of the tube (there might only be a few microlitres of the solution). For most analytical purposes, we aim to add 1 μl of the enzyme preparation. If necessary, the stock solution should be diluted to achieve a suitable concentration of enzyme (i.e. a concentration that will give the minimum amount of enzyme needed: larger amounts can, of course, be used, but it would be wasteful). Since dilution often results in loss of activity, care must be taken if it is necessary: only small aliquots of the stock should be diluted at any one time with the storage buffer containing 50% glycerol. The amount of enzyme present in preparations is quoted in Units, which are usually based on the amount of the preparation required to digest some standard DNA completely in a standard time (usually 1 h). Obviously the DNA under test is unlikely to behave exactly the same as the test DNA, but, at least initially, one can be guided by these Units in deciding how much enzyme to add. Some restriction enzymes are very stable and so remain active for many hours even under digestion conditions. The amount of such enzymes required for digestion can be reduced by extending the period of digestion (see Crouse and Amorese, 1986); indeed, it is often convenient to digest overnight.

B. Conditions

Class II restriction enzymes are not very fastidious in their requirements: the only necessary cofactor is magnesium ions. The other parameters that have to be considered are pH, ionic strength and temperature. The concentrations of the components of the buffers are often very flexible and it is our experience that most restriction enzymes work satisfactorily in 50 mM Tris-HCl, 5 mM $MgCl_2$, pH 7.5 at 37°C. (It should be noted that Tris buffers have a high temperature coefficient: for instance, a 50 mM solution of Tris-HCl which is pH 8.0 at 25°C is pH 8.6 at 5°C and pH 7.7 at 37°C. We make up our Tris buffers by mixing Tris base and Tris-HCl (from Sigma) in the proportions given in the tables supplied with the chemicals.) These conditions do not necessarily correspond to the ideal for every enzyme, and there are certainly some enzymes which would be used in different buffers. For example, *Eco*RI often gives extra bands at low ionic strength, and it is sensible to add 100 mM NaCl for this enzyme; high ionic strength is also recommended for *Sal*I; and a few commonly-used enzymes either require high pH for activity (e.g. *Sma*I), or work better under such conditions (e.g. *Bgl*I and *Bgl*II). But the use of a single buffer is very convenient, particularly when the DNA is to be digested with more than one enzyme. In the past few years, firms have started supplying suitable buffers with their enzymes: BRL, for instance, presently have eleven different standard buffers, and the requisite one is included with the enzyme preparation. This procedure is very convenient, and saves some time in preparation of buffers: we applaud it, and now generally use the buffers provided. With double digests, there may be a suitable common buffer, so that a mixture of the enzymes could be used (see above). Alternatively, one might digest first with the enzyme that requires least salt and then add more salt and the second enzyme. If neither of these procedures is satisfactory, the DNA could be precipitated (Chapter 5) after the first digestion and redissolved in the buffer for the second enzyme. The normal incubation temperature is 37°C, but a few enzymes (e.g. *Taq*I) are used at much higher temperatures. One other point to bear in mind is that glycerol will be added to the reaction mixture with the enzyme (see below) and it has been reported that some enzymes cut DNA in unexpected places when the glycerol concentration is higher than 5% (Maniatis *et al.*, 1982).

C. DNA

The rapid methods of preparing DNA (Chapter 5) normally give DNA that is satisfactory for analysis with restriction endonucleases. Any contamination of these preparations with chromosomal DNA, which would be seen as a discrete band if undigested preparations were run in a gel (see Chapter 5, Section V), will be cut into essentially random size pieces by the enzymes. Therefore, after digestion, heavy chromosomal contamination will be seen as a smear on a gel, and, as is usually the case, a small amount of contamination will no longer be visible. We sometimes find that a particular batch of DNA preparations is refractory to digestion; if so, it is usually quicker to discard the batch and start again, rather than fiddle around trying to clean the preparations up.

There are some restriction endonucleases whose recognition sites can contain the site or part of the site recognized by either the *E. coli dam* or *dcm* methylases. (These sites are GATC and CCA/TGG respectively.) Thus, these sites can be methylated in *E. coli* so that the enzymes do not act on them. If these enzymes are to be used to digest DNA isolated from *E. coli*, then the DNA has to be prepared from a *dam* or *dcm* mutant (Maniatis *et al.*, 1982). For instance, *Bcl*I recognizes the site TGATCA; this site would be methylated by the *dam* methylase and *Bcl*I does not cut the methylated site. (*Bam*HI, which recognizes GGATCG, and *Bgl*II which recognizes AGATCT, on the other hand, cut both methylated and unmethylated DNA.) Similarly, *Eco*RII recognizes the same site as the *dcm* methylase and does not act on the methylated site. The sites recognized by some enzymes sometimes overlap the *dam* or *dcm* site, so that only some of the sites that the endonuclease recognizes will be methylated and might be refractory. For instance, ClaI recognizes ATCGAT, so that, if there is a G before the site, or a C after it, a *dam* site is present and the A can be methylated; the methylated site is not cut by the enzyme.

D. Digestions

Digestions are usually carried out in small (0.4 or 1.5 ml) plastic tubes. For analytical purposes we might mix the following in a 0.4 ml tube:

1. 9 μl of DNA solution (usually in 10 mM Tris, pH 7.5, 0.1 mM EDTA);
2. 1 μl of 10 × buffer (e.g. from the supplier of the enzyme);
3. 1 μl of enzyme preparation.

or:

1. 8 μl of DNA solution as above;
2. 1 μl of 10 X buffer (e.g. from the supplier of the enzyme);
3. 1 μl of enzyme preparation.

The amount of DNA would be in the range 0.1 to 1 μg: the actual amount of any DNA preparation to add to give suitably bright bands in the gel (see below) is determined empirically. The tube is then vortexed, briefly spun in a microfuge if necessary (to force droplets to the bottom of the tube), and incubated at 37°C. After an appropriate time at the incubation temperature, 5 μl of EDTA (0.2 M, pH 8), sucrose (40% w/v), bromophenol blue (1.5 mg ml^{-1}) are added and the mixture vortexed. The sample is now ready for loading on a gel (see below). This 'stop mix' would not be added to restriction enzyme digests that are to be subjected to further enzyme treatments (e.g. ligation or labelling with radioactivity), but here it may be necessary to inactivate the restriction enzyme. This can normally be done by heating the mixture at 65°C for 10 min; there are, however, some enzymes that are resistant to this treatment (e.g. BamHI, TaqI, HindIII). Such enzymes can be inactivated with phenol: the mixture can simply be phenol-extracted, followed by precipitation of the DNA (Chapter 5), or a drop of buffered phenol can be added to the solution and, after vortexing, can be removed with buffer-saturated ether.

III. Analysis of restriction enzyme digests

A. Electrophoresis

Restriction enzyme digests can be analysed on either agarose or polyacrylamide gels. The resolution of the gels depends inversely on the size of the fragments, and very large fragments tend to comigrate, whatever their size. To some extent, this problem can be overcome by selecting the correct gel material and concentration. Thus, in general, acrylamide would be used for low molecular weight fragments (less than 1 kb), and agarose for larger fragments. (However, agarose can also be used for fragments as small as 100 bp.) Details of acrylamide gels are given in Chapter 11. As far as agarose gels go, the concentrations used vary between 0.3% (to resolve fragments of maybe 100 kb) and 2.5% to resolve fragments down to 100 bp. (Concentrations of agarose below about 0.5% are very sloppy, difficult to handle, and the gels must be used horizontally.) In general, though, it is very difficult to achieve good resolution of fragments of greater than 50 kb in a normal electrophoresis set-up. Recently, however, it has been found that periodic reversal of the polarity during electrophoresis allows resolution of linear molecules with sizes of greater than 1000 kb (see, for example, Southern et al., 1987). Electrophoresis equipment that delivers the necessary pulsed field is now commercially available and is a necessity for separation of very large fragments.

To prepare agarose gels, the agarose is weighed out in a flask, an appropriate volume of electrophoresis buffer (Tris-borate: 50 mM Tris, 50 mM boric acid, 2.5 mM EDTA, pH 8.2; this would normally contain ethidium bromide at 0.5 μg ml^{-1} — see below) is added, and the mixture heated directly over a bunsen to dissolve the agarose. The flask is frequently shaken to prevent charring (especially with agarose at 1.5% and over). Care must also be taken that the solution does not boil over. The agarose solution can then be poured into an appropriate gel former; concentrated solutions set very rapidly and it is wise to pour the solution at 100°C into prewarmed apparatus if a vertical system is being used. (It should be remembered that pouring very hot solutions onto cold glass plates cracks the plates. Pouring the gel when it is close to its setting point often results in air bubbles in the gel which are almost impossible to remove.) It is simpler to use horizontal gels. In this case, the agarose is simply poured on a perspex plate that is surrounded by sticky tape (we use 3 mm autoclave tape), that sticks above the surface, thus forming a shallow tray; the comb, which forms the wells, is suspended over the plate with bulldog clips at each end. Once poured, the gel should set in 15–20 min (this can be accelerated if the whole apparatus is put in a cold room or refrigerator). For a 'minigel', about 10 ml of the agarose solution can simply be pipetted on a dry microscope slide with a comb suspended above; surface tension is sufficient to keep the solution on the glass, and tape is not necessary. In general, we recommend horizontal gel electrophoresis because the gels are much easier to prepare, use, and process afterwards. However, one advantage of vertical gels is that the wells can be as deep as you like, which means that larger samples can be run.

After the gel has set, the comb is removed and the gel placed in the appropriate electrophoretic apparatus. If one is running a vertical gel, it will have been formed between glass plates. The whole sandwich can be transferred to the apparatus, which will have two buffer tanks containing the electrodes: one in contact with the top and the other with the bottom of the gel. If it is going to run horizontally, the simplest procedure is to use a submarine apparatus, in which the gel on its plate is placed in electrophoresis buffer with electrodes at each end. Commercial apparatus can be bought, which is relatively cheap. The buffer in the tanks is simply electrophoresis buffer (see above) without ethidium (the concentration in the gel is sufficient to stain the DNA).

When the gel has been assembled in the apparatus, the samples can be added, using a micropipette. (Note that the wells will be covered in buffer. This is why sucrose is added to the stop mix (see above): the samples drop to the bottoms of the wells because of the density of the sucrose. Other substances, such as Ficoll or glycerol, which perform the same function, can substitute for sucrose.) In general, with fragments larger than about 1 kb, if the gel is run slowly, the separation of the fragments is better. It is not, however, usually necessary to use very long running times, unless attempting to resolve large fragments: we usually aim to complete a run in about 2 h. (This requires about 100 V for a gel of 15 cm length on the apparatus that we use.) By this time, the marker dye will be close to the other end of the gel. Remember that since DNA is negatively charged at neutral pH, it will run towards the positive electrode. (If one cannot wait to make sure that the dye is moving in the right direction, a quick way of checking the polarity is the bubbles from the electrodes: there are far more from the negative one.)

B. Visualization of fragments in gels

After running, the fragments in the gel are visualized by examination in U.V. light (310 nm is best; and a transilluminator, which allows illumination of the gel from underneath, is the most satisfactory apparatus). On illumination with U.V. light, ethidium bromide (which is included in the electrophoresis buffer in the gel) fluoresces strongly when bound to DNA, so that it shows up as orange bands. It should be noted that ethidium cannot be incorporated into polyacrylamide gels (presumably it is oxidized by the perchlorate) and these have to be stained prior to exposure to U.V. light. (After electrophoresis, gels are simply immersed in electrophoresis buffer containing 1 μg ml^{-1} of ethidium bromide for 15–30 min.) It should also be noted that SDS electrophoreses with the marker dye and complexes all the ethidium it encounters. Consequently, gels loaded with samples containing SDS have to be stained even if ethidium had been included in the gel.

A photograph must be taken to analyse the pattern obtained. This should be taken through a red filter. The simplest way is to place a fixed-focus Polaroid camera on a frame that sits on top of the gel on the transilluminator. If the frame is light-proof, it is possible to visualize and to photograph in the open laboratory.

C. Determination of size of fragments

Generally, the sizes of the fragments are required. This is determined by running the samples with a set of standard fragments in the same gel. These standards are generated by digesting a well-characterized DNA such as that of λ bacteriophage with an enzyme that gives fragments with a suitable range of sizes. This can be done by oneself, but such standard 'ladders' are also commercially available (BRL, for instance supply a 'kilobase ladder' containing fourteen evenly-sized fragments between 0.5 kb and 12 kb). A standard curve of the distances travelled by these standard fragments as a function of the logarithm of their sizes should be constructed (this should be approximately linear, at least with smaller-size fragments); the distance travelled by the sample would then be measured so that its size could be determined from the graph. (Note that such determinations assume that the structural form of the DNA of the standard and the sample is the same: if one wanted to determine the size of uncut plasmid DNA, linear standards cannot be used (see Chapter 5, Section V).)

D. Southern transfer and hybridization

It is very often the case that one wants to determine the homology between the fragments in the gel and some other DNA sequence. For example, one might want to know how closely related the plasmid sequences in the gel are to another plasmid; or one might want to know the location of a particular gene whose sequence has been isolated. The general technique to determine homology is hybridization, in which single-stranded DNA reanneals with complementary strands. Single-stranded test DNA is annealed with labelled single-stranded 'probe' DNA and the mixture then assayed for the presence of double-stranded labelled hybrid DNA. The amount and the stability of this will be a measure of the degree of homology between the test and the probe. In the case of fragments in a gel, the aim is to hybridize each of them to a probe. One could isolate each of the fragments (Section V of this chapter) and then hybridize them separately, but it is much simpler to do all the hybridizations together. This is achieved by denaturing the fragments in the gel (i.e. making them single-stranded) and then transferring them to a filter to which they are irreversibly bound. This is called a Southern transfer, after the inventor of the technique. (It is also called 'blotting', because of the details of the technique — see below.) The filter is then incubated with labelled probe DNA, which hybridizes with homologous sequences; unhybridized probe is then washed away. Therefore the only label left on the filter should have hybridized to the test fragments. This can be detected by autoradiography and, since the fragments on the filter retain the same relative positions to each other as in the original gel, comparison of the position of the label on the filter and the photograph of the gel shows which fragments have homology with the probe. A good description of the practical side of nucleic acid hybridization is provided by Hames and Higgins (1985).

The following is a procedure for Southern transfer of fragments to a nitro-cellulose filter. All steps are carried out at room temperature except the final baking.

1. Electrophorese DNA fragments in agarose gel. Visualize under U.V. and photograph.
2. For optimal transfer of high molecular weight DNA, depurinate by soaking the gel in 0.25 M HCl for 7 min. (This step is not really necessary if the blotting is carried out overnight.)
3. Denature DNA by soaking the gel in 0.5 M NaOH, 1.5 M NaCl for 45 min, with constant agitation.
4. Neutralize by soaking the gel in 0.5 M Tris/HCl, 3 M NaCl, pH 7 for 45 min.
5. Set up blot with the gel inverted on top of wicks consisting of Whatman 3MM paper standing out of a container of $20 \times$ SSC (SSC is 0.15 M NaCl, 0.015 M Na$_3$ citrate). Presoak nitrocellulose in distilled water for 5 min and then $20 \times$ SSC for 1 min. Place filter on top of gel, then roll a 10 ml pipette over filter to remove excess liquid from between gel and filter. Place a piece of Whatman 3MM paper on top of filter and then a stack of paper towels on top of this, then a weight (a litre flask containing water will do) on top of these. Leave for 4 h to overnight.
6. Wash filter in $6 \times$ SSC for 5 min, then dry in a vacuum dessicator and bake at 80°C for 2 h. The dried filters can then be stored until required.

This method uses nitrocellulose filters. Recently, nylon filters have been introduced (Millipore GeneScrenPlus or Amersham Hybond-N, for example). These have some advantages over nitrocellulose: baking the DNA on the filter is not necessary and nylon filters are flexible whilst nitrocellulose is fragile when dry, so that handling is much easier.

The filters are now ready for hybridization and the labelled probe has to be prepared. This can be done by using DNA polymerase I to nick-translate the DNA, or extend the random primers; or an RNA copy can be made. Normally, ^{32}P is used so that the probe is labelled in its phosphodiester bonds, but nonradioactive methods that use biotin can also be used. Kits are available from various suppliers for these labelling procedures. (See Cunningham and Mundy (1987) for a discussion of the various methods available. Hames and Higgins (1985) also deal with the preparation of labelled probes.) The method of hybridization below assumes a ^{32}P-labelled probe and a nitrocellulose filter.

1. Place blotted filter in a sealable bag and add 20 ml prehybridization solution (6 × SSC, 5× Denhardt's solution, 0.5% SDS, 100 μg/ml calf thymus DNA; Denhardt's solution is 0.2% Ficoll (MW 400 000), 0.2% polyvinylpyrollidone, 0.2% bovine serum albumin, in 1 × SSC). Remove all air bubbles (or most) and seal bag. Place in a 65°C water bath for 2 h, occasionally moving the fluid around to shift bubbles.
2. Remove filter to a fresh bag and add 3 ml of hybridization buffer (same as prehybridization buffer with the addition of the labelled DNA probe). Seal bag and incubate at 65°C in a water bath overnight.
3. Remove filter from bag and quickly place in a bath of 2 × SSC, 0.5% SDS at 65°C. Agitate for 5 min then replace with fresh solution and agitate for a further 5 min. Wash off excess probe with 0.1–2 × SSC, 0.1% SDS at 65°C, with constant agitation (shaking water bath).
4. If you wish to reprobe, do not let the filter dry (wrap in cling film). Expose X-ray film with intensifying screens (see Chapter 11).

The prehybridization and hybridization buffers should be degassed by leaving overnight at 65°C before use. The washes can simply be carried out in small plastic boxes; indeed, so can the hybridization itself, but measures would have to be taken to reduce evaporation. The point of the prehybridization step is to cover sites other than DNA on the filter that could bind the probe; the point of the various washes afterwards (step 3) is to ensure that only stable hybrids are retained; the regime shown is a 'high-stringency' wash. If one wanted to detect hybrids that were less stable (i.e. the homology was less), these washes would be varied. (See Meinkoth and Wahl, 1984; Hames and Higgins, 1985). As implied in step 4, filters can be reused and probed again, perhaps with a different probe. To do this, the first probe has to be removed. This is simply effected by incubating the filter for 30 min in the denaturing solution (step 3 of the blotting procedure) and then washing. The filter can then be used again for hybridization. Filters can be reused many times in this way.

IV. Extraction of DNA fragments from gels

It is often desirable to isolate a specific fragment from a gel. For instance, one might want to use it as a probe against other plasmids as described above; or one might want to clone that fragment. Chapter 11 describes isolation of DNA from polyacrylamide gels; it also briefly mentions methods for isolating DNA from agarose gels. We describe here in detail the method that we routinely use; it is based on that of Dretzen et al. (1981).

1. Visualize fragment in gel by U.V. illumination. Cut slit in front of band and insert piece of DE81 paper; use two pieces if the band is heavy. (The paper can be used directly; it can be washed first by soaking in 2.5 M NaCl for 1 h, followed by three washes with water and storage in 1 mM EDTA, but the wet paper is much harder to manipulate.)

2. Replace gel in electrophoresis apparatus, turn on current and electrophorese the DNA into the paper. (Progress can be monitored periodically by placing the gel on the transilluminator.)
3. Remove paper and illuminate with U.V. Trim paper around the orange fluorescent spot of DNA.
4. Place paper in 0.4 ml Eppendorf tube and wash three times with water.
5. Add 0.1 ml of 1.5 M NaCl, 50 mM Tris-Cl, 1 mM EDTA (pH 7.5) and mash up paper with a plastic pipette tip. (Alternatively, break up paper by vortexing.)
6. Incubate at 65°C for 10 min and then cool slowly.
7. Invert tube and pierce the bottom with a red-hot needle. Place in a 1.5 ml Eppendorf tube. Centrifuge for 1 min.
8. Pierce bottom of clean 0.4 ml tube and insert a polycarbonate filter into the bottom. Add DNA solution to this tube, place in a large tube and spin for 1 min.
9. Add 100 μl of water to the solution and add 0.2 ml of buffer-saturated phenol. Vortex and then spin for 1 min. Remove upper phase and discard lower phenolic phase.
10. Add 20 μl of 4 M Na acetate and 0.12 ml of isopropanol. Mix and spin for 5 min. Discard supernatant.
11. Wash pellet with ethanol and then dissolve in 10 mM Tris-Cl (pH 7.5).

This method is quite fast and convenient and gives DNA that is suitable for labelling, cloning or further digestion.

References

Crouse, J. and Amorese, D. (1986). *Focus (BRL Technical Bulletin)* **8**, no. 3, 1–2.
Cunningham, M. W. and Mundy, C. R. (1987) *Nature (London)* **326**, 723–724.
Diver, W. P. and Grinsted, J. (1984). *Methods in Microbiol.* **17**, 157–162.
Dretzen, G., Ballard, M., Sassone-Corsi, P. and Chambon, P. (1981). *Anal. Biochem.* **112**, 295–298.
Hames, B. D. and Higgins, S. J. (Eds) 1985. "Nucleic Acid Hybridisation: A Practical Approach". IRL Press, Oxford.
Kessler, C. and Holtke, H.-J. (1986). *Gene* **47**, 1–153.
Maniatis, T., Fritsch, E. F. and Sambrook, J. (1982). "Molecular Cloning. A Laboratory Manual". Cold Spring Harbor, New York.
Meinkoth, J. and Wahl, G. (1984). *Anal. Biochem.* **138**, 267–284.
Southern, E. M., Anand, R., Brown, W. R. A. and Fletcher, D. S. (1987). *Nucleic Acids Res.* **15**, 5925–5943.

7

Electron Microscopy of Plasmid DNA

H. J. BURKARDT

Department of Microbiology and Biochemistry, University of Erlangen, Federal Republic of Germany

A. PÜHLER

Department of Genetics, University of Bielefeld, Federal Republic of Germany

I. Introduction

Since the development of a suitable preparation technique for DNA molecules, which was achieved by the introduction of spreading techniques using basic protein monolayers (Kleinschmidt and Zahn, 1959), electron microscopes (EM) have been used routinely to study plasmid molecular biology. There are several reasons for their widespread use in plasmid research, namely that most EM techniques are relatively quick in comparison to other molecular biological methods, their accuracy is only surpassed by DNA sequencing and one can actually see what happens to the DNA molecules under investigation. So electron microscopy always gives information about single individual molecules,

METHODS IN MICROBIOLOGY
VOLUME 21 ISBN 0-12-521521-5

in contrast to DNA gel techniques, for instance, which average a whole population of molecules. This individuality has two consequences: to obtain an unambiguous result from an EM study of plasmids statistical evaluation may be required; some peculiarities within a population of DNA molecules may be detected only by an EM technique. Electron microscopy can help to solve the following questions in plasmid research.

1. Visualizing of intracellular plasmid DNA.
2. Molecular weight determination by contour length measurements.
3. Measuring of DNA supertwisting.
4. Detection of inverted repeats in plasmids (IS sequences, transposons) by homoduplex experiments.
5. Plasmid-plasmid homology and therefore detection of insertions, deletions and inversions by heteroduplex analysis.
6. Gross nucleotide composition by AT-mapping and secondary structure mapping.
7. In combination with other molecular biological methods: promotor mapping by RNA polymerase binding studies and visualization of transcription complexes or transcription products by R-loop mapping.
8. Mapping restriction sites on plasmid DNA.
9. Study of basic molecular biological events in which nucleic acids are involved, such as replication, recombination, transcription and translation.

This chapter will concentrate on molecular weight determination, homo- and heteroduplexing and AT-mapping because all these methods are based on the same preparation technique for DNA visualization, are relatively simple and so can be performed in every EM laboratory without the need for complicated and expensive apparatus and chemicals.

Examples for intracellular plasmid visualization are described by Kunisada and Yamagishi (1983), for measuring of DNA supertwisting by Sperrazza et al. (1984) and for mapping of restriction sites by Moore and Griffith (1983). An excellent review of DNA electron microscopy is given by Brack (1981).

II. Technical requirements for electron microscopy of nucleic acids

A transmission electron microscope as shown in Fig. 1 is the main technical requirement. It is not necessary to have a sophisticated one, because, for all the techniques described in this chapter (and even for most other DNA techniques) the required magnification is moderate, about 10000-fold. There are no resolution problems because in plasmid research only lengths of complete DNA molecules or segments, and not DNA fine structure, are of interest. A vacuum evaporator is useful for preparing specimen support films and contrasting DNA preparations. A room with no through-pass or a special preparation box is recommended because all DNA preparations are sensitive to contamination from the environment and to draughts etc.

For DNA preparation, grids with carbon or Parlodion films may be used. The type of film preferred depends on which one is normally used in the EM laboratory. For beginners we recommend Parlodion films because they are easier to prepare, the quality is reproducible and the adsorption rate of DNA is high. A simple recipe for preparation of Parlodion films is as follows.

1. Arrange copper grids on a filter paper in a water-filled Büchner funnel.
2. Clean the water surface using a water jet pump. Place one drop of a water-free Parlodion solution in *iso*-amyl acetate (1.5–3%) onto the water surface, in the middle of the funnel. After evaporation of the solvent, which will take 5–7 min, the Parlodion will form a thin film on the surface of the water. To prepare the Parlodion solution, cut dry Parlodion into small pieces and bake them at 80°C overnight. This reduces the water content of the Parlodion. Put the pieces in a tightly sealed flask with the organic solvent and dissolve by gently shaking it over a period of a few days. Parlodion which is predissolved in ethanol or ether might be less useful because of possible water contamination which causes holes in the final film.
3. Remove this first film and discard it (it is used to clean the water surface).
4. Prepare a second film in the same manner.
5. The copper grids are covered with the film when the water is allowed to flow out of the funnel.
6. Before use, the grids, covered with a film of Parlodion, are dried while still on the filter paper. Removal of the wet grids from this paper damages the film.
7. Film thickness can be roughly estimated by its shininess. Very thin films are merely dull, thicker ones are slightly glossy. Films which are too thick for use exhibit a strong gloss or even interference colours.

III. Cytochrome *C* preparation methods: classical spreading and droplet technique

As mentioned in Section I, for EM preparation a DNA molecule has to be stabilized by a special support film. The first films which were introduced were basic protein monolayers of which cytochrome *c* (Cyt *c*) has been used most widely because of its easy availability and properties, described below.

1. The molecule is small and globular (giving a fine background grain).
2. At the water–air interface it denatures, forming a monolayer which is stable on the water surface.
3. In the neutral pH range it has a net positive charge that can interfere with the negative charges of nucleic acids which results in unfolding and binding of those molecules.

(a)

Fig. 1. View and cross-section of a transmission electron microscope.

(b)

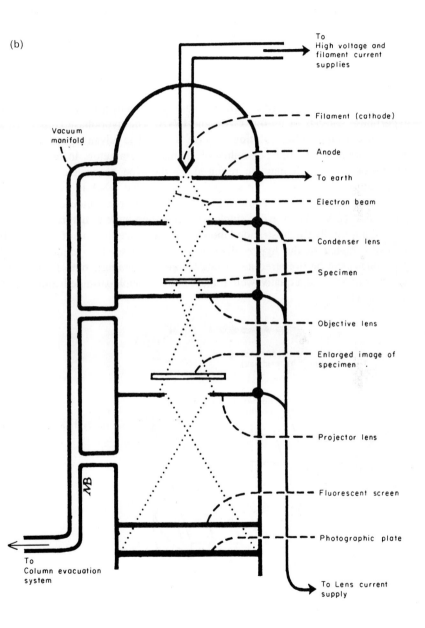

The following two Cyt c spreading methods are in use:

1. The spreading method, which is the original Kleinschmidt technique (Kleinschmidt, 1968).
2. The droplet technique, which is a simple modification of the former (Lang and Mitani, 1970).

A. Spreading method

This method uses a hyperphase which is spread over a hypophase (Fig. 2). The hyperphase consists of buffer, Cyt c and DNA. The hypophase consists of a less concentrated buffer solution or pure water. The advantage of this two-phase system lies in the possibility of changing parameters in both phases independently, which may be required by the type of nucleic acid under investigation. For example, to spread single-stranded DNA the water has to be partially replaced by denaturing agents such as formamide or dimethyl-sulphoxide (DMSO) to avoid single-stranded snap-back structures or heavy coiling of a totally single-stranded molecule. The concentration of such agents is always higher in the hyper- than in the hypophase. For routine use the hyperphase can be prepared as a stock solution, containing 60% formamide, which can be used equally well for spreading double-stranded and single-stranded DNA.

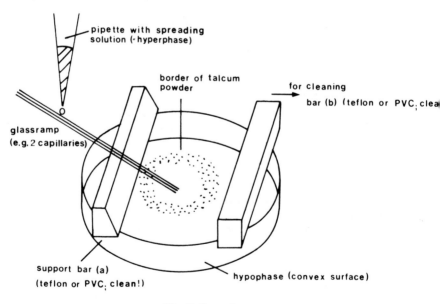

Fig. 2. Spreading apparatus.

The recipe for formamide spreading is laid out below. Note that the purity of formamide is critical for the hyperphase (not for the hypophase) and can be checked by optical density measurement. It should not exceed $0.2\,A_{270}$ (reference: water). The quality of formamide can be improved by recrystallization or distillation.

1. Preparation of hyperphase

1. Prepare a stock solution: 60% formamide (purified), 0.2 M Tris, 0.02 M EDTA adjusted with 1 N HCl to pH 8.5. The stock solution can be kept frozen at $-20°C$ for months.
2. Mix 10 μl of stock solution and 10 μl of DNA solution. This ratio is valid for DNA concentrations of about $1–10\,\mu g\,ml^{-1}$; if the DNA is more concentrated, dilute it with 1 mM EDTA.
3. Add 0.4–0.5 μl of Cyt c ($5\,mg\,ml^{-1}$) immediately before spreading.

2. Preparation of hypophase

1. Mix in a bottle 10 ml of formamide, 1 ml of 1 M Tris and 1 ml of 100 mM EDTA, fill up with double-distilled water to 100 ml and adjust pH with 1 N HCl to 8.5–8.7.
2. Use hypophase within a few hours.

3. Spreading

1. Spread 2–5 μl of the hyperphase over the hypophase as indicated in Fig. 2.
2. Before the first and between two spreadings the surface of the hypophase is cleaned by wiping it with the plastic bar.
3. Pick up parts of the Cyt c film with its adhering nucleic acid by touching the hyperphase with a copper grid with a support film. The area of the hyperphase can be made visible by sprinkling some talcum powder in front of the ramp. After spreading, the powder marks the border of the spread hyperphase.

B. Droplet method

The droplet method (Fig. 3) is simpler because it is a one-phase system; unfortunately, however, results are mostly very poor when preparing single-stranded DNA. Therefore we recommend this method only for DNA double-strand preparations.

1. Prepare in a test-tube 0.2 M ammonium acetate buffer, 1 mM EDTA, pH 7.
2. Add Cyt c from a 0.1% stock solution (in water) to a final concentration of 0.00025%.
3. Add to 0.25 ml of this mixture about 0.5–5 μl of DNA solution (depends on DNA concentration).

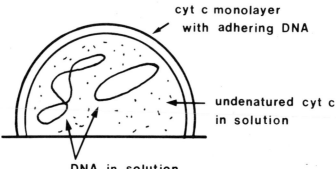

Fig. 3. Droplet preparation of plasmid DNA.

4. Pipette this volume in three to four droplets on a clean hydrophobic surface (e.g. a new plastic Petri dish).
5. Protect the droplets against dust and wait for 0.5–1 h until the Cyt *c* film has formed and DNA has adsorbed to it.
6. Pick up Cyt *c* film and adsorbed DNA by touching the droplet surface with a grid.

IV. DNA contrasting by shadowing and staining

Like most biological specimens DNA has no heavy atoms to give sufficient contrast in the bright field EM, so additional contrasting is required for visualization (unless special and complicated dark field EM methods are applied). Staining with solutions of heavy metal salts, shadowing with heavy metal or a combination of both techniques are possible. For normal preparation with Parlodion-coated grids we recommend the last technique which results in excellent contrast. The stain and shadowing also thicken the diameter of the nucleic acid and therefore improve the visibility of the fine filaments of nucleic acids.

Procedure

1. Dip grids with film and DNA immediately after preparation in staining solution, which is 10^{-5}M uranyl acetate and 10^{-4}N HCl in 90–95% ethanol. (This can be cheap alcohol, which is denatured by petroleum ether. Pure alcohol damages Parlodion film.) The staining solution is freshly prepared by diluting a 100-fold more concentrated aqueous stock solution. The stock solution is stable in the dark for several months. To reduce the amount of uranyl precipitates on the grid, filtering (pore diameter $<0.2\,\mu$m) the stain solution is recommended.
2. Remove excessive stain by submerging the grid in 90–95% ethanol without stain.
3. Dry grid in the air; a possible fluid bath in petroleum ether accelerates drying.

4. Shadow DNA with Pt, Pt/Ir or similar alloy in the shadow casting machine. To obtain an even contrast, especially of circular molecules, rotary shadowing is required.
We use the following shadow casting conditions:
  vacuum about 10^{-5} Torr
  distance (wire-specimen) about 4.5 cm
  angle about 5
  amount about 3.5 cm wire (φ 0.1 mm)

V. Evaluation of electron micrographs: configuration and contour length of plasmid DNA

A well-spread (droplet method) and contrasted plasmid molecule is shown in Fig. 4(a); Fig. 4(b) is a typical example of one which is poorly spread. The reason for the bad preparation may be contaminating agents in the DNA solution (detergents), bad Cyt c (stock solution too old) or contamination of buffer. Such a simple DNA visualization can give the following information.

1. Presence or absence of DNA in a plasmid isolation experiment.
2. Quality of isolated DNA (contamination, fragments because of contaminating nucleases).
3. Approximate concentration of DNA.
4. Form of DNA: linear, circular, supertwisted.
5. Molecular weight of DNA.

For the last point some more evaluation work is necessary. Molecular weight determination by EM is, in principle, a contour length measurement of the nucleic acid molecules (Lang, 1970). These lengths can be determined by tracing the molecule contour on an enlarged drawing or directly on an enlarged projected EM negative. Several types of instruments are available for tracing. The most simple and cheapest is a distance measurer for maps, but sophisticated measurers with computer evaluation are also available. As individual molecules are measured, a length distribution will result. We evaluate a minimum of about 20 molecules of one DNA species. Besides the mean length, the respective standard deviation should always be quoted. It indicates the quality of the length determination and should not exceed ±2%. To get a reliable absolute molecular length a calibration of the magnification of microscope and evaluation system is essential. For double-stranded nucleic acids, an external calibration is sufficient, because the molecules are relatively resistant to preparation artefacts. For this purpose, all molecules should be photographed at one microscope magnification, which is calibrated by a carbon replica grating. For single-stranded nucleic acids an internal standardization is recommended because these molecules are very sensitive to stretching and have a prominent tendency to shorten themselves by forming snap-back structures. Internal

164 H. J. BURKARDT AND A. PÜHLER

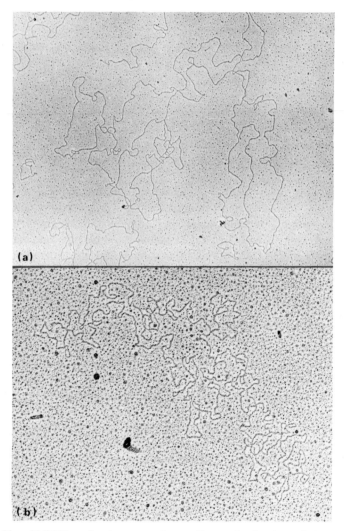

Fig. 4. Plasmid DNA molecule prepared according to the droplet technique: (a) well spread; (b) poorly spread.

calibration is achieved by adding a known single-stranded DNA, e.g. $\varphi X174$ bacteriophage DNA, to the DNA sample to be studied (Stüber and Bujard, 1977). The length can be determined in $\varphi X174$ DNA units. An analogous internal standardization is also possible with double-stranded DNA (e.g. ColE1 DNA may be used as a standard). At an ionic strength of 0.1–0.5 (which is the case for both the droplet and spreading methods) the following factor for double-stranded DNA can be used to calculate the molecular weight: 1 μm of DNA = 2.07×10^6 D (≈ 3 kb).

VI. Homoduplex experiment: detection of inverted repeats

Inverted repeats (IR) are DNA sequences which appear in two or several copies in one DNA molecule, with examples in both possible orientations. They are parts of transposons and IS sequences and, therefore, common in plasmids. They can be detected readily by a homoduplex experiment. This involves denaturing, renaturing and then spreading the DNA. Denaturation means separating the strands of double-stranded DNA; renaturation is the reverse

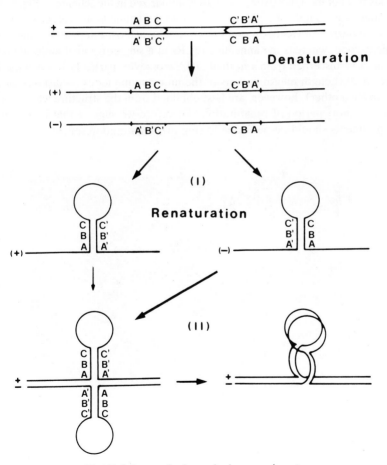

Fig. 5. Scheme of a homoduplex experiment.

process. Nevertheless, after renaturation, the parental type of molecule is not always restored. In particular, when renaturation times are short, there is not sufficient time for complementary single strands to find each other. Complementary sequences, however, within one strand will find each other quickly, as they are situated on the same molecule. Provided that the IRs are small in comparison to the total DNA molecule (as is the case for IS sequences in plasmids) then the majority of the DNA molecule remains single stranded, allowing a second annealing with a complementary single-strand DNA molecule. In this case, however, a complete restoration of the parental molecule is impossible, because the IRs on one strand have undergone intramolecular annealing. Such a process leads to a special structure which has been called an 'underwound loop' (Kleckner *et al.*, 1975; Burkardt *et al.*, 1978a). The reactions in a homoduplex experiment are summarized in the scheme of Fig. 5; Fig. 6 shows an example of a plasmid with an underwound loop. This structure is a good indicator of IRs because it has a very characteristic appearance under the microscope, whereas the annealing of IRs in a single-stranded molecule often cannot be discerned from a normal DNA crossover, particularly when the IRs are small. Length measurements of the underwound loops (= distance of IRs from each other), however, are best obtained from the structure which results from a short period of renaturation. Therefore we suggest that homoduplex experiments should be performed using two renaturation times.

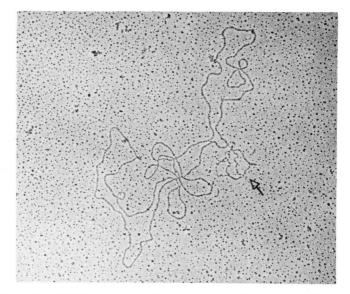

Fig. 6. Plasmid DNA molecule with an underwound loop (marked by arrow).

Procedure

1. Denaturation

    1. Mix 5 μl of purified formamide, 1 μl of 1 M phosphate buffer pH 7 and 3 μl of DNA
       (about 10 μg ml^{-1}).

    2. Boil in a water bath for 1.5 min.

2. Renaturation

    1. Add 1 μl of 2 M NaClO$_4$ and renature at 40°C for 5 min and 30 min. The short time
       mainly allows complementary sequences on the same DNA strand to anneal. The
       long time mainly results in molecules with underwound loops.

3. Spreading

    1. Dilute 2 μl of the renaturation mix in 10 μl of purified formamide and 8 μl of 1 mM
       EDTA.

    2. Add 0.4 μl of Cyt c (5 mg ml^{-1}) to complete the hyperphase and spread over 30%
       formamide, 10 mM Tris, 1 mM EDTA, pH 8.5–8.7.

The method is based on a well-balanced ratio between the different denaturing
and renaturing agents: denaturation is favoured by heat and formamide,
renaturation is favoured by high salt concentrations.

VII. Heteroduplex analysis: demonstration of homology

This is probably the most useful EM DNA technique and was, for a long time,
the only widely used method to test physically the relationship between DNA
molecules and to detect physically alterations in DNA, such as deletions,
insertions, substitutions or inversions (Davis and Davidson, 1968; Davis *et al.*,
1971; Westmoreland *et al.*, 1969). Even today when other methods such as
Southern blotting or DNA sequencing can be used to solve similar problems,
heteroduplex mapping can compete with these methods because it has the
advantages of being, in general, more exact than the former and far less time
and money consuming than the latter. As in homoduplex analysis, heteroduplex
analysis consists of a biochemical part and an EM preparation part. The
biochemical part also requires a hybridization step. This time, however, duplex
structures are formed between related, but different, DNA species. The following points should be kept in mind.

    1. Structures are only visible under the EM if their size exceeds the resolution limit
       of the DNA preparation, i.e. 50–100 base pairs.

    2. Two strands are regarded as homologous even when each third base is false
       (resolution limit of biochemical hybridization).

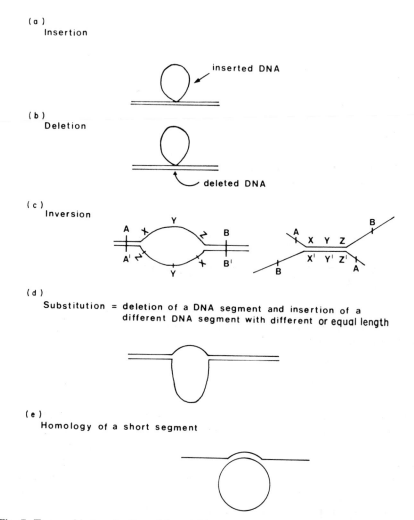

(a) Insertion

(b) Deletion

(c) Inversion

(d) Substitution = deletion of a DNA segment and insertion of a
different DNA segment with different or equal length

(e) Homology of a short segment

Fig. 7. Types of heteroduplices: (a) and (b) DNA molecule exhibiting an insertion or deletion single-stranded loop: (c) DNA fragment with an inversion bubble; (d) DNA molecule with a substitution bubble; (e) heteroduplex between a linear fragment and a plasmid DNA showing a region of homology.

In Fig. 7(a–e) all types of heteroduplex molecules are shown. For a correct interpretation it is essential that one can clearly distinguish between double-stranded and single-stranded DNA. When heteroduplex molecules are well prepared the single-stranded DNA appears slightly thinner and more kinky than the double-stranded part (cf. Fig. 6). It is also important that the DNA has as few crossovers as possible in order to determine, unambiguously, as for example, an insertion point relative to another marker. When the renaturation time is too long, especially in the presence of nicked DNA, hybridization can take place between single-stranded regions of different heteroduplex molecules leading to uninterpretable structures and a DNA network. This occurs in particular with incomplete heteroduplices which consist of a total single-stranded molecule and a single-stranded fragment partner (from nicked DNA). So only DNA preparations with a low proportion of nicked DNA should be used for heteroduplex experiments. This can be estimated by the ratio of open circular and supertwisted DNA. The ideal ratio is 1:1 because in such a DNA preparation the likelihood of an open circular DNA molecule carrying more than one nick is low. These open circular molecules, however, are required for heteroduplex formation because the denatured single strands of supertwisted DNA molecules cannot be separated from each other. They remain interlocked and are, therefore, incapable of heteroduplex formation.

The number of heteroduplex molecules that can be evaluated is reduced by the above-mentioned limitations. In addition, most of the single-stranded DNA molecules will not give rise to heteroduplices but will restore the parental molecules because the plus and minus strands are in the same reaction mixture. So, having performed a heteroduplex experiment, the number of unambiguous heteroduplex structures may be low, especially in complicated DNA hybrid molecules with more than one possible single-stranded structure. This means that patience is required when surveying the specimens under the EM.

Procedure

1. Mix 5 µl of purified formamide, 1 µl of 1 M phosphate buffer pH 7, 1.5 µl of DNA₁ (about $10 \mu g \, ml^{-1}$) and 1.5 µl of DNA₂ (about $10 \mu g \, ml^{-1}$).
2. Denature, renature and spread as in a homoduplex experiment.

VIII. Partial denaturation: gross nucleotide composition of plasmid DNA regions

Partial denaturation is a technique by which a specific physical pattern can be assigned to a double-stranded nucleic acid. In addition to this physical characterization it also yields information about the gross nucleotide composition by marking AT- and GC-rich molecule regions. So, like heteroduplexing, it can be used to identify plasmids, to test relationships between plasmids and, in addition, to study evolutionary aspects of plasmids. AT-mapping can also be used as a preliminary indicator method to look for promotors or other DNA regions which are distinguished by a high AT content.

It does not have the accuracy of DNA sequencing but surpasses this newer method by its speed and simplicity. The principle of the method is the following: a double-stranded DNA molecule can be separated into its single strands when heating it beyond its melting point (this is used in all hybridization experiments). When the temperature is raised to a point just below the melting point the strands will not be separated totally; rather denaturation (= strand separation) starts in weak DNA regions. Weak regions are defined by a high content of AT bases because these two organic bases form only two hydrogen bonds between each other (in comparison to three between G and C) and stacking forces between AT pairs are weaker than between GC pairs. Partial denaturation can also be achieved by chemical methods with denaturing agents such as alkali, DMSO or formamide (Inman, 1966; Inman and Schnös, 1970). In the microscope the unpaired (denatured) regions can be seen as bubbles on an otherwise duplex structure, if renaturation is prevented (e.g. by chemical fixation) and if a resolution limit of 50–100 bp is exceeded. An example of a partially denatured plasmid molecule is presented in Fig. 8.

The evaluation of such partially denatured molecules is the most laborious task of a partial denaturation experiment. It has to be a statistical assessment because strand separation is a statistical process and differs to some extent from one molecule to another. The first step consists of measuring separately, and in sequence, all native and denatured regions of a molecule. This, and the

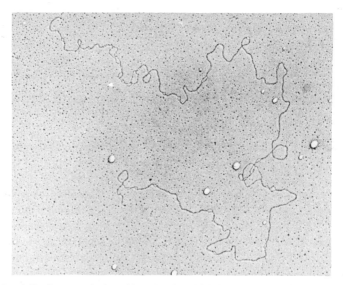

Fig. 8. Partially denatured plasmid molecule which has been linearized by restriction enzyme digestion.

further evaluation, is very much facilitated when an additional physical marker to act as a start and as an end point for the measurements is present. For circular plasmid molecules linearization, before partial denaturing, with a restriction enzyme that has a single cleavage site is useful. The extent to which further evaluation can be taken depends on whether a computer together with a suitable program are available (or not). When access to a computer is not possible the measured molecular segments are drawn to scale on a paper strip (e.g. line for native, box for denatured region) and then the paper strips are aligned so as to obtain maximum congruence of denatured and native segments.

The final diagram (AT map) emerges when all molecules are counted which are denatured at the molecular segments $x_0 \ldots x_n$. The size of one segment x_i depends on the resolution of the EM and the evaluation system. A typical system is shown in Fig. 9. The abscissa equals the total molecule length and is subdivided into the $x_0 \ldots x_n$ segments. The ordinate indicates the percentage of molecules of the whole population which are denatured in a given x_i segment. In some instances, in particular when you have fairly large plasmid molecules which are only slightly denatured, it is advantageous in our experience not only to construct diagrams based on this resolution limit, but also to enlarge

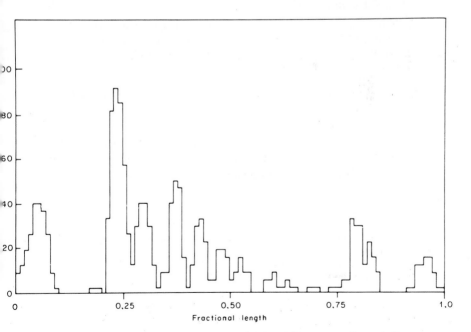

Fig. 9. Plasmid AT map. Abscissa equals total molecule length, ordinate indicates proportion of individual molecules which are denatured at a given fractional length.

artificially the denatured sites, by anything up to 3% of the total molecule length, which results in clearer AT maps that are easier to interpret than the originals (Burkardt *et al.*, 1978b). The reason for this map improvement is the following. When the denaturation bubbles are rather small and sparingly distributed along the DNA, naturally identical denaturation sites of two molecules frequently may not occupy quite the same position in the $x_0 \ldots x_n$ frame, because they are more or less displaced by preparation artefacts and false tracing and measuring. So in a system with high resolution they contribute, by error, to different denaturation peaks. However, if they are sufficiently enlarged until they overlap then they contribute to the same peak, which corresponds better to the natural conditions. The map presented in Fig. 9 was constructed using such a manipulation. When using a computer to analyse your data you can incorporate this provision into the program (including bubble enlargement). The computer will then generate the possible variations of the map of your denatured and measured molecules in a very short time.

Procedure

1. Partial denaturation

    1. Prepare denaturation stock: 0.25 ml of DMSO, 0.21 ml of 0.2 M phosphate buffer, 0.001 M EDTA, pH 7 and 0.14 ml of formaldehyde (37%).
    2. Mix 0.03 ml of denaturation stock and 0.02 ml of DNA ($\sim 10 \, \mu \mathrm{g \, ml^{-1}}$).
    3. Heat to 50–60°C for 10–20 min.

The exact denaturation conditions required depend on the (G + C)% ratio of the DNA to be studied and have to be found in a series of denaturation experiments. Formaldehyde reacts with the free amino groups of unpaired bases in DNA single strands so preventing normal reannealing of the DNA. Unbound formaldehyde disturbs, however, the Cyt *c* spreading and therefore has to be removed before spreading.

2. Formaldehyde removal

    1. Prepare in a Pasteur capillary pipette (ϕ about 7 mm) a small Sephadex G50 column (5 cm long); equilibrate with 1 mM EDTA.
    2. Apply 20 μl of denaturation mix on top of the column.
    3. Elute with about 500 μl of 1 mM EDTA.
    4. Collect three 20-μl fractions.
    5. Mix 10-μl fraction with 10 μl of hyperphase mix (Section III).
The column has to be calibrated for use. This is achieved by testing the system with radioactively-labelled DNA or using the EM to screen the fractions for DNA molecules.

IX. Molecular structure of P type plasmids: an example of DNA analysis by electron microscopy

The use of the EM in analysing plasmids will be demonstrated in this last section of the chapter. Plasmid RP4 serves as an example, some other P type plasmids are RP1, RP8, R68, R68.45 and RK2. These molecules are typical resistance plasmids, isolated from *Pseudomonas aeruginosa* (except RK2, which was isolated from *Klebsiella*), all carry resistance genes to ampicillin, kanamycin and tetracycline. They belong to the same incompatibiltiy group P. When we obtained these plasmids a few years ago their broad host range, which rendered them prominent among all other plasmids so far known, had just been discovered (Datta *et al.*, 1971). Since molecular data were sparse at that time we undertook an EM analysis of these plasmids. A Cyt *c* droplet preparation enabled a portrait of these molecules for the first time (Fig. 10) and confirmed and extended molecular weight determinations made by sucrose gradient centrifugation (Burkardt *et al.*, 1978b). A summary of the length measurement data revealed, within the range of error, identical lengths for RP1, RP4, R68 and RK2, whereas R68.45 was slightly longer (0.6 μm increase) and RP8 was 50% longer (12-μm increase) than the other plasmids (Table I). The different lengths of the latter two plasmids correspond well with their different biological properties.

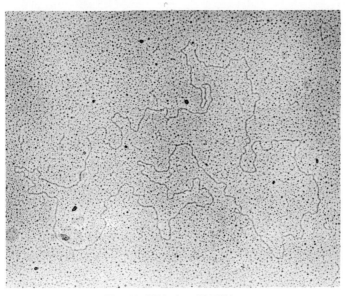

Fig. 10. RP4 plasmid DNA.

TABLE I
Lengths of P type plasmids

Plasmid	Contour length
RP1	18.9 μm
RP4	19.1 μm
RP8	31.3 μm
R68	19.1 μm
R68.45	19.7 μm
RK2	18.2 μm

1. R68.45 shows enhanced chromosomal mobilizing ability (sex factor activity) in comparison to RP1, RP4, R68 and RK2.
2. RP8 has a narrow host range in comparison to all other P type plasmids and is freely transferable only between Enterobacteriaceae and *P. aeruginosa*.

These findings, namely size and genetic characters (identical resistance genes but differences in host range and chromosome mobilizing ability), led us to question the identity and relationship of these P type plasmids. Therefore various heteroduplex experiments were performed. We started with RP1 and RP4 against RP8 because we knew from partial denaturation experiments (Burkardt *et al.*, 1978b) and biochemical hybridization studies from other groups (Holloway and Richmond, 1973) that RP8 should show extensive

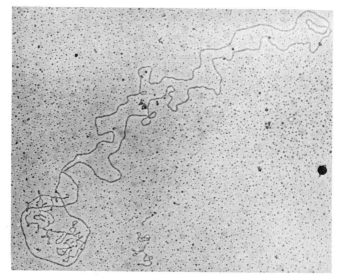

Fig. 11. RP4/RP8 heteroduplex.

homology with both RP1 and RP4. Figure 11 shows a heteroduplex between RP4 and RP8 which clearly demonstrates that the hybrid molecule is composed of a double-stranded section and a single-stranded section (which was the only single-stranded structure). Length measurements of these parts proved the following interpretation to be correct: RP8 is constituted from an entire RP4 molecule and a piece of additional DNA; it can therefore be regarded as an insertion mutant of RP4. A similar interpretation followed an examination of heteroduplices of R68.45 and RP4 (Fig. 12): R68.45 is composed of an RP4 type molecule with a 0.6 μm insertion (Riess *et al.*, 1980). Heteroduplices between RP1, RP4, R68 and RK2 exhibited no such heterologies. This series of heteroduplex experiments revealed the following relationship (at the EM resolution level): RP1 = RP4 = RK2 = R68 = R68.45—0.6 μm insertion = RP8—12 μm insertion (Burkardt *et al.*, 1979). RP1/RP8 heteroduplices displayed a characteristic single-stranded structure, which was later also found in all heteroduplex experiments with P type plasmids (Fig. 6), in addition to the single-stranded insertion loop. This structure resembled neither an insertion loop nor a substitution or inversion bubble and could not be interpreted by us at that time. The solution to the problem emerged when more was known about the genetics of P type plasmids and when we performed long-time homoduplex experiments. Genetic investigations detected the first transposon, Tn*1*, as a part of RP4 (Hedges and Jacob, 1974). This could be visualized by us as an underwound loop in a typical homoduplex experiment, as described

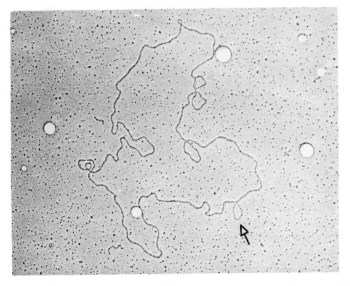

Fig. 12. RP4/R68.45 heteroduplex. R68.45 insertion is marked by an arrow.

in Section VI. As a heteroduplex experiment utilizes the same preparation techniques as a long-time homoduplex experiment, the occasional appearance of underwound loops was explained.

Partial denaturation experiments, originally performed to assign specific patterns to each P plasmid, were compared to detect relationships (Burkardt *et al.*, 1978b). The relationship question was solved more decisively by hetero-duplex experiments which are more suitable for this (Burkardt *et al.*, 1979). The AT-mapping, however, yielded information about the gross nucleotide composition of P type plasmids. By comparing these maps with genetic maps we were able to determine the relative AT content of the RP4 genes (Burkardt *et al.*, 1980). An example for RP4 is shown in Fig. 13. The correct alignment of the partial denaturation and genetic maps was facilitated by using two single restriction enzyme sites on RP4, specifically those for *Eco*RI, and *Hind*III. Without going into detail one can deduce from this figure an unusually high relative AT content for the two resistance gene regions of ampicillin (located on Tn*1*) and kanamycin. This finding may reflect evolutionary aspects of the RP4 plasmid, namely that these gene regions were not part of a primitive RP4 precursor but were acquired from different DNA sources during plasmid evolution.

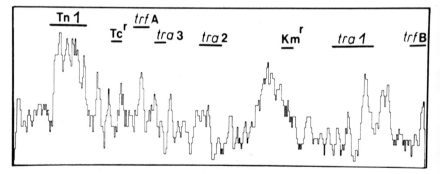

Fig. 13. Partial denaturation map of RP4 showing the relative AT content of some RP4 genes.

The EM story of IncP plasmids does not end here. Applying other EM methods, not described in this chapter, we have been able to map *E. coli* RNA polymerase binding sites on RP4 DNA and have obtained evidence of active, but as yet unknown, RP4 genes (Burkardt and Wohlleben, 1981). Electron microscopic analysis has been extended to more exotic IncP plasmids, which have been detected in the meantime and classified by their incompatibility behaviour. Electron microscopy, together with Southern blotting of restriction enzyme fragments, also in this case has revealed the relationship within this fascinating group of plasmids (Villaroel *et al.*, 1983).

References

Brack, C. (1981). *Crit. Rev. Biochem.* **10**, 113–169.
Burkardt, H. J. and Wohlleben, W. (1981). *J. Gen. Microbiol.* **125**, 301–309.
Burkardt, H. J., Mattes, R., Schmid, K. and Schmitt, R. (1978a). *Mol. Gen. Genet.* **166**, 75–84.
Burkardt, H. J., Mattes, R. E., Pühler, A. and Heumann, W. (1978b). *J. Gen. Microbiol.* **105**, 51–62.
Burkardt, H. J., Riess, G. and Pühler, A. (1979). *J. Gen. Microbiol.* **114**, 341–348.
Burkardt, H. J., Pühler, A. and Wohlleben, W. (1980) *J. Gen. Microbiol.* **117**, 135–140.
Datta, N., Hedges, R. W., Shaw, E. J., Sykes, R. B. and Richmond, M. H. (1971). *J. Bacteriol.* **108**, 1244–1249.
Davis, R. W. and Davidson, N. (1968). *Proc. Natl. Acad. Sci. U.S.A.* **60**, 243–250.
Davis, R. W., Simon, M. and Davidson, N. (1971). *Methods Enzymol.* **21**, 413–428.
Hedges, R. W. and Jacob, A. E. (1974). *Mol. Gen. Genet.* **132**, 32–40.
Holloway, B. W. and Richmond, M. H. (1973). *Genet. Res., Camb.* **21**, 103–105.
Inman, R. B. (1966). *J. Mol. Biol.* **18**, 464–476.
Inman, R. B. and Schnös, M. (1970). *J. Mol. Biol.* **49**, 93–98.
Kleckner, N., Chan, R. K., Tye, B. K. and Botstein, D. (1975). *J. Mol. Biol.* **97**, 561–575.
Kleinschmidt, A. K. (1968). *Methods Enzymol.* **12**, 361–377.
Kleinschmidt, A. K. and Zahn, R. K. (1959). *Z. Naturforsch.* **14b**, 770–779.
Kunisada, T. and Yamagishi, H. (1983). *Plasmid* **9**, 8–16.
Lang, D. (1970). *J. Mol. Biol.* **54**, 557–565.
Lang, D. and Mitani, M. (1970). *Biopolymers* **9**, 373–379.
Moore, C. and Griffith, J. (1983). *Gene* **24**, 191–198.
Riess, G., Holloway, B. W. and Pühler, A. (1980). *Genet. Res., Camb.* **36**, 99–109.
Sperrazza, J. M., Register III, J. C. and Griffith, J. (1984). *Gene* **31**, 17–22.
Stüber, D. and Bujard, H. (1977). *Mol. Gen. Genet.* **154**, 299–303.
Villaroel, R., Hedges, R. W., Maenhaut, R., Leemans, J., Engler, G., van Montagu, M. and Schell, J. (1983). *Mol. Gen. Genet.* **189**, 390–399.
Westmoreland, B. C., Szybalski, W. and Ris, H. (1969). *Science* **163**, 1343–1348.

8

Plasmid Cloning Vectors

RUSSELL THOMPSON

Institute of Virology, University of Glasgow, Church Street, Glasgow G11 5JR

I. Introduction

Concern with bacterial plasmids stems not only from an interest in the elements themselves and the functions they encode, but also from the ways in which they can be exploited. Perhaps the most widely known aspect of this exploitation is in their use as cloning vectors in genetic engineering. In this context, segments of DNA that do not replicate independently and are thus not maintained in the host cell are inserted into plasmids. The plasmids, of course, can replicate and are maintained, as are the recombinants of plasmid with foreign DNA. Thus, the foreign DNA is stably maintained in the host, under the control of the plasmid.

METHODS IN MICROBIOLOGY
VOLUME 21   ISBN 0-12-521521-5

At its simplest, a plasmid cloning vector is small and present in many copies per cell (resulting in amplification of the recombinant). It will encode one or more selectable markers, so that recombinants can be selected, and will contain single sites for one or more restriction endonucleases, so that the foreign DNA can be inserted. The insertion process involves cutting purified plasmid DNA (which is circular — Chapter 5) at a single site and then ligating the foreign DNA to the ends with DNA ligase. The resulting mixture, containing recombinant DNA molecules, is then used to transform the host bacteria to a resistance that is encoded by the plasmid, and the transformants analysed to detect bacteria containing recombinants. For an introduction to genetic engineering and the use of plasmid vectors, Old and Primrose (1985), and Brown (1986) are recommended.

These characteristics are illustrated by, perhaps, the best-known cloning vector of all, pBR322 (Balbas *et al.*, 1986). This is 4363 bp, occurs in the cell as multiple copies, encodes resistance to penicillin and tetracycline, and contains unique sites for 20 different restriction endonucleases. Plasmid pBR322 was specifically constructed as a cloning vector, and is the basis of many of the more modern vectors (see Section II). Another famous plasmid that was specifically constructed for use as a cloning vector is pACYC184 (Chang and Cohen, 1978). This is 4.2 kb, encodes resistance to tetracycline and chloramphenicol and contains unique sites for various restriction endonucleases. This plasmid is also high-copy number (although not as high as pBR322), but is compatible with pBR322, which can be useful in complementation tests (see Chapter 9). Now, this chapter shows that cloning vectors have become much more sophisticated in the last ten years; but these original vectors are still very satisfactory for many routine cloning experiments and are still often used.

This chapter is intended to give a brief overview of the present generation of plasmid cloning vectors. It is not a comprehensive listing but rather a sampling of the most recent plasmids which illustrate the general trends in vector design. The most definitive listing is contained in Pouwels *et al.* (1985) together with the supplementary update published in 1986. This is a massive tome on 'Cloning Vectors' whose loose-leaf binder design allows new descriptions to be fitted into existing chapters. Each vector has a single page entry with a map, summary of details and descriptive text. A very comprehensive listing of vectors based on the pBR322 replicon can be found in Balbas *et al.* (1986), and earlier reviews are listed in Thompson (1982).

II. General purpose vectors

The plasmid pBR322 was constructed by a series of *in vivo* and *in vitro* ligation events, with the explicit aim of generating a multi-purpose cloning vector. It has proved to be the first of a large and continuously-evolving family of vectors (Balbas *et al.*, 1986). A major trend in the later generations of pBR322-based vectors has been the increase in the number of restriction sites available for cloning, and the provision of direct means of distinguishing recombinant molecules from empty vectors.

A. pUC vectors and derivatives

At the present time, probably the most widely-used plasmid vectors are the pUC series constructed by Messing and co-workers (Vieira and Messing, 1982; Norrander *et al.*, 1983; Yanisch-Perron *et al.*, 1985). The pUC plasmids are small derivatives of pBR322. Their size is 2.69 kb and their sequences are known (Yanisch-Perron *et al.*, 1985). They contain a gene encoding resistance to penicillin. In addition, they encode the *N*-terminal α-peptide of the β-galactosidase gene, which is expressed under the control of the *lac* promoter, so that, in appropriate hosts, α-complementation occurs, with the production of β-galactosidase activity. This results in blue colonies if the bacteria are plated on isopropyl-B-D-thiogalactoside (IPTG) and the chromogenic substrate, X-gal (see Chapter 11, Section I.B.2). The DNA encoding the α-peptide contains a fragment of DNA (a polylinker) containing sites for many restriction endo-nucleases, and insertion of foreign DNA into the polylinker results in inac-tivation of the α-peptide, so that white colonies are produced, enabling immediate visual identification of recombinant clones. This is exactly analagous to the method used for cloning in bacteriophage M13mp vectors (see Chapter 11). (It should be noted, however, that there is no necessity to grow pUC-containing cells on IPTG/X-gal, or, indeed, even to use the hosts required for α-complementation: the pUC plasmids are very satisfactory vectors even without the α-complementation screen.)

The polylinkers in the pUC series of plasmids derive originally from the M13mp series of phages and, as shown in Table I, the range of sites available has been extended to 17 sites in a single polylinker, while still retaining blue/white colony screening. The arrays of cloning sites listed in Table I are also available in M13 bacteriophage vectors and in vectors containing the replication origin and packaging signals of filamentous single stranded DNA phages (see Chapter 11 and Section IV below).

Many of the vectors listed in Table I contain an ampicillin resistance gene to provide selection for those bacteria harbouring the plasmids. Subcloning DNA fragments from one of these vectors to another is complicated by the need to distinguish the desired construction from the starting plasmids; i.e. the new cloning vector should have a different selectable marker from the first. To facilitate sub-cloning procedures, Spratt *et al.* (1986) have developed a set of plasmids which have the kanamycin resistance gene from Tn*903* as the selectable

TABLE I

Polylinkers in the lac α-Peptide

Plasmid	Polylinker sites[a]	Selectable marker[b]	References
pUC8	E Sm B S P H	Ap[R]	Vieira and Messing (1982)
pUC9	reverse of pUC8	Ap[R]	Vieira and Messing (1983)
pUC18	E Ss K Sm B Xb S P Sp H	Ap[R]	Norrander et al. (1983)
pUC19	reverse of pUC18	Ap[R]	Norrander et al. (1983)
pIC-7	E C EV Xb Bg Xh Sc N H	Ap[R]	Marsh et al. (1984)
pIC-19R	E C EV Xb Bg Xh Sc N H P S B Sm E	Ap[R]	Marsh et al. (1984)
pIC-19H	H P S B Sm E C EV Xb Bg Xh Sc N H	Ap[R]	Marsh et al. (1984)
pIC-20R	E C EV Xb Bg Xh Sc N H Sp P S Xb B Sm K Sc E	Ap[R]	Marsh et al. (1984)
pIC-20H	H Sp P S Xb B Sm K Sc E C EV Xb Bg Xh Sc N	Ap[R]	Marsh et al. (1984)
pBGS8, 9, 18, 19	as pUC8, 9, 18, 19	Km[R]	Spratt et al. (1986)
pBGS130	E Sm Ss EV Sp K Xb H B S P	Km[R]	Spratt et al. (1986)
pBGS131	Bg P S B H Xb K Sp EV Ss Sm E	Km[R]	Spratt et al. (1986)
pBR328	as pUC9	Cm[R]	Quigley and Reeves (1987)

[a] Sites are listed in order of proximity to the amino terminus of the peptide. B, *BamHI*; Bg, *BglII*; C, *ClaI*; E, *EcoRI*; EV, *EcoRV*; H, *HindIII*; K, *KpnI*; N, *NruI*; P, *PstI*; S, *SalI*; Sc, *SacI*; Sm, *SmaI*; Sp, *SphI*; Ss, *SstI*; Xb, *XbaI*; Xh, *XhoI*.
[b] Ap, ampicillin; Km, kanamycin; Cm, chloramphenicol.

marker. As well as the pUC polylinkers, the set also contains vectors with the extended polylinkers constructed by Kieny *et al.* (1983). A further useful feature of this pBGS series of plasmids (Table I) is that their complete nucleotide sequence has been compiled from the sequences of their constituent parts.

Plasmids incorporating the polylinkers of pUC18 and pUC19, together with the kanamycin resistance gene from Tn*5*, have recently been described (Pridmore, 1987). Again, the complete nucleotide sequences are known and, in addition, but for reasons unknown, the plasmids consistently yield three times as much DNA as do the pUC series. Following a similar experimental line, Quigley and Reeves (1987) have constructed vectors which use chloramphenicol resistance as a selective marker and have the pUC9 or pUC19 polylinker and blue/white colony screening.

B. Restriction-site bank vectors

A second approach which has been used to increase the number of unique restriction sites available for cloning has been to construct restriction site bank vectors. These contain a large number of restriction sites collected together on a small DNA fragment inserted into a non-essential region of the vector. The plasmid pJRD158 contains 43 unique restriction sites; almost all known restriction enzymes that recognize 6 bp palindromic sequences can be used with this vector (Heusterspreute *et al.*, 1985). Of the 64 possible 6 bp palindromes, 46 are unique in pJRD184 and of these 38 are recognized by known restriction enzymes. The continuing discovery of new restriction enzymes suggests that enzymes which will recognize the remaining unique palindromes have yet to be found. Restriction site bank vectors do not have the advantage of blue/white colony screening, although the proportion of empty vectors that come through a cloning experiment can be minimized by using dephosphorylated vector DNA (Ullrich *et al.*, 1977), or by cloning between restriction sites with different sticky ends.

C. Positive-selection vectors

Several plasmids have been described which allow positive selection of transformants that contain vectors with DNA inserts (see Pouwels *et al.*, 1985 for a listing). These vectors are not widely used, partly because the small number of restriction sites available for cloning limits their flexibility. An early example of a positive selection vector is plasmid pTR262 (Roberts *et al.*, 1980). This contains the intact *c*I repressor gene from λ phage and the rightward promoter *p*R, which drives a tetracycline resistance gene. In the native vector, transcription from *p*R is repressed by the *c*I protein and cells containing the vector are sensitive to tetracycline. When fragments of DNA are cloned into the

*Hind*III or *Bcl*I sites, the *c*I coding region is interrupted and the tetracycline resistance gene is expressed. Plating on agar containing tetracycline directly identifies clones with recombinant plasmids. Nilsson *et al.* (1983) have used oligonucleotide-directed mutagenesis to make specific base changes within the coding region of the *c*I gene, which create new restriction sites while maintaining *c*I protein function. The resulting improved vector, pUN121, is smaller and has additional sites for *Eco*RI and *Sma*I/*Xma*I and insertions into these also yield tetracycline resistant transformants. The creation of new cloning sites within genes in vectors of this type is limited by the necessity of maintaining the reading frame so as to produce an active protein. (It is fortunate that the α-peptide of β-galactosidase is particularly tolerant of sequence variation at its N-terminus, thus allowing the expansion of the pUC series polylinkers (see above).)

Other positive selection systems include rescue from the lethal effects of galactose in a cell displaying a $GalK^+T^+E^-$ phenotype (Ahmed, 1984), and use of a sole carbon source (Stevis and Ho, 1987).

D. Very high-copy number vectors

Plasmid variants in which the copy number can be experimentally manipulated by altering the growth conditions have been exploited as expression vectors (see Section III). Although uncontrolled, 'runaway' plasmid replication is usually lethal to the host cell, it is still possible to get vectors with very high-copy number. Thus, a point mutant of pBR322 has been described which has a constitutive copy number of around 1000 copies per cell. The plasmid comprises 65% of total cellular DNA without altering cell viability (Boros *et al.*, 1984). Several high-copy vectors, the pHC series, have been generated from it by inserting polylinkers. The pHC plasmids encode ampicillin resistance, they are small (2 kb), their DNA sequence can be compiled from their components and, most importantly, they yield 10–20 mg of plasmid DNA from one litre of bacterial culture. However, it is worth noting that problems might arise with these plasmids if the cloned DNA encodes a product that is produced constitutively.

E. *pcn* and copy number of pBR322-based plasmids

The copy number of pBR322 derived replicons can be influenced by host cell chromosomal functions. Mutations in a recently identified *E. coli* gene, *pcnB* (plasmid copy number), have been described which reduce the copy number of pBR322 and its derivatives. Plasmid pBR322 and its derivatives are normally present at a level of about 18 copies per chromosome. This level is reduced 16-fold, making them essentially single-copy, in *pcnB* hosts (Lopilato *et al.*, 1986). The *pcnB* gene has been mapped to approximately 3 min on the *E. coli* chromosome and is distinct from other genes whose products are known to affect plasmid replication or stability. A transposon Tn*10* insertion that is closely linked to the *pcnB* mutation has been isolated and this allows the use of P1 transduction to move the *pcnB* mutation into any strain of choice.

F. Promoter probes

Another context in which considerations of plasmid copy number become important is the use of 'promoter probe' plasmids to analyse transcriptional control mechanisms. Such experiments involve the creation of transcriptional fusions between a promoterless reporter gene and the promoter under study. The pKO series of plasmids use the *gal*K gene in the role of reporter and promoter activity is monitored by simple and rapid galactokinase assays (Rosenberg *et al.*, 1986). However, the multi-copy nature of these plasmids may lead to misleading results when the control of expression of a normally single-copy chromosomal gene is examined. For this reason, Koop *et al.* (1987) have developed a low-copy number 'promoter probe' vector based on the single copy F plasmid replicon. This construct has an M13mp18 polylinker upstream of a promoterless *lacZ* gene. Reconstruction experiments using wild-type and mutant *lac* promoters in this system showed that, in contrast to the pKO vectors, the low-copy number vector faithfully reproduced the regulatory responses of a chromosomally located *lac* promoter.

G. Effect of palindromic sequences on cloning

The chromosomal background chosen as a host in cloning experiments influences the capacity to clone and stably maintain palindromic sequences. Wyman *et al.* (1985, 1986) showed that when *recBC*, *sbcB* hosts were used to propagate human DNA libraries, approximately 9% of the recombinant molecules obtained failed to propagate on Rec$^+$ hosts; many of these clones carried palindromic sequences. Leach and Lindsey (1986) extended these observations and showed that palindrome-conferred inviability is expressed in both *recA* and Rec$^+$ cells. Strains that are *recBC sbcB* are recombination proficient but lack exonucleases V and I; for reasons that are not yet understood, this appears to stabilize the long palindromic sequences found in the human and other genomes. Strains which are *recBC sbcB* are therefore, the hosts of choice for the propagation of representative libraries of these genomes.

III. Expression vectors

One of the most rapidly expanding areas of plasmid vector design and construction has concerned vectors for expressing cloned genes, either as RNA or protein. Improvements and modifications in these systems continue to appear and some of the newest vectors will be discussed under the headings of RNA, fusion protein and native protein production.

A. RNA production

The early genes of the related *E. coli* bacteriophages T3 and T7 are transcribed by the host RNA polymerase. One of the products of the phage early genes is a new RNA polymerase which initiates only at the phage late gene promoters (Chamberlin *et al.*, 1983). The phage late promoter sequence bears no resemblance to the *E. coli* promoter consensus sequence and the phage polymerases will not initiate transcription from *E. coli* promoters. The *Salmonella typhimurium* phage SP6 encodes an RNA polymerase with properties similar to the T3 and T7 enzymes (Green *et al.*, 1983). Plasmids that contain late promoters from these phages positioned upstream of an array of unique cloning sites provide a means of generating specific single-stranded RNA molecules from cloned fragments by using the cognate RNA polymerase in an *in vitro* transcription reaction. Little and Jackson (1987) have recently reviewed the use of these vectors and provide detailed experimental protocols.

A large number of vectors incorporating a T3, T7 or SP6 promoter exist and some of the newer members of this class are listed in Table II. RNA molecules are usually produced from these vectors by linearizing the purified DNA at a

TABLE II

Plasmid vectors incorporating T3, T7 or SP6 promoters

Plasmid	Promoter(s)[a]	Restriction sites[b]
pSP65[c]	SP6	R, Sc, SmB X S P H
pGEM3[d]	SP6, T7	H Sp P S X B Sm K Sc R
pGEM-5Zf	T7, SP6	A Aa Sp Ea Nc Sc2 EV Se No PS Nd Sc Bs Ns
pGEM-7Zf	T7, SP6	A Aa Sp Xb Xh E K X Av Sm As C H B Sc Bs Ns
Bluescribe[e]	T3, T7	H Sp P S X B Sm K Sc R
Bluescript	T3, T7.	K A2 Xh S C H RV R P Sm B Se Xb No Ea3 Sc2 Bs Sc

[a]Promoters direct transcription across the cloning sites from left to right. Where two promoters are present, the second directs transcription from right to left.

[b]Abbreviations are as in the legend to Table I with the addition of: A, *Apa*I; A2, *Apa*II; Aa, *Aat*II; As, *Asa*II; Av, *Ava*I; Bs, *Bst*II; Ea, *Eag*I; Ea3, *Eag*III; EV, *Eco*RV; Nc, *Nco*I; Nd, *Nde*I; No, *Not*I; Ns, *Nsi*I; Sc2, *Sac*II; Se, *Spe*I; X, *Xma*I.

[c]pSP65 and a series of related plasmids is described in Krieg and Melton (1987).

[d]The pGEM plasmids are made and marketed by Promega Biotech.

[e]Bluescribe and Bluescript are made and marketed by Stratagene.

restriction site distal to the promoter, and then using the DNA *in vitro* as template for the polymerase to synthesize run-off transcripts. Some of the vectors incorporate two different promoters arranged 'head to head' on either side of the polylinker. This allows the production of RNA from either strand of the insert according to which cognate RNA polymerase is used. DNase digestion of the reaction mixture followed by phenol extraction allows recovery of the single-stranded RNA products, which have a wide range of uses. For instance, as hybridization probes they have all of the advantages of single-stranded DNA probes and are easier to prepare, and provide a ready supply of substrate RNA. These vectors have revolutionized the study of RNA processing events.

B. Fusion protein production

Fusion proteins are generated by joining two open reading frames so that the translation product contains amino acid sequences derived from both genes. One part of the complete open reading frame is derived from the vector, the other from the cloned DNA segment. The promoter, ribosome binding site and amino-terminal region of the vector gene ensure efficient transcription and translation of the fusion product. Fusion protein production allows the expression of short gene fragments and, consequently, the dissection of functional domains within complex proteins.

The *E. coli* protein that has been most widely used in this context is β-galactosidase, the product of the *lacZ* gene. Ruther and Müller-Hill (1983) introduced a variety of restriction enzyme sites into the C-terminal region of this gene to produce a series of pUR plasmids that can be used to generate β-galactosidase fusions. The original plasmids provided *Bam*HI, *Sal*I, *Pst*I, *Xba*I and *Hin*dIII sites in all possible reading frames at the 3' end of the *lacZ* gene. To these have been added a vector with an in-frame *Nco*I site (Biernat *et al.*, 1987). *Nco*I sites (CCATGG) are often found spanning the ATG initiation codons of eukaryotic genes so this particular vector facilitates isolation of fusions. The fusion proteins produced by these recombinants have β-galactosidase activity and a simple immunoenzymatic assay can be used to screen for bacterial clones expressing any protein fragment for which an antibody is available.

Two other vector systems yield fusion to β-galactosidase with expression levels that have been increased by replacing the *lac* promoter by stronger λ phage promoters. Sisk *et al.* (1986) described a plasmid (pWS50) that is designed to select DNA fragments that contained open reading frames (ORFs) and express them as β-galactosidase fusion proteins. Transcription is driven by the λ P_L promoter and regulated by a chromosomally located cI_{857} gene which encodes a temperature-sensitive repressor of P_L. The efficient translation initiation signals of the N-terminal segment of the λ cII gene are positioned upstream and out-of-frame to lacZ. A unique NruI site is located at the cII/lacZ interface. Insertion of blunt ended DNA fragments containing ORFs at the NruI site restores the translational reading frame between cII and lacZ, and colonies containing such recombinants can be scored as Lac⁺ on indicator media. The fusion proteins produced can readily be identified on sodium dodecylsulphate (SDS) polyacrylamide gels by their abundance and large size (β-galactosidase is one of the largest polypeptides in *E. coli* extracts).

In the second case, a family of vectors constructed by Stanley and Luzio (1984) have polylinkers at the 3' end of a hybrid comprising the λ *cro* gene and the lacZ gene. Large quantities of a fusion protein are made under the control of the λ P_R promoter. The fusion proteins precipitate inside the cell and are partially protected from proteolysis. A rapid immunological screening procedure has been developed for the insoluble fusion proteins and, as a consequence, these vectors have been used to construct a human cDNA library, which was screened with antibodies to detect rare cDNAs.

The pUC plasmids can also be considered as expression vectors, since they offer the possibility of producing in-frame fusions to the lacZ α-peptide. Unregulated expression of cloned genes is not always desirable since certain fusion proteins can be toxic to the cell. Hence fragments that generate these undesirable products are not recovered and those that are recovered sustain deletions that alleviate the problem. The *lac* promoter on the pUC plasmids can be controlled by the repressor product of the lacI^q allele carried on the single copy F plasmid. In strains such as JM101, and its relatives (in which pUC plasmids are often maintained), this allele produces 10× more repressor than wild-type. However, since the *rop* (*rom*) gene of pBR322, which mediates copy number control, was deleted during the construction of pUC plasmids, they have a higher copy number than pBR322, and this elevated copy number is sufficient to titrate even the increased amount of repressor produced by the lacI^q allele. Hence expression of the lacZ α-peptide is constitutive. Thus, for example, pUC8 in JM101 produces blue colonies on Xgal, even in the absence of inducer (Stewart *et al.*, 1986b). This may cause a problem, if production of the fusion protein is lethal.

A series of plasmids (pHG) which have all of the features of pUC but which have the lower copy number of pBR322 have been described by Stewart *et al.* (1986b). For reasons which are not understood, the pHG plasmids permit an increase in the level of α-complementation, so that the cell lines are Lac⁺ on MacConkey lactose medium, unlike the case of pUC plasmids where levels of β-galactosidase are too low to be detected on this medium. This feature has important consequences for the detection of recombinant clones by the loss of α-complementation. Recombinant pHG plasmids will be Lac⁻ on MacConkey agar whereas non-recombinant parental molecules will be Lac⁺. This distinction can be made without inducing the *lac* promoter, so removing the possibility of selective loss of recombinants through lethal expression of cloned inserts. The copy number of pHG plasmids is such that the *lac* promoter is efficiently repressed, even when there is but a single copy of the wild type repressor gene in the cell.

The problem of achieving tight regulation of production of a fusion protein has also been approached by other routes. Stewart *et al.* (1986a) have used the λ phage early promoter, P_R, to drive transcription of the *lacZ* α-peptide. This promoter is tightly controlled by the repressor produced by a cI_{857} gene included on the vector, but the controlled gene can be induced by heat to yield levels of fusion protein up to 25% of the total cell protein. Similar plasmids which offer different restriction sites at which to produce fusions have been described by Windle (1986).

Yet another approach has been described by Stark (1987), who has constructed a set of expression vectors which achieve very tight control by incorporating a *lac*I^q allele on the vector. This ensures an adequate supply of *lac* repressor to regulate the expression promoter. The pTTQ series is based on the pUC plasmids with the same polylinkers and *lacZ* α-peptide. The *lac* promoter has been replaced with the stronger hybrid *tac* promoter and a strong transcriptional terminator has been placed downstream of the α-peptide. The efficiency of the *lac* repressor control was demonstrated by cloning into the vector the *lux* genes from *Vibrio harveyi* which are lethal when expressed at a high level in *E. coli*. The experiments of Stark (1987) highlight a potential problem arising from the use of pUC or pTTQ vectors. Blue/white colony screening for loss of α-complementation requires induction of the plasmid promoter; accordingly inserts which are lethal when expressed will not be recovered. Although probably not common, if it is suspected that this may be a problem, it can be circumvented by selecting recombinants under noninducing conditions.

The promoters of the ribosomal RNA genes are among the strongest in the *E. coli* genome. Boros *et al.* (1986) have constructed a hybrid promoter from the −35 region of the *rrn*BP$_2$ promoter and the −10 region of the *lac* promoter-operator, producing a strong and regulatable promoter which they have designated *rac*. A series of vectors with *Eco*RI and *Cla*I sites in all three reading frames to yield α-peptide fusions has been constructed. The pBR322 copy number derivatives appear to be efficiently regulated. When the constructions were transferred to the very high-copy number mutant plasmid pHC314 (see above) regulation was predictably reduced but the fusion protein reached more than 60% of the total cell protein.

The genes of a number of other proteins have been employed for making fusions. We will briefly discuss fusions to thymidine kinase, chloramphenicol acetyl transferase, protein A and secretable proteins.

Shapira and Casadaban (1987) have described the use of the thymidine kinase from Herpes simplex virus to form enzymatically active hybrid proteins. Although designed primarily for the study of gene regulatory elements, these plasmids also show that fusion of new sequences to the amino terminal domain of thymidine kinase does not result in the loss of enzymatic activity.

An elegant and versatile prokaryotic-eukaryotic shuttle vector has been described by Mole and Lane (1987). The vector pSEMCat$_R$I can replicate in *E. coli* cells via a plasmid replicon and in mammalian COS-1 cells via the SV40 early region. The plasmid is provided with tandem bacterial and eukaryotic promoter elements linked to the chloramphenicol acetyl transferase (CAT) gene which allow expression of this gene in both bacterial and mammalian cells. Within the CAT gene coding region is a unique *Eco*RI site, which is in frame with the *Eco*RI site at the 3′ end of the *lacZ* gene in both the pUR and λgtll expression systems. An *Eco*RI fragment containing an ORF previously recognized in one of these two systems can thus be recloned to produce a CAT fusion protein. The activity of this hybrid in bacterial and mammalian cells can be directly compared since the same construct can be used for each cell type.

Staphylococcal protein A is particularly well suited for affinity purification due to its ability to bind specifically to the Fc domain of immunoglobulins. Two plasmid vectors containing the protein A gene adapted for gene fusions have been described by Nilsson *et al.* (1985) and are marketed by Pharmacia. Fusion proteins produced by these vectors can be purified in a single step by affinity chromatography on IgG sepharose. One vector, pRIT2T, has sites for *Eco*RI, *Sma*I, *Bam*HI, *Sal*I and *Pst*I in the 3′ end of the protein A gene, and allows temperature-inducible synthesis of intracellular fusion proteins using the λ P$_R$ promoter and the temperature-sensitive regulator cI_{857}. The second plasmid, pR1T5, is a shuttle vector designed for production of proteins to be secreted. It has both an *E. coli* and a *Staphylococcus aureus* replicon and the protein A gene is expressed from its own promoter and retains its native signal sequence. Fusion proteins produced using this vector are periplasmic in *E. coli* and can be collected by osmotic shock. Alternatively, the fusion proteins are secreted from *S. aureus* cells and can be recovered from the culture supernatant. Secretion has the advantage that it allows the formation of disulphide bridges which would otherwise not form in the reducing environment of the bacterial cytoplasm.

Proteins secreted from *E. coli* cells are normally retained in the periplasmic space. Kato *et al.* (1987) have described a system whereby the outer membrane of *E. coli* is made permeable without cell lysis, thus allowing complete secretion into the growth medium. The plasmid pEAP8 carries the weakly activated *kil* gene from pMB9 and the penicillinase promoter and signal sequence from a *Bacillus*. In their test system, Kato *et al.* (1987) cloned a synthetic human growth hormone gene into a *Hin*dIII site at the carboxyl terminus of the *Bacillus* signal peptide. The signal peptide directed secretion of the growth hormone into the periplasmic space and was correctly cleaved off to yield authentic growth hormone, which was then released into the growth medium through the outer membrane which had been permeabilized by the action of the *kil* gene product. The construct yielded about 20 mg of human growth hormone per litre of culture of which 55% was free in the medium. Because this system cannot be regulated, it is not clear how useful it will be in its present form.

A vector which can be used to probe for protein export signals and analyse membrane protein topology has been described by Broome-Smith and Spratt (1986). The plasmid, pJBS633, allows the construction of translational fusions to the amino terminus of the mature form of TEM β-lactamase. In-frame fusions are identified by their capacity to allow transformants to grow when plated at high inocula on ampicillin plates. If the β-lactamase can be translocated across the inner membrane to the periplasm then the transformants form isolated colonies on agar containing ampicillin, so allowing discrimination between fusions that allow translocation and those that do not. An additional feature of this vector is the inclusion of the phage F1 origin for single-stranded DNA replication, which allows the production of single-stranded DNA for rapid DNA sequencing of recombinants (see Section IV below).

C. Native protein production

E. coli expression systems continue to provide an abundant source of bio-logically-active proteins which would be otherwise virtually impossible to obtain — human interleukin-1 and human immunodeficiency virus reverse transcriptase provide two recent examples (Huang *et al.*, 1987; Larder *et al.*, 1987). As the experiments have moved from a laboratory to an industrial scale there has been increasing interest in efficient regulation of protein production and in vector stability. Some recent developments in these areas will be discussed here.

An interesting report from Miki *et al.* (1987) concerns a variant of the vector pKK223-3 which is commercially available from Pharmacia. The original expression vector contains an array of cloning sites; most of which are not correctly spaced for the ribosome binding site. Nonetheless a cDNA for chicken lysozyme was expressed at 2.5% of the total cellular protein. Miki *et al.* (1987) report that, when the pBR322 replicon in the original cloning vector was replaced by a pUC9 replicon, the plasmid copy number became temperature dependent and increased by a factor of ten in response to a shift from 28 to 42°C. At present the authors cannot rule out the possibility that the pUC9 which they used carried a second mutation which makes the replication of the plasmid temperature sensitive. However, their construction yielded 25% of total cell protein as chicken lysozyme, suggesting that gene copy number is the major limiting factor in this case. Miki *et al.* (1987) found that *lac*I^q repression was sufficient to shut off expression from these plasmids. This is in contrast to the experience of several other authors cited above. The lysozyme forms inactive precipitates within the cells, a phenomenon often observed when disulphide-containing proteins are expressed at high levels, but techniques have been developed to purify and activate the enzyme.

A novel dual control expression system for cloning genes whose products are toxic to *E. coli* cells is described by O'Connor and Timmis (1987). Transcription of the insert is achieved in a conventional way by the strong but highly repressible λ phage P_L promoter. The additional control feature is provided by IPTG-induced synthesis of antisense RNA from a convergent *lac* promoter positioned downstream of the cloned gene. As a test of the efficiency of this system, the gene encoding the *Eco*RI restriction endonuclease was cloned in the absence of its protective methylase. It was shown that transformants containing the construct were only viable when both repression controls were operational.

The problem of ensuring plasmid stability while not compromising levels of gene expression has been addressed by Wright *et al.* (1986). They have constructed vectors which contain two replication origins. For low-copy number maintenance, the pSC101 *ori* (giving about four copies per chromosome) was used, with no segregational loss even in the absence of antibiotic selection. For the production phase, there is a pBR322-derived origin in which the primer promoter which controls the frequency of replication initiation events has been replaced with the λ P_R promoter. A cI_{857} repressor gene is also included on the plasmid and its product prevents initiation from the pBR322 *ori* at low temperatures. Heat induction causes a dramatic increase in plasmid copy number with resultant high levels of cloned gene expression yielding greater than 20% of total cell protein.

Non *E. coli* proteins vary in the extent to which they are susceptible to proteolytic degradation by the bacterial system for eliminating abnormal proteins. The product of the *lon* gene is probably involved at an early stage since *lon* mutants degrade abnormal proteins at a reduced rate. The product of the *htpR* gene, required for induction of the heat shock response, has also been implicated since improved stability of foreign proteins has been reported in *lon htpR* double mutants (Buell *et al.*, 1985). Some bacteriophages interfere with the host proteolytic system and the inhibition of protein degradation in T4-infected cells has been exploited by Duvoisin *et al.* (1986) to allow expression of unstable proteins. T4 gene *32* protein is synthesized in large amounts and its mRNA displays an unusually high stability. Plasmids pRDB8 and pRDB9 both contain the promoter region and start codon of T4 gene *32* followed by a polylinker. DNA fragments inserted in the polylinker are transcribed and translated at high levels in both uninfected and T4 infected cells. By treating the cells with rifampicin to prevent new transcription and then allowing time for all mRNAs except those containing the gene *32* stabilizing leader sequence to decay, when radiolabelled amino acids are supplied the product of the cloned gene can be uniquely labelled. In addition, T4 infection can be used to inhibit host mediated degradation. With this system, Duvoisin *et al.* (1986) successfully expressed the highly labile rabbit immunoglobulin λ light chain.

Phage T7 promoters have not been found in any DNAs apart from T7 and its relatives. This specificity has been used by Rosenberg *et al.* (1987) as the basis for a set of vectors for selective expression of cloned genes by T7 RNA polymerase. The plasmids are given a pET designation for 'plasmid for expression by T7 RNA polymerase'. Transcription is controlled by the strong T7 $\phi10$ promoter and target mRNAs can be translated from their own start signals or fused to the signals of the major capsid protein of T7. Alternatively, fusions with the 2nd, 11th or 260th codon of the T7 capsid protein can be made. The vectors can be used to produce large amounts of RNA *in vitro* or to produce RNA and protein *in vivo*, either in a strain carrying the T7 RNA polymerase gene in the chromosome or in a strain superinfected with a λ phage construct expressing the T7 polymerase.

A significant number of proteins, when expressed to high levels in the *E. coli* cytoplasm, are insoluble and aggregated, sometimes forming visible inclusion bodies. A variety of empirical techniques have been developed to solubilize, denature and refold such proteins and these have recently been reviewed by Marston (1987).

IV. Plasmid packaging in single-stranded filamentous phage particles

It is possible to prepare single-stranded DNA from a plasmid vector by cloning into it a small fragment carrying the replication and packaging signals from a filamentous single-stranded bacteriophage. These signals are normally silent but can be activated by superinfection with a helper phage. The helper then supplies, *in trans*, all of the proteins required to replicate the viral strand and package it into phage particles which are extruded into the medium. The orientation of the phage origin determines which plasmid strand is packaged. The construction and use of vectors of this type has been reviewed by Cesarini and Murray (1987) and by Geider (1986).

The pBGS plasmids as well as the pGEM Bluescribe and Bluescript vectors embody the feature of single-stranded DNA production (see Tables I and II). In addition, the pEMBL plasmids have the same polylinkers as the pBGS series but confer resistance to ampicillin rather than kanamycin (Cesarini and Murray, 1987). This allows these two sets of vectors to be used in serial subcloning experiments without resorting to gel purification of fragments. The pBGS plasmids may be unsuitable for oligonucleotide directed mutagenesis due to remnants of the inverted repeats of Tn903 which cause premature DNA polymerase termination at the ends of the kanamycin resistance gene (B. G. Spratt cited in Cesarini and Murray, 1987).

The company, Stratagene, have exploited the properties of the M13 gene 2 protein to effect a type of *in vivo* subcloning. They have developed a phage vector, λ ZAP, for cDNA cloning. When an interesting clone has been identified by nucleic acid or antibody screening of a cDNA library, it is infected with an M13 helper phage. This recognizes M13 viral strand origin signals located in the ZAP arms and causes displacement and circularization of the cDNA insert embedded in a complete Bluescript vector. These molecules are secreted as single-strand phages and can be recovered by infecting an F' strain and plating for colonies on ampicillin plates (Fernandez *et al.*, 1987). The subcloning from ZAP to Bluescript plasmid is thus achieved without any *in vitro* manipulations. The Bluescript vectors offer the full range of features: DNA and RNA sequencing, site directed mutagenesis, synthesis of RNA probes, expression of proteins and isolation of single-stranded DNA.

A feature of all of the vectors discussed so far in this section is that they can only produce one of the strands of the insert in a single-stranded form. If the other strand is required, the insert must be recloned in the opposite orientation. Peeters *et al.* (1986) have constructed the pKUN vectors to allow the separate production of either plasmid strand in a filamentous phage particle. These plasmids contain all of the features of the pUC series and the M13 (+) origin together with the origin of a second filamentous phage, Ike. M13 and Ike are specific for their parental *ori* sequences and are not interchangeable. The Ike (+) strand *ori* is in inverted orientation relative to the M13 *ori*. Thus, superinfection with an M13 helper packages one vector strand and an Ike helper packages the other strand. In order to circumvent problems which arise from the differing surface receptor requirements of M13 and Ike, Konings *et al.* (1986) have isolated a recombinant phage, Mike, which supplies the Ike-specific protein while infecting cells via the M13 receptor; a single M13 sensitive host strain can therefore be infected with either phage type with resulting production of the plasmid strand of choice.

Several *E. coli*-yeast shuttle vectors have been described which incorporate M13 *ori* fragments for single-strand production (Vernet *et al.*, 1987; Cesarini and Murray, 1987). Although in theory there is no upper limit for DNA packaging, increased phage size is associated with a high rate of insert instability caused by spontaneous deletions. This is particularly marked in phage M13 clones and can be minimized by maintaining replication in the plasmid mode. In addition, Cesarini and Murray (1987) have noted that expression plasmids with strong promoters give consistently low yields of single-stranded DNA unless the promoters are kept fully repressed. The fact that the M13 origin has no adverse effect on plasmids during normal growth, and the utility of obtaining purified single-stranded DNA will ensure that this type of vector is widely used.

V. Extended host range vectors

In general, the development of vectors from bacterial hosts other than *E. coli* has followed a similar course to that of the *E. coli* vectors themselves. Multipurpose cloning vectors, regulatable expression vectors and vectors for the isolation and study of promoter regions have been constructed for a wide range of genera and they will be briefly discussed in this section.

A broad host range plasmid vector, pNM187, based on the RSF1010 replicon and which allows regulated expression of cloned genes in Gram-negative bacteria of at least 16 different genera has been described (Mermod *et al.*, 1986). Transcription is controlled by the positively activated Pm twin promoters of a toluene-degrading plasmid and *xylS* the gene for the positive regulator. Expression is induced by micromolar quantities of benzoate or *m*-toluate. In a test case using the *xylE* gene, controlled expression was observed in most genera, with induced levels of up to 5% of total cellular protein. The level of gene expression in different genera tended to correlate with their phylogenetic distance from *Pseudomonas putida*. A set of vectors based on the broad host range replicon RSF1010 which are also useful for a wide range of Gram-negative bacteria has also been described by Chistoserdov and Tsygankow (1986).

Bacillus subtilis has several attractive features as a host for producing industrial proteins from cloned genes. It is non-pathogenic, is currently in use as an industrial microorganism and it can secrete proteins into the culture medium. Several new plasmid vectors for *B. subtilis* have been described. The set constructed by Grandi *et al.* (1986) present cloning sites downstream of the promoter and ribosome binding site of an erythromycin resistance gene. Hence, whereas the pBR322 β-lactamase gene is not normally expressed in *B. subtilis* from its *E. coli* regulatory elements, when cloned in the *Bacillus* expression plasmids, the protein is both synthesized and secreted. The finding that it is secreted indicates that the β-lactamase signal sequence is recognized by the *B. subtilis* secretory mechanism.

Promoter probe plasmids have been widely used to study gene regulation in *E. coli*. As stated earlier, they consist of a promoterless reporter gene, the product of which can easily be assayed, and upstream sites into which DNA fragments can be cloned to assay for promoter function. Drug resistance genes such as those encoding chloramphenicol acetyltransferase or aminoglycoside phosphotransferase, as well as metabolic enzymes such as galactokinase and β-

galactosidase have been used as reporter genes. This strategy has now been extended to study promoters in *Streptomyces* spp. (Ward *et al.*, 1986), *Streptococus* spp. (Achen *et al.*, 1986) and *Staphylococcus* spp. (Hudson and Stewart, 1986).

A study by Stibitz *et al.* (1986) that involved manipulation of the toxin gene of *Bordetella pertussis* affords a good example of the value of analysing genes in their normal host context and the need for vectors to accomplish this. *Bord. pertussis* is the aetiological agent of whooping cough and attempts to study virulence determinants of this organism by cloning them into *E. coli* have, until recently, not been successful. Most progress towards understanding the virulence loci has come from studying them in *Bord. pertussis* itself. In many cases it is desirable to replace a chromosomal locus with an allele that has been altered *in vitro*. Stibitz *et al.* (1986) describe a vector, pRTP1 (*return to pertussis*), which permits such allele replacement. The plasmid is designed to allow successive selection of two homologous recombination events. The first recombination is detected by selecting for acquisition of the ampicillin-resistance gene on the plasmid after transfer by conjugation. The plasmid is unable to replicate in *Bord. pertussis* cells. This stratagem selects for integration of the plasmid into the *Bord. pertussis* chromosome by recombination between the cloned gene and its chromosomal homologue. The second step uses streptomycin (Sm) to select for loss of vector sequences that encode a Sm-sensitive allele which is dominant over the chromosomal Sm-resistant allele. The DNA rearrangement requires a second recombination event within the cloned sequences and can lead to the replacement of the resident allele with the cloned version. The approach is particularly useful when studying multi-component systems such as the virulence determinants of *Bordetella*.

In terms of the application of recombinant DNA techniques, the obligatory anaerobic bacteria have been neglected until recently. The importance of this group is seen when it is considered that the vast majority of the indigenous microflora of man are obligate anaerobic bacteria. Extrachromosomal systems and gene transmission in the two major genera, *Bacteroides* and *Clostridium*, have been reviewed by Odelson *et al.* (1987). Advances in gene cloning systems in *Bacteroides* have stemmed from two advances in the ability to introduce DNA into these organisms. One advance is based on the mobilization of plasmids, by conjugation, from *E. coli* to *Bacteroides* spp. The other was the development of a transformation system to allow uptake of plasmid DNA by *Bact. fragilis*. The vectors which have been constructed are all based on shuttle plasmids, made by fusing *E. coli* and *Bacteroides* replicons. This is necessary because there are no known plasmid replicons or drug resistance determinants which can function naturally in both cell types. We can anticipate a growth in our understanding of these anaerobes as these systems are developed and exploited.

We will conclude this section on extended host range vectors by examining two shuttle vector systems for moving genes between *E. coli* and either plant or monkey cells.

Koncz and Schell (1986) describe a novel binary vector system which they have used to study the expression of a Ti-plasmid gene promoter in plants. The tumour inducing Ti-plasmids of *Agrobacterium* have been widely used for introducing foreign genes into dicotyledenous plants. The new Ti-plasmid derived binary cloning systems have two elements: a broad host range cloning vector carrying T-DNA border sequences essential for transfer and integration into the plant genome and a Ti-derived helper plasmid to provide virulence functions *in trans* (Bevan, 1984). The system described by Koncz and Schell (1986) combines all of the elements needed in plant cloning vectors in a single 'cassette' — these include selectable markers flanked by T-DNA borders, cloning sites, bacterial genes for transconjugant identification and broad host range replication and mobilization functions. The cassette can be maintained and transferred back and forth between *E. coli* and *Agrobacterium* hosts which contain plasmid RK2 replication and mobilization helper functions. The DNA from transformed plants can be reisolated in *E. coli* by virtue of the linked plasmid replicon. Two T-DNA promoters, one for gene 5 and the other for the nopaline synthase gene, are present in the cassette. By linking marker enzymes to these promoters and measuring enzyme levels it was shown that the gene 5 promoter was regulated in a tissue specific manner, whereas the nopaline synthase promoter was not.

SV40 based shuttle vectors have been widely used to analyse gene expression and DNA replication, recombination and mutagenesis. These constructs must be introduced into mammallian cells by DNA transfection, which is inefficient and may account for the high genetic instability observed (Calos *et al.*, 1983). Menck *et al.* (1987) describe new SV40 shuttle vectors which can be packaged into pseudovirions. They replicate and are transmitted as virus in monkey COS cells (which provide large T antigen) without a helper. DNA isolated from infected cells can be rescued by transformation of *E. coli*. Up to 1.4 kb of extra DNA can be accommodated in these vectors without exceeding their packaging limit. They neatly combine the convenience of analysis in bacteria with the advantages of infectious virus.

VI. Specialized vectors

A. cDNA cloning

The vector primer and linker method of Okayama and Berg (1982) has been widely used to isolate full length cDNAs from poly A^+ RNA preparations. Nakamura *et al.* (1986) constructed a cDNA library in pUC8 from potato tuber poly A$^+$ RNA. On screening this library with radiolabelled antibody against a major tuber protein, patatin, several cDNA clones were identified which expressed the precursor form, prepatatin. Translation of the cloned gene seemed to be initiated inside the cDNA sequence at the authentic start codon, despite the fact that DNA sequencing showed no convincing Shine-Dalgarno ribosome binding site homology in the vicinity of this codon. The authors concluded that, provided high levels of mRNA are produced in the bacterial cells, then the highly sensitive *in situ* radioimmunoassay can detect the levels of direct translation product produced.

Recently cDNA cloning vectors have been designed that can promote expression of the antigenicity or *in vivo* function of the protein products. The latest addition of this type is the vector pSI4001, constructed by Shigerada *et al.* (1987). This contains the *lac* promoter and operator and three sets of ribosome binding sites and associated ATG codons arranged tandemly in all possible reading frames. cDNA can be cloned downstream of these, in the appropriate orientation for translation, using the Okayama and Berg (1982) method that yields an in-frame fusion with one of the three translation starts. Expression of the cDNA reading frame and production of the protein is thus assured. Two of the three translation initiation sites exhibit high levels of expression. The third functions at a lower, but still significant rate.

New pUC-derived expression vectors to facilitate the construction of cDNA libraries, and a simplification of the original Okayama-Berg cDNA cloning protocol, have been described by Oberbäumer (1986). The plasmids contain the extended polylinker from M13tg131 (Kieny *et al.*, 1983; see pBGS131 in Table I). The incorporation of this sequence allows most of the restriction sites to be preserved when following the modified protocol (because of the different arrangement of the sites this would not have been possible had pUC18 or pUC19 been used) and, hence, easy recovery of the inserted cDNA. In the first step the vector is linearized at a site near the centre of the polylinker and short dC tails are added. The termini are then blocked by addition of a dideoxy C residue so that they will be unable to participate in subsequent tailing reactions. One tail is then removed by cleavage at a second polylinker restriction site and DNA purified by gel filtration. Then a dT tail is added to the new terminus. At this point the vector is ready for the cDNA synthesis step which follows conventional routes. mRNAs are annealed to the vector by pairing of their 3' poly A tract and the oligo dT tail. Reverse transcriptase is used to make the first DNA strand which is then 'G-tailed' to allow circularization. The molecule is then repaired in the normal way. Note that there is no need for an additional linker fragment as required by the original method. A library of 200 000–450 000 transformants was constructed from 2 µg of human placental poly A$^+$ RNA (Oberbäumer, 1986). These numbers are large enough to ensure a complete library.

B. DNA sequencing and mutagenesis

Sequencing of long DNA fragments is considerably simplified if a nested set of deletions can be generated and several enzymatic techniques are available to isolate such sets *in vitro*. An *in vivo* method, which is technically much simpler, uses transposon-induced deletions (Ahmed, 1984, 1985). This employs a plasmid containing a portion of the Tn9 transposon and a truncated *gal* operon. Growth in the presence of galactose causes accumulation of UDP-galactose which is lethal to the cell. Galactose-resistant mutants arise by deletions that extend from a fixed site at the transposon terminus through the *gal* genes to variable sites in the adjoining DNA. Peng and Wu (1986) improved this system by cloning into the vector an M13 origin fragment that allows the production of single-stranded DNA for dideoxy sequencing by M13K07 superinfection. Gold Biotechnology market a kit which contains a pair of plasmids that have been further improved by incorporating the *lacZ* α-peptide and polylinkers from pUC18 and 19, so permitting blue/white colony screening for the initial cloning step. Hundreds of galactose-resistant colonies are pooled, and their plasmid DNA extracted. The required deletion size range is selected after separation on agarose gels and these molecules are used to transform an M13-sensitive host. The deletions in the recovered recombinant plasmids are ordered by restriction digestion. Single-stranded DNA is prepared by M13K07 super-infection and sequenced.

Eckert (1987) has described a pair of new vectors which facilitate rapid sequencing using the Maxam and Gilbert chemical degradation method. They allow a single defined end to be labelled, without resort to a gel purification step. The plasmids contain a pair of *Tth*111I sites which flank a *Sma*I site and all are embedded in a polylinker. The DNA sequence recognized by *Tth*111I is 5'GACTNAGTC3' where N can be any base. The enzyme cuts within this sequence to leave the central base as an unpaired 5' overhang. The two vector sites each have a different central base. Thus, for a blunt ended fragment cloned into the *Sma*I site *Tth*111I digestion will generate four ends, each with a unique 5' unpaired base. By selecting an individual [α-^{32}P] labelled deoxynucleoside triphosphate, either of the two ends flanking the insert can be uniquely labelled by DNA polymerase I and sequenced directly. Both strands of the insert can, therefore, be sequenced. For fragments cloned with one blunt end in the *Sma*I site and a second end in one of the polylinker sites (*Hind*III, *Pst*I, *Sal*I, *Xba*I, *Bam*HI, *Sac*I, *Bst*EII or *Eco*RI) a single *Tth*111I site will remain in the molecule allowing direct sequencing of one strand. The plasmid yield from a 1.5 ml miniprep is sufficient for four separate labelling experiments each yielding over 2×10^6 cpm which is sufficient for five or more sequence runs.

In vitro mutagenesis of cloned genes is an important route towards understanding protein function. Barany (1985) has developed a technique which permits insertion of single-stranded hexameric linkers into pre-existing cohesive end restriction sites such that a new restriction site is generated and two new amino acids are inserted into the protein. Such mutations when introduced into the pBR322 β-lactamase often generated a temperature-sensitive phenotype and the mutant enzyme demonstrated altered substrate specificity. A range of TAB linkers (*two amino acid B*arany) are commercially available from Pharmacia.

C. *In vivo* cloning

E. coli phage Mu is especially suitable for *in vivo* cloning because it transposes hundreds of times as it replicates when derepressed for its lytic functions. Temperature-sensitive alleles of the Mu repressor allow synchronous derepression of replication. Groisman and Casadaban (1986) have described mini-Mu phage with plasmid replicons for *in vivo* cloning and the production of *lac*Z fusions. The mini-Mu replicons can be used to clone DNA by growing them with a complementing Mu phage and using the resulting lysate to transduce Mu-lysogenic cells. The vectors have selectable genes for resistance to antibiotics and low- or high-copy number plasmid replicons. Packaging constraints limit the size of clonable fragments to 17–31 kb, which is more than adequate for the isolation of bacterial genes. The high frequency with which clones can be obtained suggests that it may be possible to select directly for traits that require two unlinked genetic segments (Groisman and Casadaban, 1986). The presence of selectable antibiotic resistance markers, plasmid replicons and the transfer origin of the broad host range plasmid RK2 allows the introduction of these vectors into a variety of Gram-negative species with a view to cloning genes back into *E. coli*.

Mini-Mu plasmids that contain cosmid replicons and can be used for *in vivo* cloning have also been described (Gramajo and Mendoza, 1987; Mendoza *et al.*, 1986). In this system derepression of mini-Mu transposition results in bacterial DNA sequences becoming flanked by integrated mini-Mu elements. These segments can then be packaged into λ phage heads by superinfection with a λ helper phage. This produces a cosmid-transducing lysate from which clones of particular bacterial genes can be recovered by selection after infection of an appropriate host strain(s). Although limited to *E. coli* and related Gram-negative bacteria this technique is still of value, particularly for the analysis of multi-factorial traits.

D. Cosmids

Of presently available cloning vectors cosmid vectors can accommodate the largest amount of DNA. They are technically more difficult to use than λ phage vectors and remain something of a specialist area. A recent review of these vectors and experimental protocols for their use have been provided by Little (1987). For successful packaging into λ phage particles there must be two *cos* sites on the same DNA molecule separated by 38–52 kb. Ligation reactions which favour the formation of contatemers of cosmid/30–45 kb passenger DNA/cosmid will provide substrate molecules that are efficiently packaged *in vitro*. Cosmid transduction frequencies are of the order of 10^6 colony forming units/µg (Little, 1987).

Cross and Little (1986) report anecdotal information of instability of cosmid clones, overgrowth of libraries by isolates containing small molecules and unclonable sequences. Some of these problems can be overcome by propagation of the recombinant DNA molecules in *recA* or *recA recBC sbcB* strains, as discussed in Section II and by use of the λ based Lorist B vector which appears to be significantly different from cosmids based on the ColEl replicon.

Lorist·B contains SP6 and phage T7 RNA polymerase promoters adjacent to the cloning sites. Cross and Little (1986) describe techniques for isolating RNA probes that correspond to the extreme ends of any cloned fragment from these vectors. The isolation requires neither DNA fragment preparation nor any knowledge of the restriction map and yields DNA probes which can be used for systematic chromosome walking.

References

Achen, M. G., Davidson, B. E. and Hillier, A. J. (1986). *Gene* **45**, 45–49.

Ahmed, A. (1984). *Gene* **28**, 37–43.

Ahmed, A. (1985). *Gene* **39**, 305–310.

Balbas, P., Soberon, X., Merino, E., Zurita, M., Lomeli, H., Valle, F., Flores, N. and Bolivar, F. (1986). *Gene* **50**, 3–40.

Barany, F. (1985). *Proc. Natl. Acad. Sci. U.S.A.* **82**, 4202–4206.

Bevan, M. (1984). *Nucl. Acids Res.* **12**, 8711–8721.

Biernat, J., Hasselmann, H., Hofer, B., Kennedy, N. and Köster, H. (1987). *Protein Eng.* **1**, 345–351.

Boros, I., Posfai, G. and Venetianer, P. (1984). *Gene* **30**, 257–260.

Boros, I., Lukacsovich, T., Baliko, G. and Venetianer, P. (1986). *Gene* **42**, 97–100.

Broome-Smith, J. K. and Spratt, G. (1986). *Gene* **49**, 341–349.

Brown, T. A. (1985). "Gene Cloning: An Introduction". Van Nostrand Reinhold (UK).

Buell, G., Schulz, M. F., Selzer, G., Chollet, A., Movva, N. R., Semon, D., Escanez, S. and Kawashima, E. (1985). *Nucl. Acids Res.* **13**, 1923–1938.

Calos, M. P., Lebowski, J. S. and Botchan, M. R. (1983). *Proc. Natl. Acad. Sci. U.S.A.* **80**, 3015–3019.

Cesareni, G. and Murray, J. A. H. (1987). *In* "Genetic Engineering", Vol. 9, pp. 135–154 (J. K. Setlow, Ed.), Plenum Press, New York.

Chamberlin, M., Kingston, R., Gilman, M., Wiggs, J. and de Vera, A. (1983). *Meth. Enzymol.* **101**, 540–568.

Chang, A. C. Y. and Cohen, S. N. (1978). *J. Bacteriol.* **134**, 1141–1156.

Chistoserdov, A. Y. and Tsygankov, Y. D. (1986). *Plasmid* **16**, 161–167.

Choo, K. H., Filby, G., Greco, S., Lau, Y-F, and Kan, Y. W. (1986). *Gene* **46**, 277–286.

Cross, S. H. and Little, P. F. R. (1986). *Gene* **49**, 9–22.

de Mendoza, D., Gramajo, H. C. and Rosa, A. L. (1986). *Mol. Gen. Genet.* **205**, 546–549.

Duvoisin, R. M., Belin, D. and Krisch, H. M. (1986). *Gene* **45**, 193–201.

Eckert, R. L. (1987). *Gene* **51**, 247–254.

Fernandez, J. M., Short, J. M., Renshaw, M., Huse, W. D. and Sorge, J. (1987). *Gene* (submitted).

Geider, K. (1986). *J. Gen. Virol.* **67**, 2287–2303.

Gramajo, H. C. and de Mendoza, D. (1987). *Gene* **51**, 85–90.

Grandi, G., del Bue, M., Palla, E., Mele, A., Colletti, E. and Toma, S. (1986). *Plasmid* **16**, 1–14.

Green, M. R., Maniatis, T. and Melton, D. A. (1983). *Cell* **32**, 681–694.

Groisman, E. A. and Casadaban, M. J. (1986). *J. Bacteriol.* **168**, 357–364.

Hadfield, C., Cashmore, A. M. and Meacock, P. A. (1987). *Gene* **52**, 59–70.

Heusterspreute, M., Thi, V. H., Emery, S., Tournis-Gamble, S., Kennedy, N. and Davison, J. (1985). *Gene* **39**, 299–304.

Huang, J. J., Newton, R. C., Pezzella, K., Covington, M., Tamblyn, T., Rutlege, S. J., Gray, J., Kelley, M. and Lin, Y. (1987). *Mol. Biol. Med.* **4**, 169–181.

Hudson, M. C. and Stewart, G. C. (1986). *Gene* **48**, 93–100.

Kato, C., Kobayashi, T., Kudo, T., Furusato, T., Murakami, Y., Tanaka, T., Baba, H., Oishi, T., Ohtsuka, E., Ikehara, M., Yanagida, T., Kato, H., Moriyami, S. and Horikoshi, K. (1987). *Gene* **54**, 197–202.

Kieny, M. P., Lathe, R. and Lecocq, J. P. (1983). *Gene* **26**, 91–99.

Koncz, C. and Schell, J. (1986). *Mol. Gen. Genet.* **204**, 383–396.

Konings, R. N. H., Luiten, R. G. M. and Peeters, B. P. H. (1986). *Gene* **46**, 269–276.

Koop, A. H., Hartley, M. E. and Bourgeois, S. (1987). *Gene* **52**, 245–256.

Krieg, P. A. and Melton, D. A. (1987). *Meth. Enzymol.* (in press).

Larder, B., Purifoy, D., Powell, K. and Darby, G. (1987). *EMBO J.* **6**, 3133–3137.

Lau, Y-F. and Kan, Y. W. (1984). *Proc. Natl. Acad. Sci. U.S.A.* **81**, 414–418.

Leach, D. and Lindsey, J. (1986). *Mol. Gen. Genet.* **204**, 322–327.

Little, P. F. R. (1987). In "DNA Cloning", Vol. III (D.M. Glover, Ed.), pp. 19–42, IRL Press, Oxford.

Little, P. F. R. and Jackson, I. J. (1987). In "DNA Cloning", Vol. III (D. M. Glover, Ed.), pp. 1–8, IRL Press, Oxford.

Marsh, J. L., Erfle, M. and Wykes, E. J. (1984). *Gene* **32**, 481–485.

Marston, F. A. O. (1987). In "DNA Cloning", Vol. III (D. M. Glover, Ed.), pp. 59–88, IRL Press, Oxford.

Menck, C. F. M., Sarasin, A. and James, M. R. (1987). *Gene* **53**, 21–29.

Mermod, N., Ramos, J. L., Lehrbach, P. R. and Timmis, K. N. (1986). *J. Bacteriol.* **167**, 447–454.

Miki, T., Yasukoci, T., Nagatani, H., Furuno, M., Orita, T., Yamada, H., Imoto, T. and Horiuchi, T. (1987). *Protein Eng.* **1**, 327–332.

Mole, S. E. and Lane, D. P. (1987). *Nucl. Acids Res.* **15**, 9090.

Monod, M., Denoya, C. and Dubnau, D. (1986). *J. Bacteriol.* **167**, 138–147.

Nakamura, K., Hattori, T. and Asahi, T. (1986). *FEBS Letts.* **198**, 16–20.

Nilsson, B., Abrahmsén, L. and Uhlen, M. (1985). *EMBO J.* **4**, 1075–1080.

Nilsson, B., Uhlen, M., Josephson, S., Gatenbeck, S. and Philipson, L. (1983). *Nucl. Acids Res.* **11**, 8019–8030.

Norrander, J., Kempe, T. and Messing, J. (1983). *Gene* **26**, 101–106.

Oberbäumer, I. (1986). *Gene* **49**, 81–91.

O'Connor, C. D. and Timmis, K. N. (1987). *J. Bacteriol.* **169**, 4457–4462.

Odelson, D. A., Rasmussen, J. L., Smith, C. J. and Macrina, F. L. (1987). *Plasmid* **17**, 87–109.

Okayama, H. and Berg, P. (1982). *Mol. Cell. Biol.* **2**, 161–170.

Old, R. W. and Primrose, S. B. (1985). "Principles of Gene Manipulation", 3rd edn. Blackwell, Oxford.

Peeters, B. P. H., Schoenmakers, J. G. G. and Konings, R. N. H. (1986). *Gene* **41**, 39–46.

Peng, Z-G. and Wu, R. (1986). *Gene* **45**, 247–252.

Pouwels, P. H., Enger-Volk, B. E. and Brammar, W. J. (Eds) (1985). "Cloning Vectors". Elsevier, Amsterdam.

Pridmore, R. D. (1987). *Gene* **56**, 309–312.

Quigley, N. B. and Reeves, P. R. (1987). *Plasmid* **17**, 54–57.

Roberts, T. M., Swanberg, S. L., Poteete, A., Riedel, G. and Backman, K. (1980). *Gene* **12**, 123–127.

Rosenberg, M., Brawner, M., Gorman, J., Reff, M. (1986). *In* "Genetic Engineering; Principles and Methods", Vol. 8 (J. K. Setlow and A. Hollaender, Eds), pp. 151–180, Plenum Press, New York.

Rosenberg, A. H., Lade, B. N., Chui, D-S., Lin, S-W., Dunn, J. J. and Studier, F. W. (1987). *Gene* **56**, 125–135.

Rüther, U. and Müller-Hill, B. (1983). *EMBO J.* **2**, 1791–1794.

Shapira, S. K. and Casadaban, M. J. (1987). *Gene* **52**, 83–94.

Shigesada, K., Itamura, S., Kato, M., Hatanaka, M., Imai, M., Tanaka, M., Masuda, N., Nagai, J. and Nakashima, K. (1987). *Gene* **53**, 163–172.

Shoemaker, N. B., Getty, C., Guthrie, E. P. and Salyers, A. A. (1986). *J. Bacteriol.* **166**, 959–965.

Sisk, W. P., Chirikjian, J. G., Lautenberger, J., Jorcyk, C., Papas, T. S., Berman, M. L., Zagursky, R. and Court, D. L. (1986). *Gene* **48**, 183–193.

Spratt, B. G., Hedge, P. J., te Heesen, S., Edelman, A. and Broome-Smith, J. K. (1986). *Gene* **41**, 337–342.

Stanley, K. K. and Luzio, J. P. (1984). *EMBO J.* **3**, 1429–1434.

Stark, M. J. R. (1987). *Gene* **51**, 255–267.

Stevis, P. E. and Ho, N. W. Y. (1987). *Gene* **55**, 67–74.

Stewart, G. S. A. B., Lubinsky-Mink, S. and Kuhn, J. (1986a). *Plasmid* **15**, 182–190.

Stewart, G. S. A. B., Lubinsky- Mink, S., Jackson, C. G., Cassel, A. and Kuhn, J. (1986b). *Plasmid* **15**, 172–181.

Stibitz, S., Black, W. and Falkow, S. (1986). *Gene* **50**, 133–140.

Thompson, R. (1982). *In* "Genetic Engineering", Vol. 3 (R. Williamson, Ed.), pp. 1–52, Academic Press, London.

Ullrich, A., Shine, J., Chirgwin, J., Pictet, R., Tischer, E., Rutter, W. J. and Goodman, H. M. (1977). *Science* **196**, 1313–1319.

Vernet, T., Dignard, D. and Thomas, D. Y. (1987). *Gene* **52**, 225–233.

Vieira, J. and Messing, J. (1982). *Gene* **19**, 259–268.

Ward, J. M., Janssen, G. R., Kieser, T., Bibb, M. J., Buttner, M. J. and Bibb, M. J. (1986). *Mol. Gen. Genet.* **203**, 468–478.

Windle, B. E. (1986). *Gene* **45**, 95–99.

Wright, E. M., Humphreys, G. O. and Yarranton, G. T. (1986). *Gene* **49**, 311–321.
Wyman, A. R., Wolfe, L. B., Botstein, D. (1985). *Proc. Natl. Acad. Sci. U.S.A.* **82**, 2880–2884.
Wyman, A. R., Wertman, K. F., Barker, D., Helms, C. and Petri, W. H. (1986). *Gene* **49**, 263–271.
Yanisch-Perron, C., Vieira, J. and Messing, J. (1985). *Gene* **33**, 103–119.
Yarranton, G. T., Wright, E., Robinson, M. K. and Humphreys, G. O. (1984). *Gene* **28**, 293–300.

9

Detection and use of transposons

P. M. BENNETT AND J. GRINSTED

Department of Microbiology and Unit of Molecular Genetics, University of Bristol, Medical School, Bristol, UK

AND

T. J. FOSTER

Microbiology Department, Trinity College, Dublin, Ireland

I. Introduction

Transposable elements are discrete sequences of DNA that can move from one genetic locus to another, on the same or on a different replicon, by a process of recombination called transposition (or translocation). They are ubiquitous among bacteria and many different ones have been discovered. The size of these elements can vary markedly from one to another, and many different properties can be transposable. When a transposable element contains an accessory gene, encoding some marker, it is called a transposon: the terms transposable element and transposon are often used interchangeably. The transposable markers may be antibiotic resistance determinants, toxins and other virulence factors, or metabolic functions. Transposable elements have been discovered in many species of bacteria, both Gram-positive and Gram-negative bacteria and in archaebacteria. The properties of many of these elements have been reviewed extensively elsewhere (Cohen, 1976; Kleckner, 1977 & 1981; Nevers and Saedler, 1977; Calos and Miller, 1980; Starlinger, 1980; Grindley and Reed, 1985).

Transposable elements retain their structural integrity during transposition, and the recombination does not require DNA base sequence homology between the donor and recipient genetic loci. Accordingly, transposition occurs independently of *recA*-controlled homologous recombination. A transposable element, in general, can be assigned to one of two main classes, designated classes I and II. Assignment is essentially on the basis of structure (Fig. 1). Class I accommodates IS elements (which are 0.7–2 kb in size) and IS-based transposons called composite transposons, while Class II (at least 3.5 kb in size) accommodates a set of ancestrally-related elements called compound transposons. Both classes contain representatives from Gram-positive as well as Gram-negative bacteria. Although the majority of transposons can be so categorized, some cannot. The most obvious examples of transposable elements that belong neither to class I nor to class II are Tn7, which has the structure of a compound transposon but is unrelated to the transposons grouped in class II (Lichtenstein and Brenner, 1982), the Gram-negative bacteriophage, Mu, which replicates by a process of transposition (Toussaint and Resibois, 1983) and the Gram-positive conjugative transposon, Tn916, and related elements (Franke and Clewell, 1981; Clewell *et al.*, 1985). However, it seems reasonable to predict that as more transposable elements are found and analysed, the number of exceptions will grow.

Irrespective of type or class, most transposable elements consist of a unique DNA sequence flanked by short (8–40 bp) inverted nucleotide sequence repeats (IR sequences). These serve to delineate the element. As noted, two main types of transposon have been described (Fig. 1). One type, exemplified by Tn3, has short IRs (38 bp) flanking a unique DNA sequence that accommodates

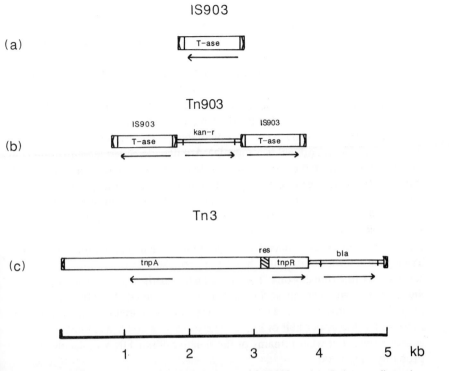

Fig. 1. Structures of some transposable elements. (a) IS*903*, a class I element (insertion sequence); T-ase, gene for transposase. (b) Tn*903*, a composite transposon based on IS903, and encoding resistance to kanamycin (kan-r) (Grindley and Joyce, 1981). (c) Tn*3*, a Class II element (compound transposon) encoding resistance to β-lactam antibiotics (Grindley and Reed, 1985); *tnpA*, gene for transposase; *tnpR*, gene for resolvase; *bla*, gene for TEM β-lactamase; the resolution site (*res*) is in the intergenic region indicated by the hatched box.

Transposition functions are represented by wide open boxes, and other functions by narrow open boxes; the shallow arrow heads at the ends of each transposable element represent the terminal inverted repeats (IRs). (IR of IS*903* is 18 bp, and of Tn*3* is 38 bp.) Arrows underneath genes indicate the directions of transcription.

All elements are drawn to the same scale, indicated by the scale at the bottom.

three genes, two of which (*tnpA* and *tnpR*) are involved in transposition, while the third (*bla*) encodes resistance to certain β-lactam antibiotics e.g. ampicillin. The sequence transposes as an entity and, for the purposes of transposition, cannot be subdivided. The second type has larger terminal repeats, sometimes directly repeated but more often inverted, flanking a unique central sequence. These terminal repeats comprise two copies of an IS element which act in concert to transpose the intervening DNA sequence. This central sequence is responsible for the cell phenotype conferred by the transposon e.g. resistance to an antibiotic, but contributes no transposition functions (i.e. it is a passenger transported by the flanking IS elements, which, in many cases, have been demonstrated to be capable of transposition independently of the rest of the structure). Hence class I transposons have an obvious modular structure.

The process of transposition is illustrated in Fig. 2. Replicon A, which carries the transposon, interacts with replicon B via the transposable sequence in one of two ways. In the case of class II elements, such as Tn*3*, the reaction is a replicative recombination event. It is mediated by the product of the *tnpA* gene of the transposon, and yields a joint DNA molecule called a cointegrate (Fig. 2) comprising both the donor and recipient replicons with a copy of the transposable sequence at each replicon junction. The two copies of the transposon are carried as direct repeats. A site-specific recombination between the two copies of the transposon releases the transposition product, replicon B::Tn*3*, and the transposon donor. The enzyme that mediates the site-specific recombination is usually encoded by the transposon itself (by *tnpR*; see above). Transposition via cointegrates has been proposed as the major pathway for several Tn*3*-like transposons (Tn*1*, Tn*3*, Tn*21*, Tn*501*, Tn*1721*) (but see below), and as the form of replication used by bacteriophage Mu. In contrast, the majority of transposition events involving class I elements occur directly, as indicated in the upper half of Fig. 2. The process is nonreplicative and, hence, conservative, and what remains of the deleted donor DNA molecule is, most probably, degraded.

Among both sets of elements are those that can transpose by more than one mechanism. Some class I elements e.g. IS1, have been shown to transpose via cointegrate formation, although the direct pathway is most favoured, and since these elements encode neither a specific site for recombination, nor a site-specific recombinase, then transposition via the cointegrate pathway yields a relatively stable cointegrate molecule (even in a Rec⁺ host, where, in principle, normal recombination between homologous sequences could effect resolution). Among class II elements, Tn*1*, Tn*21* and Tn*1000* have been shown to transpose directly, and this form of transposition may account for 5% or more of all transpositions with a particular element (Bennett *et al.*, 1983; Tsai *et al.*, 1987).

As a consequence of a transposition event, a short DNA sequence at the target site (about 5 or 9 bp) is duplicated. The resultant direct repeats are positioned on either side of the transposon, adjacent to the IR sequences at its new location. This finding suggests that cleavage of the recipient DNA at the target is not at the same point on both strands of the DNA (i.e. a flush cut) but rather is staggered, and that duplication is generated by a repair process after the transposon has been inserted (Arthur and Sherratt, 1979; Shapiro, 1979). The presence of short direct repeats flanking a DNA insertion can, therefore, be used as evidence of transposition. Several models have been

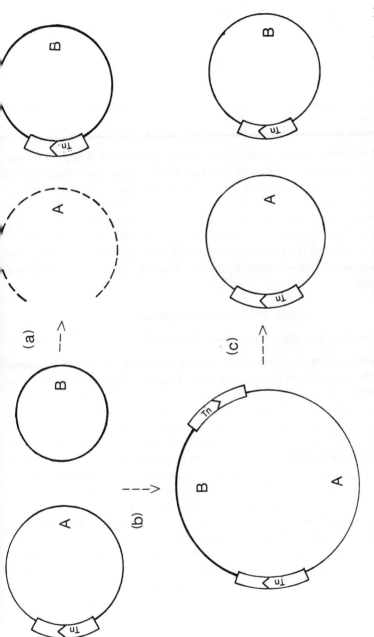

Fig. 2. Pathways of transposition. Replicon A carries a transposable element (Tn); replicon B is the target molecule. (a) Conservative transposition; the end result is that the element (Tn) has been cut from replicon A and inserted into replicon B to give the new recombinant (B::Tn). The remains of A will be degraded. (b), (c) Replicative transposition; the first step, (b), generates a cointegrate with directly-repeated copies of the element at the junctions of the replicons; the cointegrate is then resolved, (c) to the new recombinant (B::Tn) plus the original donor (A).

proposed to explain transposition (see Grindley and Reed, 1985), but it is not necessary to elaborate them any further in this article.

Transposon insertion can have several consequences. Among the most useful are:

1. A DNA sequence of characteristic structure is integrated into the target replicon. This can be analysed by physical methods e.g. restriction enzyme mapping (Chapter 6).
2. The new sequence (i.e. the transposon) has a selectable marker that automatically becomes associated with each site of insertion.
3. The point of insertion may be within a gene and so cause a mutation.

Bacterial plasmids confer a rich variety of cell phenotypes (Chapter 2), and many of the genes that encode these functions are carried on transposons. It follows that the characterization of a newly isolated plasmid should include tests to discover if transposable elements comprise part of the plasmid. All such tests employ the same basic strategy, namely, transposition of a DNA sequence from one replicon to another. Various combinations have been used (plasmid to plasmid, plasmid to chromosome, plasmid to phage); which combination will be most suitable in a particular situation is usually a case of trial and, very often, error.

II. Detection of transposons

A. Transposition to a self-transmissible plasmid

The simplest tests for transposition involve plasmid-plasmid combinations, because the experiments usually rely on technically simple bacterial matings (Chapter 3) to separate the transposition products from the original donor and recipient plasmid molecules. Indeed, the first bacterial transposon, Tn*1* (formerly TnA), was discovered in this way (Hedges and Jacob, 1974). Many plasmids isolated from nature transfer poorly, or not at all by conjugation. This deficiency can be exploited in the search for transposable sequences carried by these plasmids. The plasmid of interest is transferred into a suitable *Escherichia coli* K12 *recA* strain (by conjugation, transformation or transduction, as convenient, see the appropriate chapters). A Rec⁻ strain, preferably a *recA* strain, is used to eliminate the possibility of host-cell mediated recombination which may confuse the issue. A second plasmid, which transfers by self-mediated conjugation at a respectable frequency (10^{-3} to 1 transconjugants per donor), and which also is unable to comobilize the test plasmid (Chapter 3), or does so poorly, is then transferred (by conjugation) into the same cell. We routinely use the IncW plasmid R388 (markers TpSu, Ward and Grinsted, 1981) or the IncP plasmid pUB307 (markers KmTc, Bennett *et al.* 1977). Once the potential donor and recipient plasmids are resident in the same cell, we store the culture on Dorset Egg slopes for up to two weeks at room temperature, after initial

incubation at 30°C for 24 h. This strategy is followed because for some transposons e.g. Tn*1* (Bennett, unpublished results), transposition events have been
shown to accumulate over a period of days. Hence, the delay may optimize the
chance of discovery. We have found that it is not necessary to follow this
procedure with all transposons, but we would advise it in the case of an
uncharacterized one, and particularly when no transposition can be detected
soon after strain construction i.e. within 24–36 h of obtaining the dual plasmid
starter strain.

Protocol 1
1. Streak a loopful of the plasmidless recipient strain on nutrient agar. We use, for
   example, JC6310 (Bennett and Richmond, 1976), which is Str^r, *recA* and requires
   histidine, lysine and tryptophan for growth.
2. Streak the donor culture (cells which contain both the test plasmid, and R388, for
   example) at right angles across the first streak.
3. Incubate the plate at 37°C for 18 h.
4. Collect the bacterial growth where the streaks intersect (with a wire loop) and
   suspend it in 1 ml of nutrient broth.
5. Prepare a 1 in 10 dilution series (to 10^{-4}) of the suspension in nutrient broth.
6. Spread 0.1 ml aliquots of the undiluted suspension and the dilution series on an
   appropriate selective agar: when JC6310 is used as the recipient, this could be
   nutrient agar containing streptomycin (100–200 μg ml^{-1}) and an antibiotic appropriate for the putative transposable resistance determinant. Alternatively, an
   appropriate minimal medium could be used, if auxotrophic counterselection is
   possible (i.e. if the donor and recipient bacteria were different auxotrophs, or if
   the donor was an auxotroph and the recipient was a prototroph). For example,
   with JC6310 as recipient, a minimal salts agar containing glucose (0.2% w/v),
   tryptophan, histidine and lysine (all at 20 μg ml^{-1}) and the antibiotic selective for
   the putative transposon would be appropriate.
7. Incubate plates for 24–36 h at 37°C (n.b. cells growing on minimal agar grow more
   slowly than those on nutrient agar).
8. Transconjugants are tested to establish that they contain transposon-carrying
   derivatives of the chosen recipient plasmid (R388 in the example). This can be
   determined by demonstrating genetic linkage of the putative transposon marker
   to the plasmid markers; e.g. the putative recombinant plasmid is transferred to
   another strain and transconjugants are selected for acquisition of one of the original
   plasmid markers (Tp or Su for R388). These transconjugants are then tested to
   see if they have also acquired the putative transposon marker. If this unselected
   marker is transferred, together with the selected marker on all occasions then
   the two markers have become genetically linked and transposition has probably
   occurred. Physical analysis of the plasmid DNA is then undertaken to confirm the
   genetic analysis (Chapter 6).

This form of experiment is straightforward and variations can be introduced
to enhance the rigour of the selection. One such variation is to use an interspecies cross, instead of a cross between strains of the same bacterial species.
Many plasmids display a relatively narrow host range; in contrast, plasmids
such as pUB307 (Bennett *et al.*, 1977) have an extended host range. This

difference can be successfully exploited to isolate new transposons. In summary, a plasmid such as pUB307 is transferred (by conjugation) to the strain carrying the test plasmid. The resulting strain is then mated with another bacterial strain to which the test plasmid will not transfer (because of lack of establishment or maintenance, rather than a conjugal defect), but to which the pUB307 will transfer (previously established). Transconjugants are selected on an agar supplemented appropriately for growth of the recipient bacterium and containing, in addition, a reagent selective for the putative transposon-encoded phenotype. Any transconjugants obtained are examined genetically, as above, to establish linkage of the test marker to the mobilizing plasmid (e.g. pUB307), and then physically (e.g. restriction enzyme analysis — Chapter 6) to confirm the presence on the plasmid of a newly acquired DNA sequence.

B. Transposition indicated by cointegrate formation

During the experiment outlined in the previous section, it may be found that all markers of both plasmids (test and recipient/target) have been transferred to the recipient strain. Thus, although the design of the experiment assumes that the test plasmid is not transferred during the mating, transposable elements can mediate the transfer of the entire DNA sequence of a nonconjugative plasmid by a conjugative one. This form of mobilization results from physical linkage of the test and conjugative plasmids. Such fusions are often formed by transposition (see above), mediated by a transposon on one of the plasmids. They represent cointegrates that have not resolved in the donor, either because the transposable element lacks a resolution system or transfer occurred before the cointegrate could be resolved. In the latter case resolution will occur in the transconjugant, which will then contain the test plasmid and the target plasmid with a transposon insertion (Fig. 2); in the former case the transconjugant will contain a cointegrate plasmid species. Mobilization of what is normally a non-conjugative plasmid is *prima facie* evidence for the presence of a transposable element; but it must be demonstrated that transconjugants carry either cointegrates of the form illustrated in Fig. 2, or products derived from such a structure. It is not sufficient simply to demonstrate cotransfer of both sets of plasmid markers. It should be noted that mobilization by cointegrate formation can be used to detect an element, such as an IS element that does not carry a selectable phenotypic marker.

In *E. coli*, a simple experiment can be performed to exploit cointegrate formation and the following example illustrates the point. Plasmid pACYC184 is a small nonconjugative plasmid that encodes resistance to chloramphenicol and tetracycline (Chang and Cohen, 1978). It was constructed *in vitro* and during this process lost the genetic information that would permit comobilization by a conjugative plasmid (Chapter 3). Hence, conjugal transfer of this plasmid requires, except in rare circumstances, covalent linkage of pACYC184 and the mobilizing plasmid at the time of transfer, i.e. it must be part of a cointegrate.

Protocol 2
1. Plasmids pUB307::TnA (KmTcAp and pACYC184 (CmTc) are introduced into the same cell e.g. JC6310, the former by conjugation and the latter by transformation.
2. Strain JC6310 (pUB307::TnA, pACYC184) is crossed with an appropriate recipient, and chloramphenicol resistant transconjugants are selected (i.e. mobilization of pACYC184 is selected).
3. Chloramphenicol resistant transconjugants are found to be, in general, resistant also to kanamycin, tetracycline and ampicillin, i.e. have acquired all the markers of both plasmids.
4. Plasmid DNA is isolated from the transconjugants and used to transform another strain of *E. coli* to resistance to chloramphenicol.
5. Transformants are found to be, in general, resistant to chloramphenicol, tetracycline (pACYC184 markers) and ampicillin (a pUB307 marker).
6. A physical examination of the plasmids in the transformants demonstrates that they are pACYC184::TnA recombinant plasmids, i.e. TnA has transposed from pUB307::TnA to pACYC184.

In this example, pACYC184 is transferred in the form of a transient cointegrate intermediate, generated in the first stage of TnA transposition. The cointegrate is resolved in the recipient cell, via a site-specific recombination mechanism encoded by the element itself (see above), to regenerate the original transposon donor (pUB307::TnA) and a derivative of pACYC184 carrying TnA. It is this latter derivative that is then recovered by the transformation step. The efficiency of separation achieved by the transformation step is aided by the fact that transformation with small plasmids is somewhat more efficient than with larger plasmids.

However, as mentioned earlier, not all transposable elements encode a site-specific resolution system, in which case the cointegrate may survive, particularly in a *recA* cell line. Resolution can usually be demonstrated, in these cases, in a Rec$^+$ background, although the process may be slow.

The example illustrates how to detect a transposon on a conjugative plasmid. The strategy can also be employed to identify transposons on nonconjugative plasmids. In this case pACYC184 is replaced by the test plasmid and pUB307::TnA is replaced by, say, R388 (or even pUB307, although this plasmid does carry other transposable elements).

C. Transposition detected by an increase in gene copy number

The following strategy may be appropriate if all others fail. As the number of copies of a particular gene in a cell increases, the overall level of gene expression increases, often more or less proportionately. In terms of drug resistance, this can mean that a particular determinant confers a higher level of resistance when the gene(s) is carried on a high copy number plasmid than when it is on a low copy number plasmid. This difference can be used, in these cases, to select for transposition.

Protocol 3
1. A small multicopy plasmid such as ColE1, pBR322 or pACYC184 is introduced (usually by transformation) into the same cell as the test plasmid.
2. The cell line with both plasmids is spread on agar containing growth requirements and the antibiotic to which the putative transposon confers resistance, at a concentration two to three times that needed to prevent colony formation of the strain that carries only the test plasmid. An inoculum of 10^8–10^9 cells per plate is used.
3. Plates are incubated under normal conditions.
4. Colonies that are obtained on the selective agar comprise cells that putatively contain transposon-carrying derivatives of the multicopy plasmid. These can be recovered following plasmid isolation and transformation, as indicated in Protocol 2. The transformants are selected with normal concentrations of the antibiotic, not the increased concentrations used in Step 2.
5. To maximize the chances of recovering plasmid-transposon recombinants, the colonies obtained after Step 3 can be pooled. Plasmid DNA can be isolated from the mixture prior to transformation, rather than the plasmid DNA being isolated from cultures of individual colonies. (The object is to demonstrate transposition and not to estimate transposition frequency, nor to obtain an insertion in a particular locus.) These elaborations, if necessary, can be done at a later stage, when the existence of the transposon has been established.

This strategy has been used successfully to demonstrate transposition of TnA elements (Wallace *et al.*, 1981; Dodd and Bennett, 1983). With some transposons, the approach would be inappropriate because phenotypic expression is not proportional to gene copy number (e.g. Tn*10*: in the case of this element, the higher the number of copies of Tn*10* in the cell, the lower the resistance to tetracycline conferred). Therefore, although in principle this approach can be successful, in particular instances it may be inappropriate. However, in most cases the level of resistance would be expected to rise as the transposon copy number increases. Unfortunately, we cannot offer much advice as to how to judge if this type of approach is likely to be successful. From the above examples (TnA, which confers ampicillin resistance and Tn*10*, which confers tetracycline resistance), amplification of a gene(s) which specify products altering the architecture of the cell, particularly the envelope, is likely to be more damaging than amplification of a gene, the product of which is an enzyme that destroys or modifies an antibiotic.

D. Bacteriophages as transposon targets

Temperate bacteriophages can also be used as the target DNA to detect transposons: e.g. λ (Kleckner *et al.*, 1977, 1978), P22 (Kleckner *et al.*, 1977), P1 (Gottesman and Rosner, 1975; Kuner and Kaiser, 1981) and fd (Herrmann *et al.*, 1978) have all been used for this purpose. However, it should be remembered that enlargement of the bacteriophage genome beyond a certain limit may prevent subsequent packaging. For example, the genome of bacteriophage λ can be increased by about 10% (5 kb) and still be packaged to yield infective particles. Larger insertions can only be accommodated if there is a compensating deletion elsewhere on the phage genome. Several deletion derivatives of phage λ are available and have been used successfully (see later), but a description of them is beyond the scope of this text (see Kleckner *et al.*, 1977, 1978; Shapiro and Sporn, 1977). The use of phage vectors for transposon detection is, generally technically more demanding than the other procedures dealt with, and in our view should only be contemplated when other methods have failed.

E. Transposon detection: final thoughts

The various techniques outlined in this part of the chapter, together with others discussed later in a different context, should provide sufficient variety for application in most experimental situations. It may be necessary to change the identity of the potential recipient replicon to take account of the marker on the potential transposon. The examples given above are of transposons encoding resistance to an antibiotic in *E. coli*, because it is the most familiar and extensively used system. Furthermore, the system is not limited to transposons encoding antibiotic resistance. Other plasmid markers could also be examined given a suitable selection procedure. The protocols are applicable to Gram-positive systems too, although they will almost certainly require some modification to accommodate the particular system under investigation. All the techniques depend on an efficient transfer system; conjugation, transformation or transduction, to separate transposition products from the donor plasmid, the target replicon and suitable recipient replicons. In our experience, finding the most suitable system is a matter for trial and error. A method that works effectively for one transposon may not work for a different element.

Finally, when a new transposable element has been discovered (or been constructed), it is given an IS number (for an insertion sequence) or a Tn number (for a transposon). To ensure that the same number is not given to two different elements, assignments are coordinated by Dr. E. Lederberg (1987). Unassigned numbers can be obtained from her upon request.

III. Use of transposons in genetic analysis

The application of drug resistance transposons to microbial genetic analysis has been reviewed elsewhere (Kleckner *et al.*, 1977) and practical details have also been published (Davis *et al.*, 1980; de Bruijn and Lupski, 1984). It is not intended to duplicate this material in this chapter. However, by exploiting transposons, it is now possible to perform genetic manipulations in organisms where no genetic system was previously available. In terms of the bacterial chromosome, these include the isolation of auxotrophic mutations (arising from transposon insertion in a gene the product of which is necessary for biosynthesis of required metabolite), provision of portable regions of homology for chromosome transfer (i.e. the generation of transient or stable Hfr strains (Hayes, 1968) by transposon mediated integration of a self-transmissible plasmid into the bacterial chromosome) and the generation of deletions (see Kleckner, 1981).

In plasmid analysis, transposons can be used to detect genes and define their limits and to facilitate gene cloning (Bennett, 1985), as well as provide easily selected markers for cryptic plasmids (or those which, for manipulative purposes, might be assumed to be cryptic because they carry markers which are difficult to demonstrate like pathogenicity determinants). The basis of all transposon use in plasmid analysis is the random (or almost random) nature of the insertions. (In principle, any sequence can be disrupted.) This means that a collection of plasmid-transposon recombinants can be built up on the linkage of the transposon and plasmid markers. The set can then be screened to identify those recombinants that do not confer a property normally encoded by the plasmid. This subset contains plasmid recombinants in which the transposon insertion has disrupted the original plasmid region responsible for the property of interest. Mutation sites (i.e. of transposon insertion) are easily determined because of the physical presence of a specific nucleotide sequence with a distinctive restriction enzyme pattern (Chapter 6). Hence, a physical and genetic map of the plasmid can be constructed.

A. Techniques for selecting transposon insertions

The methods that have been devised to select transposon-carrying derivatives of the test plasmid are of two general types:

(a) enrichment for plasmid::transposon recombinants by selecting cells that have retained the transposon but lost the remainder of the transposon donor (among these, some cells will contain the desired plasmid derivatives), prior to the recovery of recombinants by plasmid transfer to a suitable recipient as indicated in Protocol 3; and,

(b) selection of the desired plasmid derivatives directly from the strain containing the target plasmid and the transposon donor by transfer (conjugation or transformation) into a new host without prior enrichment.

B. Transposon donors

Transposon donors are of three types:
(a) the bacterial chromosome,
(b) temperate bacteriophages such as λ, and,
(c) plasmids, often nonconjugal or mutants unable to replicate at temperatures above 40°C.

1. The bacterial chromosome

This system is very convenient if the plasmid to be mutated is a conjugal one. E. coli derivatives carrying one of a variety of transposons integrated into their bacterial chromosomes have been described. The most useful of these would appear to be the drug resistance transposons Tn1/Tn3 (Bennett and Richmond, 1976), Tn7 (Barth and Grinter, 1977; Barth et al., 1978) and a set of transposon derivatives generated from Tn1721 (Ubben and Schmitt, 1986). The protocol is straightforward.

1. Transfer the test plasmid (usually by conjugation) into the transposon donor strain (Chapter 3).
2. Cross the donor strain carrying the test plasmid with a suitable recipient strain of E. coli or another Gram-negative recipient.
3. Spread aliquots of the mating mixture on agar which contains the antibiotic to which the transposon confers resistance thereby selecting for the recipient bacterium.
4. Transconjugants will contain transposon derivatives of the test plasmid. These can be analysed to determine the phenotypic and genetic effects of the insertions, which can then be located on the plasmid by restriction enzyme mapping. Insertions in a particular plasmid gene may be recognized by their mutational effects. However, it should not be forgotten that transposon insertions can be very polar (i.e. expression of genes downstream from that in which the insertion is located will also be adversely affected, if they are transcribed from a common promoter). This must be taken into account when analysing coordinately controlled gene systems like genes necessary for conjugal transfer of the plasmid.

Use of the bacterial chromosome as a transposon donor is not restricted to generating derivatives of conjugal plasmids. With minor modification, the system may prove effective for nonconjugal plasmids. If the frequency of transposition to the plasmid is of the order of 10^{-4} or higher, then total plasmid DNA can be isolated from a derivative of the donor strain carrying the test plasmid and used to transform a suitable recipient strain. Transformants are selected on medium containing the antibiotic to which the transposon confers resistance. Most, if not all the transformants recovered, will carry a transposon derivative of the test plasmid. However, in some instances the frequency of transposition is so low that transposon-carrying derivatives cannot be recovered. In these cases it may be possible to enrich for the desired plasmid species prior to plasmid isolation and transformation. This exploits the fact that, in general, the level of resistance increases as the resistance gene copy number increases (see Protocol 3). Therefore, in a population of cells, the few cells that not only have a copy of the transposon on the chromosome but also have additional copies on the plasmids they carry, are likely to display a higher level of drug resistance than the rest.

In practice, aliquots of a culture of the transposon donor containing the test plasmid are spread on a series of nutrient agar plates containing increasing concentrations of antibiotic to which the transposon confers resistance. We would normally use a set of drug concentrations based on the minimum inhibitory concentration (MIC) of the donor strain itself e.g. 0.5, 1.0, 1.5 and 2.0 times the MIC. Colonies growing in the presence of the drug at concentrations above the normal MIC can be examined individually for derivative plasmids by isolation and transformation to a suitable recipient, or alternatively the colonies can be pooled prior to plasmid isolation. The latter is our normal procedure. Plates with up to 200 colonies are flooded with 5 ml of growth medium (or a suitable buffer) and a glass spreader is used to form a bacterial suspension. The suspension is dispensed into a sterile centrifuge tube after which the cells are harvested and their plasmids isolated. The plasmid preparation is then used to transform a suitable recipient, selecting transformants that have acquired antibiotic resistance to which the transposon confers resistance.

Most systems of this type i.e. transposition from the bacterial chromosome to the plasmid, have been elaborated in *E. coli*. However, similar systems do exist in other bacteria, particularly in selected Gram-positive organisms such as *Staphylococcus aureus*, *Bacillus subtilis* and *Streptococcus faecalis* (Khan and Novick, 1980; Franke and Clewell, 1981; Clewell *et al.*, 1985).

2. Temperate bacteriophages

The principle of methods which use phage vectors to donate transposons is to infect a recipient cell with a specialized transducing phage (Hayes, 1968) which carries the transposon, under conditions where the phage cannot lysogenize or replicate lytically (e.g. in principle, superinfection of a lysogen would suffice). Ideally, each drug-resistant transductant colony is derived from an independent transposition event into one of the replicons in the recipient cell (either the plasmid or the chromosome). The proportion of plasmid-linked insertions relative to those in the chromosome will depend on the size and copy number of the test plasmid and the ease of transposition into each of the two replicons. This can only be determined experimentally. If a particular plasmid proves to be refractory to a particular transposon, then alternative elements will have to be tried.

If a plasmid-mediated phenotype has been lost simultaneously with the acquisition of the transposon, it is a reasonable assumption that the element has been inserted into the plasmid. Nonetheless, the test plasmid should be transferred into another bacterial strain. If the test plasmid is conjugal then transfer by conjugation would be appropriate (unless the transposon insertion has mutated the transfer system), otherwise plasmid isolation followed by transformation (Chapter 4) is usually the most convenient recovery procedure. Transconjugants or transformants are selected for acquisition of an original plasmid marker and then screened for nonselected acquisition of the transposon marker. This will only be transferred at the same frequency as the plasmid proper if it has physically become a part of the plasmid.

Wild-type temperate bacteriophages are not always suitable transposon delivery vehicles. Accordingly, several of them have been adapted to serve the particular purpose of transposon carriers. The most notable of these are derivatives of phage λ, which carry transposons Tn5, Tn9 or Tn10 (Rosner and Gottesman, 1977; Kleckner et al., 1978). The best derivatives for donating Tn5 and Tn10 have mutations which have removed the phage attachment site (att) and the integration function (int) (e.g. the b221 deletion). The extra advantage gained from such phage derivatives is that the deleted phage genome can accommodate more foreign DNA and still be packaged to form an infective phage particle (Brammer, 1982). They may also have a temperature-sensitive repressor, cI857, which prevents repressed phage genomes being maintained after infection. Another benefit to be gained from phage delivery of these particular transposons is that they appear to transpose at higher frequencies immediately after entering a cell than when established. This is probably because of temporary escape from normal control systems that act to minimize the frequency of transposition (since unrestricted transposition is undesirable, given its mutagenic potential).

3. Plasmids temperature-sensitive for replication

The elimination of the transposon donor plasmid by exploiting the temperature-sensitive (ts) character (natural or generated by mutation, Ubben and Schmitt, 1986) of a gene product needed for replication can be used. The ts plasmid is maintained in the cell together with the test plasmid by growth at the permissive temperature (usually 30°C) with appropriate selection, if necessary (established experimentally). The culture is then diluted (1/100) into fresh growth medium lacking selective agents and growth continued at the nonpermissive temperature (usually 42°C). After several hours (6–18 h) incubation, the culture is again diluted (1/100) with fresh growth medium containing the antibiotic to which the transposon confers resistance and, if necessary, an agent selective for the test plasmid and incubation at the nonpermissive temperature continued (usually 18 h). Dilutions of the resulting culture are spread on agar incorporating the same antibiotics and the plates are incubated at the nonpermissive temperature. Colonies which form are tested to verify loss of markers, other than that of the transposon, associated with the *ts* plasmid (because stable inheritance of the transposon could be the result of reversion of the mutation conferring temperature sensitivity, or by integration of the entire plasmid into one of the other replicons in the cell). Plasmid derivatives are purified by transfer to another bacterial host, by whatever transfer system is convenient. Transformation is often the best since this proceeds *via* a pure DNA stage.

Several *ts* plasmid vectors have been described: F'*ts1141ac*::Tn*10* delivers Tn*10* (Kleckner *et al.*, 1977); pMR5, a *ts* mutant of RP1 (Robinson *et al.*, 1980) and pSC304, a derivative of pSC101 (Kretschmer and Cohen, 1979) both donate Tn*1*/Tn*3*; pRN3091, a derivative of the penicillinase plasmid pI258 of *Staphylococcus aureus*, donates the erythromycin resistance transposon Tn*551* (Novick *et al.*, 1979); pTV1 accommodates the Class II Gram-positive transposon Tn*917* (Youngman *et al.*, 1983). Plasmid pMR5 is of particular interest because of its wide host range. In principle, transpositions of Tn*1* can be selected in any of the diverse hosts in which IncP plasmids can be propagated, provided that the cells grow at the restrictive temperature. Plasmid pRN3091 could possibly be transferred into other Gram-positive species and used to generate mutations. The range of elements transposed by this method could be extended by constructing *ts* plasmid derivatives incorporating other transposons. Indeed, Ubben and Schmitt (1986) have generated a series of derivatives of the transposon Tn*1721*; each of which has a different drug resistance marker and is carried on a *ts* plasmid. Hence, Tn*1725* encodes chloramphenicol resistance, Tn*1731* encodes tetracycline resistance and Tn*1732* encodes kanamycin resistance; this is a suitable range of drug resistance determinants for most purposes.

Plasmids which are *ts* for replication are a type of suicide donor (i.e. under the particular experimental conditions chosen, the transposon delivery system cannot be maintained). Hence, persistence of the transposon marker probably indicates that the element has transposed. Simple tests, indicated above, will determine whether transposition has been into the plasmid of interest or into the bacterial chromosome. It is beyond the scope of this text to detail other suicide vectors, but reference to journals such as the *Journal of Bacteriology*, *Plasmid* or *Gene* will rapidly indicate what is available.

4. Nonconjugal plasmids

Small nonconjugal plasmids may make good transposon donors, particularly if, in the system used, they are not comobilized (Chapter 3). Examples of such plasmids that can be used in *E. coli* are the laboratory-derived plasmids pBR322, which confers ampicillin and tetracycline resistance, and pACYC184, which confers chloramphenicol and tetracycline resistance. A variety of transposon derivatives of these plasmids has been constructed which can act as donors. The following experimental procedure can be adopted when the test plasmid is self-transmissible.

1. Transfer the test plasmid, by conjugation, into the same host as the transposon donor, preferably a *recA* strain.
2. Cross the constructed strain with a suitable recipient. The choice of mating procedure will depend on the particular test plasmid and the frequency of transposition. If in doubt, filter or plate matings are recommended (Chapter 3).
3. Spread the mating mixture on an agar selective for both the recipient host and the transposon. Colonies which appear comprise recipient cells carrying transposon derivatives of the test plasmid.

It is worth noting that the two plasmids mentioned require DNA polymerase I to replicate, unlike the bacterial chromosome and the majority of bacterial plasmids. As a consequence, these small plasmids cannot replicate in a *polA* mutant of *E. coli*. Hence, if the recipient is *polA*, the selection is much tighter. If, however, transposon donor markers other than that of the transposon are transferred, some form of cointegrate will almost certainly have been recovered. Unless these form all, or the majority of the transposition products, they can be ignored when using transposons for genetic analysis. With the commonly-used transposons, this is not likely to be a common occurrence.

5. Use of an incompatible plasmid

The phenomenon of plasmid incompatibility may also be exploited when selecting for transposon insertions (Foster *et al.*, 1975). Two incompatible plasmids, one being the test plasmid, are introduced into the same cell, preferably a Rec⁻ strain in order to prevent plasmid-plasmid recombination. Selection is imposed for the antibiotic resistance of the transposon on the first plasmid and for a resistance marker on the second (test) plasmid. Derivatives are then selected in which the two markers are stably inherited. This can be achieved by alternate cycles of growth on antibiotic-containing and drug-free media. Once the element has transposed to another replicon in the cell, the donor plasmid is likely to be lost because of incompatibility coupled with selection of the test plasmid. Among the survivors will be those carrying transposon insertion in the bacterial chromosome and those carrying insertions in the test plasmid. The plasmid derivatives can be isolated by transfer to another bacterial host. Cotransfer of the transposon and test plasmid markers is selected.

IV. Factors affecting the frequency of transposition

An important factor to be considered when planning plasmid mutagenesis or designing experiments to tag a cryptic plasmid with a transposon is the frequency at which the element transposes to the plasmid concerned. This frequency is usually expressed as the ratio of the number of plasmid::transposon recombinants to the total number of target plasmids. This value may range from almost 1 to less than 10^{-8}, depending on the components of the experimental system. The factors discussed below may influence the successful outcome of such experiments.

A. The host strain

The frequency of transposition may vary from one species to another and from strain to strain within the same species. For instance, transposon Tn*10* transposes from a vector into the chromosome of derivatives of *E. coli* C at a 10- to 100-fold higher frequency than into the chromosome of *E. coli* K12 (Beacham and Garrett, 1979).

B. Selection of antibiotic resistance

When transposing Tn*10* into the bacterial chromosome, it has been found that 12.5 μg ml^{-1} of Tc is the optimal concentration for selecting resistant colonies rather than the commonly-used concentrations of 20–25 μg ml^{-1} (Beacham and Garrett, 1979). We would assume that the lower concentration is also appropriate for selecting transpositions into a plasmid. Incorporation of sodium pyrophosphate (2.5 mM) into the Tc agar reduces the background growth of Tc^s cells and also prevents those cells, in which transposition events have occurred, being infected with phages on the surface of the agar (Kleckner *et al.*, 1978). If Tn*10* is inserted into a multicopy plasmid, it should be noted that the level of Tc^r expressed by the transposon in the multicopy state is reduced by a factor of 10–20 (Coleman and Foster, 1981). In this case, a concentration of tetracycline no greater than 5 μg ml^{-1} should be used. We are not aware of any reports of other transposon drug resistance markers behaving anomalously.

C. The method of detection

Transposition of Tn*5* occurs at a much higher frequency immediately after the transposon enters a new cell (Biek and Roth, 1980). However, if the cell already carries a copy of the transposon, the frequency is lower. This is interpreted to mean that Tn*5* transposition is repressed in cells in which the transposon is established, but can be derepressed by a form of zygotic induction. This is supported by the finding that transposition of Tn5 from the chromosome to a small multicopy plasmid occurred at a 10- to 100-fold lower frequency than transposition from a λ::Tn5 donor phage (T. J. Foster, unpublished data). It is possible that other transposons behave in a similar manner, but the point has not been tested rigorously.

In addition, the nature of the donor and recipient replicons may influence the transposition efficiency. Thus, it has been reported that the ampicillin resistance transposon Tn1 transposes from the bacterial chromosome to different plasmids at different frequencies under otherwise identical conditions (Bennett and Richmond, 1976). Furthermore, we have found it almost impossible to transpose Tn7 to small ColE1-like plasmids, while transposition to a variety of conjugative plasmids was readily demonstrated (P. M. Bennett, unpublished data). The basis of the distinction is unknown, but indicates that several transposons may have to be tried before the most appropriate is found. In this respect, Tn5 is reputed to insert into any replicon in a random manner.

D. Temperature

The incubation temperature of cultures may be important in determining the frequency of transposition. Thus, Tn3 transposes at a much higher frequency when the cells are incubated at 30°C than at 37°C, while at 42°C, transposition of Tn3 is completely inhibited. Transposition of Tn21, Tn501, Tn1721 and Tn2501 displays similar temperature dependence (A. K. Turner and J. Grinsted, unpublished). Interestingly, transposition of Tn1000 (γδ), which is related to Tn3, is not so sensitive (Tsai et al., 1987). The temperature characteristics of other transposons have not been reported. In practice, we overcome potential problems of this type by storing the strain in which transposition occurs on Dorset egg slopes at room temperature or 30°C for a few days, prior to recovery of the transposition products.

E. Specificity of insertion

One of the factors that might influence the choice of a transposon as a mutagen is its insertion site specificity. Transposon insertion may appear to be relatively nonspecific when the distribution of insertions in a large replicon is examined, but for a particular combination of transposon and plasmid, the distribution of insertions may not be random. Thus, it has been reported that insertion of a Tn1 analogue (Tn802) and Tn501 into the 54 kb plasmid pUB307 was not random but, rather, showed distinct regional preference. Some regions of the plasmid sustained frequent insertions while others had few or none at all (Grinsted et al., 1978). A detailed study involving fine-structure genetic and physical analysis of Tn*3* insertions in the small plasmid pTU4 showed that the transposon appeared to insert preferentially into AT-rich regions of the replicon (Tu and Cohen, 1980). Similar findings have been reported for other transposons (Miller *et al.*, 1980; Galas *et al.*, 1980). Insertions into a particular plasmid gene by a particular transposon may be rare due to this selectivity. Accordingly, many plasmid derivatives may have to be screened before the desired one is found. Alternatively, it may be necessary to try several different transposons before the desired recombinants are recovered. Unfortunately, there is no substitute for experimentation.

V. Analysis of plasmid structure and function

The analysis of transposon insertions is a powerful method of probing plasmid structure and function. A number of independent insertions in the plasmid are collected and scored for defects in known plasmid-encoded functions. These insertions are then mapped physically using restriction enzyme analysis (Chapter 6). This task is usually straightforward because the restriction maps of the test plasmid and the transposon are known. From the data, the positions of various functions can be placed on the plasmid map. Further, if a sufficient number of insertions in a particular gene are examined, the size of the gene may be determined (Bennett, 1985). Silent regions of the plasmid may be identified by mapping insertions which do not change any known plasmid-encoded property. Regions in which no insertions occur may be provisionally identified as those essential for plasmid maintenance (remembering the conditions given above in Section IV.E). These could represent genes for plasmid replication and partition functions, where insertions would prevent or jeopardize plasmid inheritance.

There are numerous examples of plasmid structure and function being elucidated by a combination of transposon and deletion analysis. Thus, Tn*3* has been used to analyse ColE1 (Dougan and Sherratt, 1977) and RSF1010 (Heffron *et al.*, 1975). Tn*7* was used in the structural analysis of RP4 (Barth and Grinter, 1977) and a Ti plasmid of *Agrobacterium tumifaciens* (van Vliet *et al.*, 1978). Similarly, Tn*5* was used to probe pKM101 (Langer *et al.*, 1981).

Transposon analysis can also help in studies of the genetic structure of a particular plasmid-encoded property (including cloned genes). A number of independent insertion mutations in the gene(s) of interest are isolated. Physical mapping of these insertions provides a minimum estimate of the gene coordinates. Examination of proteins expressed by these mutant plasmids in minicells (or a similar system) (see Chapter 10) will enable gene products to be identified. The insertion will disrupt the gene and so destroy the structure of the protein encoded by the gene. The insertion may also eliminate expression of distal genes in an operon (polar insertion). Thus, if a single protein band is missing from the profile of the mutant plasmid, it will be a good candidate for the gene product. The direction of transcription may be elucidated by examining different insertions in the same gene and comparing the sizes of the truncated protein fragments. The more distal the insertion in the gene, the larger the peptide produced. In studies of this type, it is wise to use only those insertions which have the same orientation. The peptides formed will usually be hybrid species, since the translation termination signal will be provided by the transposon. How soon such a sequence will be encountered will depend on the reading frame entering the transposon sequence. Similarly, an operon might be identified if a number of missing proteins in a set of mutants can be correlated with the physical map positions of the insertions. This type of logic was used to postulate an operon of three genes required for the synthesis of the K88 adhesion antigen of enteropathogenic E. coli (Kehoe et al., 1981). It should be appreciated that such studies are more easily performed on small multicopy plasmids than on large low copy ones because,

(a) the small plasmids are, in general, more easily segregated into minicells and,

(b) the protein profiles obtained are much less complex. Indeed, to analyse protein synthesis directed by large plasmids it may be more appropriate to use the maxicell technique (see Chapter 10).

Given that transposon insertions behave generally like single point insertions, they can be analysed by similar means, such as complementation analysis. Most of the problems that occur when attempting complementation tests with plasmid mutants involve the construction of heterozygous cells. Two mutant derivatives of the same plasmid will be incompatible (Chapter 2) and may not possess differential selective markers. This will prevent the construction of a stable heterozygous cell. In practice, the problem may be overcome by the following approaches.

(a) *Transient heterozygotes*. The construction of transient heterozygotes (i.e. cells carrying the two mutant plasmids unstably), followed by an assessment of the degree of complementation before segregation could occur, was used to analyse *tra* mutants of *F'lac* (Achtman et al., 1972).

(b) *Selection for the phenotype generated by complementation*. If the phenotype formed by complementation generates a selective marker, this can be used to select for the heterozygote. For example, selecton of mercuric-ion-resistant colonies allowed the detection of complementation between different Tn*1* insertion mutations in the *mer* region of otherwise isogenic R100*drd*1 plasmid derivatives (Foster et al., 1979).

(c) *Construction of stable heterozygous cells*. This may apply to the complementation analysis of functions which cannot be used as selective markers.

In these cases, it is better to attempt complementation analysis with two distinguishable compatible plasmids that can be used in strain construction. However, complementation tests between naturally-occurring plasmids in different incompatibility groups may be impracticable because the genes concerned may have diverged too much to allow complementation between mutants. Hence, the genes of interest may have to be transferred to another compatible plasmid by gene cloning techniques. Mutations can then be generated both in the parent plasmid and in the cloned genes, and complementation tests performed on these. Mutant alleles may also be switched from the parent plasmid into the cloned genes before attempting complementation. One practical point to remember when analysing the results is that a difference in the copy numbers of the two plasmids may affect the quantitative result obtained. This may be important, particularly when analysing control mutants generated by transposon insertion.

VI. Other uses of transposons

A. Tagging cryptic plasmids

It is not uncommon to isolate bacterial strains that carry one or more plasmids to which no phenotype can be assigned. The absence of an easily-selected marker makes plasmid recovery and, as a consequence, characterization difficult. One solution is to generate plasmid derivatives by transposon tagging. The drug resistance determinant of the transposon then provides a convenient selectable marker. In these experiments, there is no real alternative to transferring the transposon into the cell carrying the cryptic plasmid(s). This is most conveniently done on a conjugative plasmid that is *ts* for replication e.g. pMR5 (see above, Section III.B.3). The procedure is essentially as described above: pMR5 (or a suitable alternative) is transferred into the test strain; acquisition of the transposon is selected (Ap^r for Tn*1* on pMR5) at the permissive temperature (30°C for pMR5); single colonies are cultured at the permissive temperature in the presence of the selective agent; subcultures are then grown in the absence of the drug at the non-permissive temperature (42°C for pMR5); these cultures are then subcultured into fresh growth medium with the selective drug, incubated at the nonpermissive temperature for 18 h, and then appropriate dilutions are spread onto nutrient agar containing the selective drug; colonies are allowed to develop at the nonpermissive temperature. These colonies are expected to carry the transposon on one of the indigenous replicons of the test strain. Plasmid derivatives can be recovered by isolation and then transformation of a suitable laboratory strain.

B. Generation of deletions *in vitro*

A powerful method of plasmid analysis that complements transposon insertion analysis is deletion analysis. In this technique, portions of the test plasmid are removed by restriction enzyme digestion and then ligated. The derivative plasmids are recovered by transformation and analysed by restriction enzyme mapping (Chapter 6) to determine (or confirm) the extent of the deletion. Cells carrying the derivatives are then tested to determine which plasmid-encoded traits have been lost, so placing the genes for these characteristics within the deleted segment. The successful application of deletion analysis may be hampered by a lack of suitable restriction enzyme sites with which to generate the deletions. However, a suitable transposon may rectify the situation. For example, the kanamycin resistance transposon Tn5 has been used as just such a mobile set of restriction enzyme sites to generate deletions of the plasmid pKM101 (Langer *et al.*, 1981) and a set of overlapping deletions within the cloned R100*drd*1 *tet* genes have been used in fine structure genetic analysis (Coleman *et al.*, 1983). Again, the procedure is straightforward: a series of transposon insertions in different sites in the region of interest is obtained; a restriction enzyme that cleaves in the transposon at least once and also cleaves once in the adjacent plasmid sequence is chosen and plasmid DNA cleaved, ligated and used to transform a suitable laboratory strain. If the restriction regime has been chosen such that derivatives will not retain the transposon marker, they are easily identified. If, however, the transposon marker is retained, deletion derivatives can usually be identified by comparing the sizes of the plasmids in the transformants with the transposition recombinant; deletion derivatives will be smaller. Analysis of uncut plasmid minipreps usually suffices to identify promising candidates.

The power of this technique lies in the fact that the end points of the deletions can be precisely mapped by physical parameters, and they can be used in marker rescue recombination experiments with point mutations, so placing point mutations on the genetic map of the region.

C. Generation of non-polar derivatives of transposon insertion mutations *in vitro*

A transposon insertion mutation which exerts strong polarity on a distal gene may be converted to a nonpolar, single site mutation by restriction enzyme cleavage and ligation (Pannekoek *et al.*, 1980). The plasmid DNA is cleaved with an enzyme that cuts the transposon inverted repeats but not anywhere else in the plasmid. The cleaved DNA is ligated and used to transform a suitable laboratory strain. Derivatives which express distal gene function(s) are sought. These will be mostly nonpolar derivatives which retain a mutational lesion at the original transposition site. Some precise excision revertants may also be recovered, depending on the transposon used in the experiment.

One set of transposons that is ideally suited to this type of experiment was created by Ubben and Schmitt (1986) (see Section III.B.3). Each one of these transposons has *Eco*R1 sites which start just 13 bp from each end. Hence, restriction with *Eco*R1 and ligation removes all but 30 bp of each of the transposons. However, the remaining insert will be 35 bp since the transposition would have inserted the transposon sequence and also generated a 5 bp direct repeat at the point of insertion which would have remained.

D. Generation of gene fusions and provision of alternative promoters

Recently, Ubben and Schmitt (1987) reported the construction of two new sets of transposons based on Tn*1721*. The first set comprises transposons that have a promoterless β-galactosidase gene at one end, so that when these elements insert in the appropriate orientation within a gene, the promoter of that gene can be used to drive transcription of the β-galactosidase gene. Each element also carries a selectable drug resistance marker, so these elements are ideal tools for transcriptional analysis. Transposon insertions can be selected using the drug resistance marker and then analysed to determine the expression of β-galactosidase. This can be correlated with the orientation of insertion. These elements permit the determination of both the direction of transcription of a gene under investigation and a measure of its relative transcription activity. It should be noted that the techniques of gene fusion using a transposable element were originally developed using bacteriophage Mu and mini-Mus derived therefrom (see Chapter 8; also Castilho *et al.*, 1984).

The second set of elements reported by Ubben and Schmitt (1987), comprises transposons that have the *tac* promoter at one end, complete with *lac* repressor binding site. Each transposon is also provided with the *lacI* allele, *lacI^q*, which produces more repressor than the wild-type allele, so the promoter is normally strongly repressed. When active, the promoters direct transcription away from the transposon, so that if one inserts the gene of interest upstream and in the appropriate orientation, that gene comes under the control of the mobile *tac* promoter, and can be activated by the addition of isopropyl-β-D-thiogalactoside (IPTG) to the culture medium. These elements may have utility when analysing genes that are naturally expressed at low levels.

The transposons in both these sets also have the *Eco*R1 sites within their short inverted repeats and so the inserted sequences can be removed freely.

VII. Concluding remarks

This chapter has offered advice on how to detect transposable elements on plasmids and how to use transposons to analyse plasmids. It is not comprehensive, but rather pragmatic and introductory. The examples chosen have invariably been drug resistance determinants, since these are the elements we know. However, it should be remembered that drug resistance is not the only form of marker to be found on transposable elements. Indeed, the perceived importance of drug resistance transposons best reflects the human interest in this class of transposon and the ease with which they can be detected and studied, rather than their abundance. Much more investigation is needed before a decision on this point can be made. However, the point is not trivial, because transposons give to bacterial DNA and in particular, to bacterial plasmids an enormous genetic flexibility that is reflected in the adaptability displayed by bacteria to alterations in their environment. Hence, this area of research can be expected to continue productively for the foreseeable future, both in terms of the discovery of novel transposons and with respect to our understanding of their transposition mechanisms and how these contribute to bacterial evolution.

References

Achtman, M., Willetts, N. S. and Clark, A. J. (1972). *J. Bacteriol.* **10**, 831–842.
Arthur, A. and Sherratt, D. (1979). *Mol. Gen. Genet.* **175**, 267–274.
Barth, P. T. and Grinter, N. J. (1977). *J. Mol. Biol.* **113**, 455–474.
Barth, P. T., Grinter, N. J. and Bradley, D. E. (1978). *J. Bacteriol.* **133**, 43–52.
Beacham, I. R. and Garrett, S. (1979). *FEMS Microbiol. Lett.* **6**, 341–342.
Bennett, P. M. (1985). *In* "Genetics of Bacteria" (J. Scaife, D. Leach and A. Galizzi, Eds), pp. 97–115. Academic Press, New York.
Bennett, P. M. and Richmond, M. H. (1976). *J. Bacteriol.* **126**, 1–6.
Bennett, P. M., de la Cruz, F. and Grinsted, J. (1983). *Nature (London)* **305**, 743–744.
Bennett, P. M., Grinsted, J. and Richmond, M. H. (1977). *Mol. Gen. Genet.* **154**, 205–211.
Biek, D. and Roth, J. R. (1980). *Proc. Natl. Acad. Sci. U.S.A.* **77**, 6047–6051.
Brammar, W. J. (1982). *In* "Genetic Engineering 3" (R. Williamson, Ed.), pp. 53–81. Academic Press Limited.
Calos, M. P. and Miller, J. H. (1980). *Cell* **20**, 579–595.
Castilho, B. A., Olfson, P. and Casadaban, M. J. (1984). *J. Bacteriol.* **158**, 488–495.
Chang, A. C. Y. and Cohen, S. N. (1978). *J. Bacteriol.* **134**, 1141–1156.
Clewell, D. B., An, F. Y., White, B. A. and Gawron-Burke, C. (1985). *J. Bacteriol.* **162**, 1212–1220.
Cohen, S. N. (1976). *Nature (London)* **263**, 731–738.
Coleman, D. C. and Foster, T. J. (1981). *Mol. Gen. Genet.* **182**, 171–177.
Coleman, D. C., Chopra, I., Shales, S. W., Howe, T. G. B. and Foster, T. J. (1983). *J. Bacteriol.* **153**, 921–929.
Davis, R. W., Botstein, D. and Roth, J. R. (1980). "A Manual for Genetic Engineering. Advanced Bacterial Genetics". Cold Spring Harbor, New York.
de Bruijn, F. J. and Lupski, J. R. (1984). *Gene* **27**, 131–149.
Dodd, H. M. and Bennett, P. M. (1983). *Plasmid* **9**, 247–261.
Dougan, G. and Sherratt, D. (1977). *Mol. Gen. Genet.* **151**, 151–160.

Foster, T. J., Howe, T. G. B. and Richmond, K. M. V. (1975). *J. Bacteriol.* **124**, 1153–1158.
Foster, T. J., Nakahara, H., Weiss, A. A. and Silver, S. (1979). *J. Bacteriol.* **140**, 167–181.
Franke, A. E. and Clewell, D. B. (1981). *J. Bacteriol.* **145**, 494–502.
Galas, D. J., Calos, M. P. and Miller, J. H. (1980). *J. Mol. Biol.* **144**, 19–41.
Gottesman, M. M. and Rosner, J. L. (1975). *Proc. Natl. Acad. Sci. U.S.A.* **72**, 5041–5045.
Grindley, N. D. F. and Joyce, C. M. (1981). *Cold Spring Harbor Symp. Quant. Biol.* **50**, 125–133.
Grindley, N. D. F. and Reed, R. R. (1985). *Annu. Rev. Biochem.* **54**, 863–896.
Grinsted, J., Bennett, P. M., Higginson, S. and Richmond, M. H. (1978). *Mol. Gen. Genet.* **166**, 313–320.
Hayes, W. (1968). "The Genetics of Bacteria and their Viruses". Blackwell, Oxford.
Hedges, R. W. and Jacob, A. E. (1974). *Mol. Gen. Genet.* **132**, 31–40.
Heffron, F., Rubens, C. and Falkow, S. (1975). *Proc. Natl. Acad. Sci. U.S.A.* **72**, 3623–3627.
Herrmann, R., Neugebauer, K., Zentgraf, H. and Schaller, H. (1978). *Mol. Gen. Genet.* **159**, 171–178.
Kehoe, M. A., Sellwood, R., Shipley, P. and Dougan, G. (1981). *Nature (London)* **291**, 122–126.
Khan, S. A. and Novick, R. P. (1980). *Plasmid* **4**, 148–154.
Kleckner, N. (1977). *Cell* **11**, 11–23.
Kleckner, N. (1981). *Annu. Rev. Genet.* **15**, 341–404.
Kleckner, N., Roth, J. and Botstein, D. (1977). *J. Mol. Biol.* **116**, 125–159.
Kleckner, N., Barker, D. F., Ross, D. G. and Botstein, D. (1978). *Genetics* **90**, 427–461.
Kleckner, N., Roth, J. and Botstein, D. (1977). *J. Mol. Biol.* **116**, 125–159.
Kuner, J. M. and Kaiser, D. (1981). *Proc. Natl. Acad. Sci. U.S.A.* **78**, 425–429.
Langer, P. J., Shanabruch, W. G. and Walker, G. C. (1981). *J. Bacteriol.* **145**, 1310–1316.
Lederberg, E. M. (1987). *Gene* **51**, 115–118.
Lichtenstein, C. and Brenner, S. (1982). *Nature (London)* **297**, 601–603.
Miller, J. H., Calos, M. P., Galas, D., Hofer, M., Buchel, D. E. and Muller-Hill, B. (1980). *J. Mol. Biol.* **144**, 1–18.
Novick, R. P., Edelman I., Schwesinger, M. D., Gruss, A. D., Swanson, E. C. and Patee, P. A. (1979). *Proc. Natl. Acad. Sci. U.S.A.* **76**, 400–404.
Nevers, P. and Saedler, H. (1977). *Nature (London)* **268**, 109–114.
Pannekock, H., Hille, J. and Noordermeer, I. (1980). *Gene* **12**, 51–61.
Perkins, J. B. and Youngman, P. J. (1986). *Proc. Natl. Acad. Sci. U.S.A.* **83**, 140–144.
Robinson, M., Bennett, P. M., Falkow, S. and Dodd, H. M. (1980). *Plasmid* **3**, 343–347.
Rosner, J. L. and Gottesman, M. M. (1977). *In* "DNA Insertion Elements, Plasmids and Episomes" (A. I. Bukhari, J. A. Shapiro, and L. Adhya, Eds), pp. 213–218. Cold Spring Harbor, New York.
Shapiro, J. A. (1979). *Proc. Natl. Acad. Sci. U.S.A.* **75**, 1933–1937.
Shapiro, J. A. and Sporn, P. (1977). *J. Bacteriol.* **129**, 1632–1635.
Starlinger, P. (1980). *Plasmid* **3**, 241–259.
Toussaint, A. and Resibois, A. (1983). *In* "Mobile Genetic Elements" (J. A. Shapiro, Ed.), pp. 105–158. Academic Press Limited.

Tsia, M.-M., Wong, R. Y.-P., Hoang, A. T. and Deonier, R. C. (1987). *J. Bacteriol.*
169, 5556–5562.
Tu, C.-P. D. and Cohen, S. N. (1980). *Cell* **19**, 151–160.
Ubben, D. and Schmitt, R. (1986). *Gene* **41**, 145–152.
Ubben, D. and Schmitt, R. (1987). *Gene* **53**, 127–134.
van Vliet, F., Silva, B., van Montagu, M. and Schell, J. (1978). *Plasmid* **1**, 446–455.
Wallace, L. J., Ward, J. M. and Richmond, M. H. (1981). *Mol. Gen. Genet.* **184**, 87–
91.
Ward, J. M. and Grinsted, J. (1981). *Gene* **3**, 87–95.
Youngman, P. J., Perkins, J. B. and Losick, R. (1983). *Proc. Natl. Acad. Sci. U.S.A.*
80, 2305–2309.

10

Detection of Gene Products Expressed from Plasmids

G. DOUGAN AND N. F. FAIRWEATHER

Department of Molecular Biology, Wellcome Research Laboratories, Langley Court, Beckenham, Kent BR3 3BS, U.K.

I. Introduction

A major difficulty encountered when analysing plasmid or phage-encoded mRNA and polypeptides in whole (normal) bacterial cells is that the majority of these products are masked by those encoded by the host cell's chromosome. This difficulty often remains even when the genes of interest have been amplified by cloning into multicopy vectors. One approach to overcoming this problem is to use expression cloning vectors to increase the levels of gene expression using powerful bacterial promoters. This approach is used for expressing high levels of specific polypeptides which have to be purified for biological or biochemical analysis. The biotechnology industry has exploited high level expression systems to produce commercially important eukaryotic proteins such as insulin and interferons. However, in some research projects it is more convenient to take alternative approaches to detect plasmid-encoded polypeptides expressed at low levels. It is often desirable to see if a particular open-reading frame, inferred using DNA sequencing, is expressed under normal circumstances using a natural promoter. It may be convenient to identify the

METHODS IN MICROBIOLOGY
VOLUME 21   ISBN 0-12-521521-5

gene products of a multicistronic operon using a simple experiment. If a relatively large piece of DNA has been cloned, it is sometimes desirable to use a rapid screen to detect polypeptides or mRNA expressed from the fragment without DNA sequencing. In certain situations, immunological or enzymatic approaches can be useful aids to identifying proteins. A number of techniques have been developed in recent years which allow the detection of gene products expressed from cloned DNA. Some of the techniques to be described can also be used for the detection of chromosomally-encoded gene products. In this review, we shall cover techniques including the use of bacterial minicells, maxicells, cell free (*in vitro*) protein synthesis, and we shall also discuss RNA detection systems including hybridization and S1 nuclease analysis.

II. Methods for detecting polypeptides expressed from plasmids

A. The bacterial minicell system

1. Introduction

Minicells are small (1.0 μm), spherically shaped, anucleated bodies that are produced at the polar ends of certain mutant strains of rod shaped bacteria. Minicell-producing mutants have been isolated from *E. coli* (Adler *et al.*, 1967), *B. subtilis* (Van Alstyne and Simon, 1971; Reeve *et al.*, 1973), *Vibrio cholera* (Gardner, 1930), *Haemophilus influenzae* (Setlow *et al.*, 1973), *Erwina amylovora* (Gemski and Griffin, 1980) and a variety of *Salmonella* species (Epps and Idziak, 1970; Tankersley, 1970; Tankersley and Woodward, 1973). In addition to normal cell division, these mutants can undergo an aberrant cell division where a septum is produced close to either pole of the cell, resulting in the formation of a minicell. These minicells contain little or no chromosomal DNA and cannot divide. However, they do remain metabolically active for long periods after their formation (Black, 1976) and can transcribe and translate DNA.

Most minicell-producing mutants were isolated fortuitously by microscopic screening of mutagenized or irradiated cells. The original *E. coli* K12 minicell-producing mutant, P678-54, was isolated by Adler *et al.* (1967). A single colony isolate (named X925) of P678-54 was subsequently selected for its high minicell yield. These original mutants have a complex and incompletely defined genotype, possessing a number of mutations affecting galactose utilization and possibly a number of suppressor and hidden mutations (Frazer and Curtiss, 1975). Most *E. coli* K12 minicell-producing strains in use today are derivatives of these original Adler mutants. The minicell strain which we use, termed DS410, was isolated by selecting Gal$^+$ Sm^r recombinants after mating P678-54 with the *Hfr* donor CSH74 (Dougan and Sherratt, 1977a). Strain DS410 is nonsuppressing and gives a high-minicell yield. The genetic basis of minicell production is poorly understood. Mutations at both of two independent loci are probably required for minicell production in *E. coli* K12. These are *min*A, which maps between the *lac* and *pur*E at 9 to 12 min, and *min*B, which maps between *pyr*C and *trp* at 24 to 27 min on the map of Bachmann and Lowe (1980).

It was recognized early on that if plasmids were introduced into a minicell-producing strain, the plasmids could segregate into the minicells (Inselberg, 1970; Roozen *et al.*, 1970). Segregation into minicells is usually more efficient for small multicopy plasmids, such as ColE1, ColE2 and CloDF13 (Dougan and Sherratt, 1977a and b; Hallewell and Sherratt, 1976; Kool *et al.*, 1972) than for large plasmids. However, many low-copy number conjugative plasmids, including the F-factor, R1, R64 and ColV, segregate into minicells (Roozen *et al.*, 1971; Cohen *et al.*, 1971a; Levy and McMurray, 1974). Cosmids and other large recombinant plasmids readily segregate into minicells.

The nucleic acid of a wide range of double-stranded DNA phage can be introduced into minicells by directly infecting purified minicells with the phage (Reeve *et al.*, 1980). Attempts to infect purified minicells with the single-stranded filamentous phage M13 resulted in no detectable disassembly of the virion or conversion of the single-stranded DNA to the double-stranded replicative form (RF) (Smits *et al.*, 1978). However, when the minicell-producing strain is infected with M13, the RF form of M13 DNA can segregate into minicells in the same manner as a plasmid (Smits *et al.*, 1978; Reeve *et al.*, 1980). When minicells are infected with virulent phage, there is no burst of phage-directed lysis as there is in whole cells. Instead, phage-encoded polypeptides continue to be produced for several hours after infection (Reeve *et al.* 1980). This suggests that the normal regulation of phage expression may be disrupted in minicells and that caution should be exercised when attempting to draw conclusions on regulation based on minicell data alone. This observation also applies to the regulation of plasmid encoded products.

Although replication of plasmids and phage in minicells does not occur or is very inefficient, transcription and translation do occur with reasonable efficiency. However, a number of observations have suggested that the rate of gene expression in minicells is slower than in whole cells. Minicells actively synthesizing protein have been shown to accumulate guanosine tetraphosphate and guanosine pentaphosphate, suggesting that ribosome stalling may occur during protein synthesis (Nothling and Reeve, 1980); also the rate of removal of the signal sequence from a number of exported polypeptides is slower in minicells (Dougan *et al.*, 1979).

Since minicells contain little or no chromosomal DNA, but may transcribe and translate for long periods after their formation, they provide an excellent system in which plasmid or phage-encoded products can be labelled in the absence of significant background expression. The plasmid is first introduced into the minicell-producing strain by conjugation or transformation. The minicells harbouring the plasmid, which has segregated into them, are then separated (purified) from the whole bacterial cells (Fig. 1). If phage-encoded products are to be analysed, the minicells are purified first and then infected directly with the phage. The purified minicells, harbouring the plasmid or phage, are then incubated in a suitable medium and the plasmid or phage products are labelled with radioactive precursors, usually ^{35}S-methionine or ^{14}C-labelled amino acid hydrolysate for polypeptides and ^{3}H-uridine for mRNA. After cell lysis and electrophoresis to separate total minicell polypeptides or mRNA, the plasmid-encoded products are detected and identified by autoradiography or fluorography. In this system only the plasmid or phage-encoded products should be strongly labelled.

Introduction of plasmid DNA into minicell producing strain, purification of single colony
↓
Growth overnight for 12–15 h in rich medium (200 ml)
↓
Low speed differential centrifugation to remove large whole cells
↓
Sedimentation through two or three sucrose gradients to purify minicells
↓
Labelling of minicells with radioactive precursor
↓
Analysis by autoradiography of SDS-polyacrylamide gels of whole minicells
or fractionated extracts

Fig. 1. Experimental steps involved in minicell analysis of expression from plasmid DNA.

2. Methods

The detailed methodology for isolating minicells has been reviewed in detail elsewhere and therefore will not be repeated here. The procedure which is most frequently used is outlined in schematic form in Fig. 1. The minicells are separated from whole cells using differential centrifugation. The main points to remember are that minicells are viable and as such they must be separated under controlled conditions. In addition, the level of transcription and translation in minicells is very low therefore every effort should be made to reduce the level of contaminating whole bacteria to a minimum. It is recommended that the protocol discussed in detail in Volume 17 of this series is adhered to closely when the system is being set up for the first time, although a number of potentially time-saving variations can be introduced into the protocol when handling the system is routine. The main problems encountered using the minicell system include heavy background labelling from host gene products, low minicell yields and inefficient radioactive label incorporation. Potential solutions to these problems were discussed in Volume 17. The minicell system has been somewhat superseded in recent years by other powerful techniques which are less labour intensive. Nevertheless, minicells can offer some specific advantages over some of these other approaches. Once the system is working, clean and reproducible results can usually be obtained. The system is also useful for studying the processing of polypeptides and their transport to different compartments of the bacterial cell. Because metabolism in minicells is apparently "slower" than in whole cells, it is sometimes possible to identify a primary polypeptide product before post-translational modification occurs and therefore get a realistic assessment of the actual size of a cistron.

Almost all of the published studies of plasmid or phage-encoded products in minicells have involved the *E. coli* minicell system and, to a lesser extent, the *B. subtilis* minicell system. No attempt at a complete review of the literature will be made here, but selected examples, which should demonstrate the value of the minicell system, will be described.

3. Analysis of RNA in minicells

Early work on ^{3}H-uridine incorporation into purified minicells demonstrated that RNA can readily be synthesized in minicells harbouring plasmid or phage DNA (Frazer and Curtiss, 1975). Practically all of this RNA is encoded by the plasmid or phage. Kool *et al.* (1974) showed that less than 2% of the RNA labelled in minicells harbouring the plasmid CloDF13 could hybridize to *E. coli* chromosomal DNA and Smits *et al.* (1978) could detect no hybridization between chromosomal DNA and RNA synthesized in minicells carrying the RF form of the M13 phage. Thus, minicells provide an excellent system in which to study plasmid or phage-encoded mRNA molecules.

A good example of transcriptional analysis in minicells has been published by van den Elzen *et al.* (1980). These workers detected 25 different RNA species after gel electrophoresis of labelled RNA isolated from minicells harbouring the cloacinogenic plasmid CloDF13. Three of these RNAs, of 2400, 2200 and 100 bases in length were produced in large amounts. By analysing the RNA molecules produced by a number of transposon insertion mutants of CloDF13 and combining these data with information from hybridization and RNA polymerase binding experiments, a transcriptional map of CloDF13 was constructed. The 2400 and 2200 base RNA species were shown to be overlapping transcripts which originate from the same promoter. These were assigned to the region of CloDF13 encoding the cloacin and cloacin immunity determinants.

Very little recent work has described the use of minicells for analysing RNA expression.

4. Analysis of polypeptides in minicells

The minicell system has been used extensively to study polypeptides encoded by a variety of plasmids and phage. Large conjugative plasmids such as R100, R64 and R6-3drd12 (Levy, 1973; Levy and McMurray, 1974; Cohen et al., 1971b) and small nonconjugative plasmids like ColE1 and ColE2 (Dougan and Sherratt, 1977a; Hallewell and Sherratt, 1976) have been studied in minicells. Large conjugative plasmids are difficult to study because their segregation into minicells is inefficient and because they encode a large number of gene products. Nevertheless, a number of polypeptides encoded by large plasmids have been identified in minicells. For example, the 36 kilodalton (kD) molecular weight (MW) Tet protein was first identified by the fact that its expression could be induced in minicells harbouring tetracycline-resistance plasmids (Levy and McMurray, 1974).

Small plasmids are easier to study because of their efficient segregation into minicells and their genetic simplicity. The polypeptides encoded by a number of commonly-used vector plasmids have been very well characterized in minicells. These include ColE1 (Dougan and Sherratt, 1977a), pSC101 (Tait and Boyer, 1978), pACYC184 (Chang and Cohen, 1978), pBR322 and pBR325 (Covarrubias et al., 1981) and ColE1-Trp plasmids (Hallewell and Emtage, 1980). TEM β-lactamase encoded by pBR322 and related plasmids is expressed in minicells as three different related polypeptides of MW 30 kD, 28 kD and 26 kD. The 28 kD MW polypeptide is the active β-lactamase protein (Dougan and Sherratt, 1977a), while the 30 kD MW polypeptide is a precursor form carrying the signal sequence (Achtman et al., 1979), and the 27 kD MW polypeptide is a breakdown product which cross-reacts with anti-β-lactamase sera (Dougan et al., 1979). The precursor form of β-lactamase is unusually stable in minicells. This is also the case with the precursor forms of a number of other exported polypeptides. Thus, minicell studies may help to identify precursor polypeptides.

The products of bacterial genes encoded by large plasmids or by the chromosome can readily be studied in minicells if the genes of interest are first cloned into a small vector plasmid. By isolating specific mutants of the cloned determinants, the polypeptides encoded by individual cistrons can be identified in minicells, even in the absence of any other assay. Deletion mutants or transposon insertion mutants are particularly useful because they may result in truncated polypeptide fragments which can be detected in minicells. This often aids the determination of the transcriptional orientation of a particular gene. This approach has been successfully used to identify gene products which would otherwise be difficult or impossible to identify. The power of this approach should be apparent from the examples described below.

Polypeptides involved in the transposition of Tn3 have been identified by studying mutants of Tn3 in minicells (Fig. 2). Transposon Tn3 encodes a 19 kD MW polypeptide which represses transposition of Tn3 and is involved in resolving cointegrate intermediates during Tn3 transposition (Dougan et al., 1979; Gill et al., 1979; Arthur and Sherratt, 1979). In addition, a 100 kD MW Tn3 polypeptide, which is detected only in minicells harbouring repressor-defective mutants (tnpR), is required for transposition and is referred to as 'transposase' (Gill et al., 1979; Chou et al., 1979). Similar studies on mutants

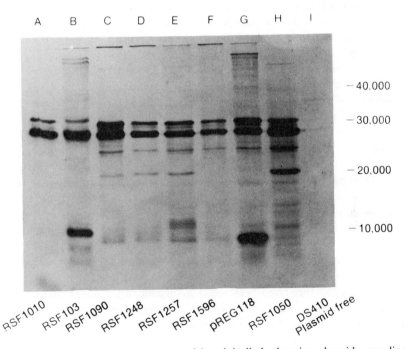

Fig. 2. Patterns of polypeptides expressed in minicells harbouring plasmids encoding wild type and mutant Tn3 elements. Track H shows expression from wild type Tn3 (three forms of β-lactamase and 19000 mol. wt repressor visible): tracks C, D and E show expression from Tn3 element defective in transposase (108000 mol. wt); tracks B and G show expression from Tn3 defective in repressor (transposase visible); track F shows expression from Tn3 defective in transposase and repressor. The Tn3 element of RSF103 contains a deletion in the β-lactamase gene and is carried on RSF1010. All other transposon mutants (tracks C to G) are derivatives of RSF1050 (pMB8::Tn3). The polypeptides of below 10000 mol. wt visible in tracks B and G are truncated forms of the repressor which normally autoregulates its own expression (from Dougan et al., 1979).

of Tn5 have identified four polypeptides, which are encoded by the inverted repeats (Rothstein *et al.*, 1980).

A number of virulence determinants of enterotoxigenic *E. coli* (ETEC) have been analysed by studying mutants of cloned determinants in minicells. The structure of the heat-labile (LT) toxin, which had proved very difficult to purify by conventional means, was elucidated with the aid of cloning and minicell analysis (Dallas and Falkow, 1979; Dallas *et al.*, 1979). The LT toxin could be precipitated from lysed minicells with anticholera-toxin sera, showing that the two toxins are structurally related. Subsequently, DNA sequencing studies confirmed the data obtained from earlier minicell studies (Spicer *et al.*, 1981). Another virulence factor of ETEC strains which has been studied with the aid of minicell analysis is the K88 adhesion antigen (Kehoe *et al.*, 1981). By studying both deletion and Tn5 insertion mutants of a cloned K88 determinant in minicells, five cistrons involved in the expression of K88 have been identified and mapped and the polypeptide products of four of these cistrons have been identified. Only one of these polypeptides, the 23 500 MW K88 antigen subunit, can be detected after staining SDS-polyacrylamide gels of total whole cell protein or of total minicell protein. The other three polypeptides can only be detected by labelling minicells (Shipley *et al.*, 1981; Kehoe *et al.*, 1981; Dougan *et al.*, 1983). The product of the fifth cistron whose expression is required for assembly of K88 fimbriae was not detected in minicells harbouring K88 recombinant plasmids. This cistron encodes a polypeptide whose sequence resembles that of pilin subunit and it may well be a minor component of the K88 fimbriae (Moori *et al.*, 1984). Quite clearly the protein is expressed at a level too low even to be detected in minicells.

E. coli minicells will apparently transcribe and translate most genes normally expressed by whole *E. coli* cells. Thus, expression of polypeptides from a wide range of bacteria including salmonellae (Koduri *et al.*, 1980), streptococci (Russell *et al.*, 1987), staphylococci (Coleman *et al.*, 1986), pseudomonads (Coleman *et al.*, 1983) and even some eukaryotic organisms (Gatenby, 1981) has been detected in minicells. Since some foreign DNA is not well transcribed or translated in *E. coli*, such DNA would be difficult to use in the minicell system without modifying expression, using expression vectors.

5. Fractionation of minicells

As mentioned above, one potential advantage of using the minicell system is in studies on secreted proteins. Signal sequences are probably cleaved more slowly in minicells from secreted proteins and precursors can often be readily visualized in gels especially if pulse-labelling is used. The stability of precursor proteins can also be readily increased by treating minicells with ethanol, which inhibits the signal sequence removal (Dougan *et al.*, 1983). Once it has been established that a particular protein is processed and is thus potentially a secreted protein, it is possible to localize the protein to a particular cellular compartment. Minicells can be lysed by sonication or by using a French press, but they are more resistant to both of these treatments than are whole cells (Frazer and Curtiss, 1975). Whitholt *et al.* (1976) described a lysis procedure designed for stationary phase whole cells of *E. coli*, involving the initial formation of spheroplasts and lysis of the spheroplasts by osmotic shock. We have found that minicells are efficiently lysed by this procedure. Thus, it is a relatively simple task to localize a polypeptide to either the cytoplasmic or envelope fraction of minicells (Kennedy *et al.*, 1977; Levy and McMurray, 1974).

A number of attempts have been made to prepare inner and outer membrane fractions of minicells (Goodall *et al.*, 1974; Levy, 1975). The best-characterized and most successful to date is the attempt to localize *tra* cistron products (Achtman *et al.*, 1979), involving an adaptation of the technique of Osborn *et al.* (1972). Achtman *et al.* (1979) separated different fractions of minicell envelopes on sucrose gradients and obtained five main peaks. The two heaviest peaks were enriched for outer membrane polypeptides, the central peak appeared to consist of a mixture of both inner and outer membranes (possibly membrane adhesion sites) and the two lighter peaks were enriched for inner membranes. Although the separation was not as efficient as can be obtained with whole cell envelopes, these workers were able to assign a number of the *tra* cistron products to either the inner or outer membrane fractions.

Selective solubilization of cell or minicell envelopes with Triton X-100 and Sarkosyl has often been used to obtain preliminary data on membrane localization of polypeptides, but this technique is not reliable when used alone. Although it is considered that 1% (w/v) Sarkosyl solubilizes only inner membrane proteins, it has been reported that proteins which have been shown to be located predominantly in the outer membrane by other methods were up to 70% soluble in Sarkosyl (Ferrazza and Levy, 1980).

Periplasmic polypeptides can readily be located in minicells using the cold shock technique developed by Hazelbauer and Harayama (1979). The labelled minicells are resuspended in 20% (w/v) sucrose and incubated on ice for 10 min before pelleting. The pellet is resuspended in cold distilled water and after a further 10 min incubation on ice, the minicells are pelleted. The periplasmic proteins which have been released into the supernatant are concentrated by precipitation with trichloroacetic acid and detected by SDS-polyacrylamide gel electrophoresis and autoradiography. We have used this procedure to locate the product of one of the K88 adhesion cistrons (the 17 000 MW *adh*C product) in the periplasmic fraction of minicells. In this case, β-lactamase, which is expressed by the vector part of the K88 hybrid plasmid, acts as a useful internal control (Dougan *et al.*, 1983).

B. The maxicell system

Sancar et al. (1979) described a simple system, commonly called the maxicell system, whereby plasmid products expressed in whole bacterial cells can be selectively labelled. The multicopy plasmid (or phage) is introduced into an E. coli strain defective in repairing DNA which has been damaged by ultra-violet light (U.V.). Commonly used strains include AD2480 (recA13, uvrA6, pro, thi) and CSR600 (recA1, uvrA6, phr-1) and SE5000. CNR600 offers the advantage of being defective in the photoreactivation repair system. The strain harbouring the plasmid is grown to late logarithmic phase and then exposed to a carefully controlled dose of U.V. irradiation. This treatment should effectively destroy the ability of the chromosomal DNA to direct protein synthesis, but, because the plasmid is smaller and present in multiple copies, should result in a percentage of plasmid genes escaping U.V. damage. To achieve the correct balance between chromosome damage and plasmid damage, the U.V. dose must be very carefully controlled. The irradiated culture is then incubated overnight at 37°C in the dark in the presence of cycloserine, which leads to lysis of cells which are still able to divide. Unlysed cells are recovered, labelled with radioactive precursors and the labelled products are analysed by SDS-polyacrylamide gel electrophoresis and autoradiography.

The level of background polypeptide synthesis can vary using this technique and it is critical to control carefully the U.V. dose if good expression from plasmid DNA is to be obtained. The dose of U.V. required for optimal results will obviously vary greatly for different E. coli strains. The maxicell system is now becoming increasingly popular for use in identifying plasmid encoded gene products. Jenkins and Nunn (1987) used the system to identify four polypeptides encoded by an E. coli determinant involved in short chain fatty acid metabolism. Brahansha and Greenberg (1987) used maxicells to identify a 59 kD polypeptide encoded by Spirochaeta aurantia DNA which complemented the E. coli trpE gene. Isberg et al. (1987) used maxicells together with an in vitro transcription/translation system to identify a 103 kD polypeptide encoded by Yersinia pseudotuberculosis which promoted the invasion of eukaryotic cells. These are only a few of the many examples of the use of maxicells now in the literature. Since maxicells are whole bacteria, they can readily be fractionated using conventional techniques for membrane analysis (Spooner et al., 1987).

C. Protein synthesis in vitro

The cell-free protein-synthesizing system first described by Zubay et al. (1970) involves the preparation of an E. coli cell extract and the addition of a complex mix of components which allows RNA and polypeptides to be synthesized in vitro using purified DNA as a template. The system has been extensively used for a number of years to study expression from both wild-type and hybrid plasmids (Kennedy *et al.*, 1977; Collins, 1979; Villafane *et al.*, 1987; Hallenback and Kaplan, 1987). In the past, the major disadvantage of the system was that it is expensive and difficult to set up initially in a new laboratory, often requiring advice from someone already familiar with the system.

However, this problem is not as serious as it once was, since a number of companies now sell *E. coli*-derived *in vitro* transcription/translation systems at a price that can be afforded by many research groups. The main advantage of the *in vitro* system is simplicity. Once the DNA template has been purified, a number of coupled transcription/translation reactions can be carried out simply by adding the DNA to the mix and incubating in the presence of label. The DNA template may be prepared from restriction-enzyme-generated DNA fragments excised from agarose gels, and this can aid in the localization of gene products to defined regions of DNA (Pratt *et al.*, 1981). The reaction also takes place in a solution which is free of whole bacteria; thus exogenous components can be readily added to reaction mixtures to study their effect on expression. This is well illustrated by the work of Glass *et al.* (1987) who used an *in vitro* transcription system to study a promoter under stringent control. Transcription and translation *in vitro* is normally accurate, using the expected start and stop signals used *in vivo*. However, there are a number of important differences from *in vivo* studies. The cell-free system is normally prepared from plasmid-free *E. coli*, and many of the regulatory proteins that might act on a piece of cloned DNA will be absent from the mixture. Thus, if a regulatory protein is required to induce expression of a cloned gene, the gene will not be expressed *in vitro* unless the regulator or its gene is added. This can, of course, be an advantage for studying gene expression control. Polypeptides which are normally processed after translation, such as many exported polypeptides, normally accumulate as precursors in the cell-free system. The signal sequences are not cleaved off because there are few membranes in the reaction mixtures. Again this can be useful for study processing. It is possible to induce cleavage of signal sequences from accumulated precursors by adding membrane vesicles from whole cells (Pratt *et al.*, 1986).

Compared to whole cells, the amount of gene product produced *in vitro* is usually very small. For example, quantities of polypeptide accumulated are usually too low to be detected in immunological assays such as immunoblotting. In addition, many polypeptides are unstable when expressed *in vitro* and often a significant level of protein degradation can be detected when the polypeptides are separated onto gels (Fig. 3). This can complicate attempts to identify plasmid-encoded polypeptides. Finally, *E. coli*-derived *in vitro* translation systems will only express genes which are able to be expressed by whole *E. coli* cells. Thus, in some cases, for expression of non-*E. coli* genes, cell-free systems have been prepared using the organisms from which the foreign DNA was derived (Chang and Kaplan, 1982).

Fig. 3. Autoradiograph showing *in vitro* transcription/translation (tracks 1–3) and minicells (tracks 4 and 5) of recombinant plasmids expressing the alpha toxin of *Clostridium perfringens* and vector controls. Track 1, no-DNA control; 2, pBR322; 3, pBR322-alpha toxin; 4, pACYC184-alpha toxin; 5, pBR322-alpha toxin. The arrows indicate the alpha toxin polypeptide. Note the precursor form is accumulated in the *in vitro* transcription/translation samples and the lower molecular weight processed form accumulates in the minicell samples.

D. Other methods for detecting specific polypeptides

The aim of a cloning experiment is often to move DNA encoding a particular protein of interest into a cloning vector in order to identify or characterize a gene product of interest. The protein often has specific properties which can be monitored by the investigator. For example, the protein may be an enzyme, or it may complement mutations which occur in *E. coli* strains, or an antibody may be available which recognizes the protein. Obviously, properties like these can be used to identify recombinant plasmids encoding the proteins and can be used to detect both chromosome and plasmid-encoded proteins. Many of the techniques involving SDS-polyacrylamide gel electrophoresis (SDS-PAGE), which can be used to identify polypeptides, are discussed below.

1. Assays for enzymatic activity in situ on SDS-polyacrylamide gels

Although SDS-PAGE normally partially denatures proteins, some separated polypeptides can retain their original biological or enzymatic activities, especially if SDS is removed from the gel on completion of electrophoresis. A number of ingenious techniques have been devised for detecting polypeptide-associated activities *in situ* in gels, and many examples have been reported in the literature. Two examples will be discussed here. *Streptococcus mutans* produces several high molecular weight glucosyltransferases, which are responsible for the polymerization of sucrose to high molecular weight polysaccharide. The glucosyltransferases are structurally related but each is responsible for synthesizing a particular type of polymer. The determinants for at least two of the glucosyltransferases have been cloned in plasmids in *E. coli* (Russell *et al.*, 1987). The polypeptides encoded by *S. mutans* glucosyltransferase genes were detected in *E. coli* cells harbouring recombinant plasmids following SDS-PAGE of whole recombinant *E. coli* cell lysates. Washed SDS-PAGE gels were incubated in sucrose solutions and an opaque deposit of freshly polymerized polysaccharide was visualized in the region of the gel corresponding to the position of the glucosyltransferase polypeptide (MW 170000 daltons) thus facilitating its identification.

A similar method was used to identify a polypeptide with phospholipase C activity cloned into *E. coli* from *Pseudomonas aeruginosa*. Filter paper was soaked in a phospholipase C sensitive chromogenic substrate and was laid on the gels to detect activity (Coleman *et al.*, 1983).

2. Immunoblotting and other immunological assays

Immunoblotting (Western blotting) is a very powerful technique to be used in conjunction with SDS-PAGE to detect polypeptides, provided suitable antisera are available. It is relatively easy to perform and small quantities of antibody can be used to detect extremely low levels of antigen. Separated polypeptides are transferred to a solid phase on which they are immobilized, usually nitrocellulose, by transverse electrophoresis (Towbin *et al.*, 1979; Burnette, 1981). Thus, the nitrocellulose has a copy of the polyacrylamide gel with the polypeptides attached by strong noncovalent interactions. Antibody can then be used to probe the filter for antigen using standard protocols. Initially, excess protein binding sites on the filter are blocked by washing in either excess proteins or detergent. The blocked filter is then treated with antibody probe which binds to the protein under investigation. After washing, a second anti-antibody is added to detect antibody bound to antigen on the filter. Alternatively, bound antibody can be detected using protein A. The second antibody or protein A is usually tagged with either ^{125}I, gold or with an enzyme such as peroxidase or alkaline phosphatase. The filter is then washed a second time and bound antibody is detected using either autoradiography or a chromogenic substrate, whichever is appropriate.

Like any immunological technique the sensitivity of immunoblotting is very dependent on the quality of antibody used. There is also a great variation in the quality of tagged second antibody offered by different companies and it is best to 'shop around' to find the most suitable reagent. It should be remembered that some proteins lose antigenicity after SDS-treatment and may not be detected using immunoblotting. One other precaution is that not all monoclonal antibodies recognize proteins after Western blotting. Nevertheless, the technique is very powerful and has been used successfully to detect many polypeptides.

Often, if a protein is detected using minicells, maxicells or cell-free systems, no antibody is available that will recognize the protein. A novel approach can be used to raise specific antibody if the DNA sequence of the relevant gene is known. Peptides can be synthesized which correspond to predicted amino acid sequences and such peptides can be used to vaccinate animals. In this way, antibodies can be raised which will recognize the native whole protein (Fujita et al., 1987).

An alternative immunological procedure commonly used to detect expressed proteins is immunoprecipitation. Proteins are normally precipitated as antibody complexes from lysed whole cells or cell supernatants. The antibody-antigen complexes are normally brought down using protein A bound either to an inert matrix or as part of whole *Staphylococcus aureus* cells. Immunoprecipitation has been used successfully in combination with minicell analysis (Dougan et al., 1979). It has the advantage over immunoblotting that native protein complexes can be identified rather than partially denatured polypeptides. A large number of other immunological techniques can be used to detect antigens, some of which have been reviewed elsewhere (Leinonen, 1985).

III. Detection of RNA in cells using cloned DNA fragments

In the previous sections, methods have been discussed where proteins expressed from cloned genes can be detected. As more reliable techniques become available, the study of the transcription from cloned genes is receiving more attention. In this section some of the information that may be obtained about the products of transcription from cloned genes will be described. The techniques will not be described in detail, but reference will be made to papers describing particular techniques and their application to the study of transcription.

A. Isolation and handling of bacterial RNA

Bacterial RNA is inherently more difficult to work with than DNA, primarily because of the extremely short half-lives of some mRNAs (Mangiarotti and Schlessinger, 1967). In addition, large amounts of RNase are synthesized by some bacterial species, and this must be inactivated quickly upon lysis of the cells to avoid rapid degradation of the RNA. RNase is a very stable enzyme (e.g. it is resistant to boiling) and therefore every precaution must be taken to avoid the accidental introduction of RNase from laboratory glassware and hands. The handling of RNA and the preparation of RNase-free solutions and glassware is outlined by Maniatis et al. (1982) and it is recommended that this be read before any RNA isolation is attempted.

Bacterial RNA is most often prepared by treating lysed cells with hot acid-phenol (von Gabain et al., 1983). Alternatively, the RNA may be purified from lysed cells by caesium chloride centrifugation. Since RNA has a greater buoyant density in caesium chloride than either DNA or protein, it forms a pellet on the bottom of the tube after centrifugation. The latter method has been used to purify RNA from Bordetella pertussis (Nicosia and Rappouli, 1987). Contaminating DNA may be removed from RNA preparations by commercially available RNase-free DNase.

B. Quantitation of mRNA using cloned DNA probes

The amount of mRNA transcribed from a cistron can be measured accurately by hybridization using a DNA probe which contains a part or all of the gene of interest. The amount of a particular mRNA species may be measured from cells grown under different conditions, or from a variety of strains differing in the amounts of gene product synthesized. Newbury et al. (1987) studied the levels of mRNA expressed from the histidine operon of S. typhimurium by growing cells in the presence of (^{3}H)-uridine to label total cellular RNA. Purified RNA was hybridized to a single-stranded DNA probe encoding part of the operon which was previously bound to a nitrocellulose filter. After washing the filters, the ^{3}H-RNA retained by the filter could be measured by scintillation counting. Instead of using radioactive RNA, total cellular RNA may be hybridized to a radiolabelled DNA or RNA probe. A popular method is to spot various concentrations of RNA on to a filter and to hybridize against the probe (dot-blotting).

Single-stranded RNA probes with high specific activities (greater than 10^9 cpm/μg) are easily synthesized using commercially-available plasmid vectors containing powerful promoters specific for particular RNA polymerases (see Chapter 8). Upon addition of the purified RNA polymerase and radiolabelled nucleotides, transcription of part or all of the cloned gene occurs. RNA molecules of different sizes may be obtained by cleaving the DNA with suitable restriction enzymes prior to the *in vitro* transcription. This method is similar to that for generating high specific activity single-stranded DNA probes (Burke, 1984) for use in S1 nuclease analysis (see below). Using RNA probes made in this way, Arico and Rappuoli (1987) measured the levels of mRNA expressed from the *Bordetella pertussis* toxin gene, and they were able to demonstrate the absence of toxin transcripts in strains of *B. parapertussis* and *B. bronchiseptica* which, although they have a toxin gene, do not produce toxin.

C. Sizing of RNA transcripts

The size of a particular mRNA species is most often assessed by 'Northern blotting'. This procedure, like 'Southern blotting' for DNA fragments (Southern, 1975, see Chapter 6), involves electrophoresis of RNA samples through agarose. This is followed by the transfer of the RNA to a nitrocellulose or nylon filter and hybridization of the immobilized RNA fragments to a radiolabelled DNA or RNA probe. The RNA must be fully denatured in order to obtain migration in linear proportion to the log of its molecular weight. This may be achieved by denaturing the RNA by treatment with glyoxal and dimethylsulphoxide followed by agarose gel electrophoresis. Alternatively the RNA may be electrophoresed through gels containing formaldehyde. Both these techniques are described in detail by Maniatis *et al.* (1982). RNA may also be electrophoresed through gels containing methyl mercuric hydroxide, although because of its extreme toxicity, this technique is not recommended.

Where a gene is part of a polycistonic operon, the size of the mRNA which hybridizes to the DNA probe indicates the length of the operon. The histidine transport operon contains four genes, *his*J, *his*Q, *his*M and *his*P, and Northern blot analysis has confirmed that the largest mRNA species is 3310 nucleotides (Newbury *et al.*, 1987). By using DNA probes originating from the 5' and 3' ends of the operon, these authors could locate the ends of the large mRNA species outside the *his*J and *his*P genes. When an operon gives rise to more than one transcript, these are each visualized on the blot. The relative intensities of these bands give an indication of the strengths of the promoters within the operon (see for example Newbury *et al.*, 1987).

D. Location of transcriptional start sites

The precise site of initiation of transcription may be located by S1 nuclease mapping. This method, first described by Berk and Sharp (1977), involves hybridization of RNA with a radiolabelled DNA probe and the subsequent digestion of the overhanging single-stranded DNA by S1 nuclease. (This nuclease is specific for single-stranded DNA.) The reaction products can be electrophoresed on a DNA sequencing gel alongside a DNA sequencing ladder. The size of the DNA protected from S1 is then determined from the known size of the sequencing reaction products and, as the 5' end of the DNA has not been degraded, the extent of hybridization to the mRNA is known. Similarly, the 3' end of the mRNA may be determined by using a DNA probe encoding this region of the gene. The S1 nuclease technique is described fully in Maniatis *et al.* (1982) and by Favalora *et al.* (1980). It has been used extensively in the analysis of mRNAs from eukaryotics and prokaryotes, and can be used, for example, in the construction of a transcriptional map of a plasmid (Brosius *et al.*, 1982).

An alternative method of mapping the 5' ends of mRNA molecules is primer extension. Here a synthetic oligonucleotide primer is hybridized to the RNA. This is then used together with reverse transcriptase to prime synthesis of a complementary DNA molecule which will extend until it reaches the end of the RNA template. It is often convenient to use the same oligonucleotide to prime a DNA sequencing reaction which runs alongside the primer-extended fragment on a denaturing acrylamide gel. Thus, the position of the base at the 5' end of the mRNA molecule can be read directly from the DNA sequencing reaction. In practice, it is advisable to use both S1 mapping and primer extension to map the 5' ends of transcripts (see for example Tunnacliffe *et al.*, 1986).

References

Achtman, M., Manning, P. A., Edelbluth, C. and Herrlich, P. (1979) *Proc. Natl. Acad. Sci. U.S.A.* **76**, 4837–4841.

Adler, H. I., Fisher, W. D., Cohen, A. and Hardigree, A. A. (1967). *Proc. Natl. Acad. Sci. U.S.A.* **57**, 321–326.

Arico, B. and Rappouli, R. (1987). *J. Bacteriol.* **169**, 2847–2853.

Arthur, A. and Sherratt, D. J. (1979). *Mol. Gen. Genet.* **175**, 267–274.

Bachmann, B. J. and Lowe, K. B. (1980). *Bacteriol. Rev.* **44**, 1–56.

Berk, A. J. and Sharp, P. A. (1977). *Cell* **12**, 721–732.

Black, J. W. (1976). Masters Thesis, University of Tennessee.

Brahanasha, B. and Greenberg, E. P. (1987). *J. Bacteriol.* **168**, 3764–3769.

Brosius, J., Cate, R. L. and Perlmutter, A. P. (1982). *J. Biol. Chem.* **257**, 9205–9210.

Burke, J. F. (1984). *Gene* **30**, 63–68.

Burnette, W. N. (1981). *Anal. Biochem.* **112**, 195–203.

Chang, A. C. Y. and Cohen, S. N. (1978). *J. Bacteriol.* **134**, 1141–1156.

Chang, J. and Kaplan, S. (1982). *J. Biol. Chem.* **257**, 15110–15121.
Chou, J., Lannaux, P. G., Casadaban, M. J. and Cohen, S. N. (1979). *Nature (London)* **282**, 801–806.
Cohen, S. N., Silver, R. P., McCoubrey, A. E. and Sharp, P. A. (1971a). *Nature (London) New Biol.* **231**, 249–252.
Cohen, S. N., Silver, R. P., McCoubrey, A. E. and Sharp, P. A. (1971b). *Ann. N.Y. Acad. Sci.* **182**, 172–187.
Coleman, K., Dougan, G. and Arbuthnott, J. P. (1983). *J. Bacteriol.* **153**, 909–915.
Coleman, D. C., Arbuthnott, J. P., Pomery, H. M. and Birkbeck, T. H. (1986). *Microbiol. Pathag.* **1**, 549–564.
Collins, J. (1979). *Gene* **6**, 29–42.
Covarrubias, L., Cervantes, L., Covarrubias, A., Soberon, X., Vichido, I., Blanco, A., Kupertztoch-Portnoy, Y. N. and Bolivar, F. (1981). *Gene* **13**, 25–35.
Dallas, W. S. and Falkow, S. (1979). *Nature (London)* **277**, 406–408.
Dallas, W. S., Gill, D. M. and Falkow, S. (1979), *J. Bacteriol.* **139**, 850–858.
Dougan, G. and Sherratt, D. J. (1977a). *Mol. Gen. Genet.* **151**, 151–160.
Dougan, G. and Sherratt, D. J. (1977b). *J. Bacteriol.* **130**, 846–851.
Dougan, G., Dowd, G. and Kehoe, M. (1983). *J. Bacteriol.* **153**, 364–370.
Dougan, G., Saul, M., Twigg, A., Gill, R. and Sherratt, D. J. (1979). *J. Bacteriol.* **138**, 48–54.
Epps, N. A. and Idziak, E. S. (1970). *Appl. Microbiol.* **19**, 338–344.
Favalora, J., Treisman, R. and Kamen, R. (1980). *Methods in Enzymology* (L. Grossman and K. Moldare, eds.) **65**, 718–749. Academic Press, London.
Ferrazza, D. and Levy, S. (1980). *J. Bacteriol.* **144**, 149–151.
Frazer, A. C. and Curtiss, R. (1975). *Curr. Top. Microbiol. Immunol.* **69**, 1–84.
Fujita, N., Ishikama, A., Nagasawa, Y. and Veda, S. (1987). *Mol. Gen. Genet.* **210**, 5–9.
Gardner, A. D. (1930). *In* "A System of Bacteriology in Relation to Medicine" (P. Fildes and J. C. G. Ledingham, Eds), Vol. 1. pp. 159–170. H.M.S.O., London.
Gatenby, A. A., Castleton, J. A. and Saul, M. W. (1981). *Nature (London)* **291**, 117–121.
Gemski, P. and Griffin, D. E. (1980). *Infect. Immun.* **30**, 297–302.
Gill, R., Heffron, F. and Falkow, S. (1979). *Nature (London)* **282**, 797–801.
Glass, R. E., Jones, S. T., Norman, T. and Ishihana, A. (1987). *Mol. Gen. Genet.* **210**, 1–4.
Goodall, E. W., Schwartz, V. and Teather, R. M. (1974). *Eur. J. Biochem.* **47**, 567–572.
Hallenback, P. L. and Kaplan, S. (1987). *J. Bacteriol.* **169**, 3669–3678.
Hallewell, R. A. and Emtage, S. (1980). *Gene* **9**, 27–46.
Hallewell, R. A. and Sherratt, D. J. (1976). *Mol. Gen. Genet.* **146**, 239–245.
Hazelbauer, G. L. and Harayama, S. (1979). *Cell* **16**, 617–625.
Inselberg, J. (1970). *J. Bacteriol.* **102**, 642–647.
Isberg, I. R. R., Voorhis, D. L. and Falkow, S. (1987). *Cell* **50**, 769–778.
Jenkins, I. S. and Nunn, W. D. (1987). *J. Bacteriol.* **169**, 42–52.
Kehoe, M., Sellwood, R., Shipley, P. and Dougan, G. (1981). *Nature (London)* **291**, 122–126.
Kennedy, N., Beutin, L., Achtman, M., Skurray, R., Rahmsdorf, V. and Herrlich, P. (1977). *Nature (London)* **270**, 580–585.
Koduri, R. K., Bedwell, D. M. and Brenchley, J. E. (1980). *Gene* **11**, 227–237.
Kool, A. J., Pranger, H. and Nijkamp, N. J. J. (1972). *Mol. Gen. Genet.* **115**, 314–323.

Kool, A. J., Van Zeben, M. S. and Nijkamp, J. J. (1974). *J. Bacteriol.* **118**, 213–224.
Leinonen, M. (1985). In "Enterobacterial surface antigens" (T. K. Korhonen, E. A. Dawes and P. H. Makala, Eds), Chapt. 14, pp. 179–206. Elsevier Press, Amsterdam.
Levy, S. B. (1973). *Bacteriol. Proc.* **62**, 134–140.
Levy, S. B. (1975). *Proc. Natl. Acad. Sci. U.S.A.* **72**, 2900–2904.
Levy, S. B. and McMurray, L. (1974). *Biochem. Biophys. Res. Commun.* **56**, 1060–1068.
Mangiarotti, D. and Schlessinger, D. (1967). *J. Mol. Biol.* **29**, 395–418.
Maniatis, T., Fritsch, E. F. and Sambrook, J. (1982). "Molecular Cloning. A Laboratory Manual." Cold Spring Harbor, New York.
Moori, F. R., van Buvien, M., Koopman, G., Roosendaal, B. and De Graaf, F. K. (1984). *J. Bacteriol.* **154**, 482–487.
Newbury, S. F., Smith, N. H., Robinson, E. C., Hiles, I. D. and Higgins, C. F. (1987). *Cell* **48**, 297–310.
Nicosia, A. and Rappouli, R. (1987). *J. Bacteriol.* **169**, 2843–2846.
Nothling, R. and Reeve, J. N. (1980). *J. Bacteriol.* **143**, 1060–1062.
Osborn, M. J., Gander, J. E., Parisi, E. and Carson, J. (1972). *J. Biol. Chem.* **247**, 3962–3972.
Pratt, J. M., Jackson, M. E. and Holland, I. B. (1986). *EMBO J.* **5**, 2399–2405.
Pratt, J. M., Bulnois, G. J., Darby, V., Orr, E., Wahle, E. and Holland, I. B. (1981). *Nucleic Acids Res.* **9**, 4459–4474.
Reeve, J. N., Lanka, E. and Schuster, H. (1980). *Mol. Gen. Genet.* **177**, 193–197.
Reeve, J. N., Mendelson, N. H., Coyne, S. I., Hallock, L. L. and Cole, R. M. (1973). *J. Bacteriol.* **114**, 860–866.
Roozen, K. J., Fenwick, R. G. and Curtiss, R. (1971). In "Informative Molecules in Biological Systems" (L. G. H. Ledoux, Ed.), pp. 249–264. North-Holland Publ., Amsterdam.
Roozen, K. J., Fenwick, R. G., Levy, S. B. and Curtiss, R. (1970). *Genetics* **64**, 554–556.
Rothstein, S. J., Jorgenson, R. A., Postler, K. and Reznikoff, W. S. (1980). *Cell* **19**, 795–805.
Russell, R. R. R., Gilmin, M. L., Mukasa, H. and Dougan, G. (1987). *J. Gen. Microbiol.* **133**, 935–944.
Sancar, A., Hack, A. M. and Rupp, W. D. (1979). *J. Bacteriol.* **137**, 692–693.
Setlow, J. K., Boling, M. E., Allison, D. P. and Beattie, K. L. (1973). *J. Bacteriol.* **115**, 153–161.
Shipley, P., Dougan, G. and Falkow, S. (1981). *J. Bacteriol.* **145**, 920–925.
Smits, M. A., Simons, G., Konings, R. N. Y. and Schoenmakers, J. G. G. (1978). *Biochem. Biophys. Acta* **521**, 27–44.
Southern, E. M. (1975). *J. Mol. Biol.* **89**, 503–517.
Spicer, E. K., Kavanaugh, W. M., Dallas, W. S., Falkow, S., Konigsberg, W. H. and Shafer, D. E. (1981). *Proc. Natl. Acad. Sci. U.S.A.* **78**, 50–54.
Spooner, R. A., Bagelasassan, M. and Franklin, F. L. D. (1987). *J. Bacteriol.* **169**, 3581–3586.
Tait, R. C. and Boyer, H. W. (1978). *Cell* **13**, 73–81.
Tankersley, W. G. (1970). Masters Thesis, University of Tennessee.
Tankersley, W. G. and Woodward, J. M. (1973). *Bacteriol. Proc.* 97.
Towbin, H., Staehelin, T. and Gordon, J. (1979). *Proc. Natl. Acad. Sci. U.S.A.* **76**, 4350–4354.
Tunnacliffe, A., Sims, J. E. and Rabbits, T. H. (1986). *EMBO J.* **1**, 1245–1252.

Van Alstyne, D. and Simon, M. I. (1971), *J. Bacteriol.* **108**, 1366–1379.
van den Elzen, P. J. M., Konings, B. N. H., Veltkamp, E. and Nijkamp, N. J. N. (1980). *J. Bacteriol.* **144**, 579–591.
Villafane, R., Bechhofer, D. H., Narayanan, L. S. and Dubnaw, D. (1987). *J. Bacteriol.* **169**, 4822–4829.
von Gabain, A., Belasco, J. G., Schottel, J. L., Chang, A. C. Y. and Cohen, S. N. (1983). *Proc. Natl. Acad. Sci. U.S.A.* **80**, 653–657.
Whitholt, B., Boekhout, M., Brock, M., Kingman, J., Van Heerikhuisen, H. and de Ley, L. (1976). *Anal. Biochem.* **4**, 160–170.
Zubay, G., Chambers, D. A. and Cheong, L. C. (1970). *In* "The Lactose Operon", pp. 375–391. Cold Spring Harbor, New York.

11

DNA Sequencing

NIGEL L. BROWN AND PETER A. LUND*

*Department of Biochemistry and Unit of Molecular Genetics, University of Bristol, Bristol
BS8 1TD, U.K.*

1. Introduction

Originally, DNA sequencing was a specialized technique. However now it is a standard and routine procedure in many laboratories; indeed, DNA sequencing may be the first analytical procedure applied to a newly isolated and not well understood genetic element. Since the first description of the rapid DNA

* Present address: Advanced Genetic Sciences, Oakland, CA 94608 CA.

METHODS IN MICROBIOLOGY
VOLUME 21 ISBN 0-12-521521-5

sequencing methods, which allowed several hundred base pairs of DNA to be sequenced (Sanger and Coulson, 1975; Maxam and Gilbert, 1977), the methods have developed so that it is now feasible to sequence the entire human genome.

This chapter is not intended to be a comprehensive review of methods and potential pitfalls, as new examples of both are continually being discovered. However, working methods are described, and the reader is encouraged to look to the original literature for other methods. Many changes in methodology have been made since the earlier version of this chapter (Brown, 1984) was written, and new methods will undoubtedly appear. The first apparatus for automated DNA sequencing is now coming into use, and will become much more reliable, cheaper and popular over the coming years.

Whatever method is used, it should be remembered that a DNA sequence is either correct or not, and it is nearly always easier to obtain the wrong DNA sequence than the correct one. A single mistake can completely alter the biological interpretation of the data.

A. Principle of rapid gel methods

All rapid DNA sequencing methods (and the analogous, but rarely used, rapid RNA sequencing methods) follow the same principle. A number (usually four) of nested series of radioactive single-stranded DNA fragments is generated, such that one end is common to all fragments, and the other end is generated in a base-specific manner (Fig. 1). The different base-specific series of fragments are analysed in parallel by slab-gel electrophoresis, under conditions in which the DNA is denatured and its migration determined primarily by the length of the oligonucleotide chain. The gel system is capable of resolving oligonucleotides differing in length by one nucleotide. The radioactive fragments are detected by autoradiography, and the relative order of these fragments on the autoradiogram gives the relative order of the specific bases from the common end, that is, the DNA sequence. The smallest fragments run farthest on the gel system, and it follows that, if the common end is a 5' end, the sequence will be read up the gel in the 5' to 3' direction, and *vice versa*. Depending on the exact conditions used, it is possible to read sequences up to about 500 nucleotides from the common end. Thus, the rapid accumulation of sequence data is possible.

B. Outline of the methods

1. Chemical method

Maxam and Gilbert (1977) devised the most widely used of the DNA sequencing methods. It is applicable to single as well as double-stranded DNA, relies on

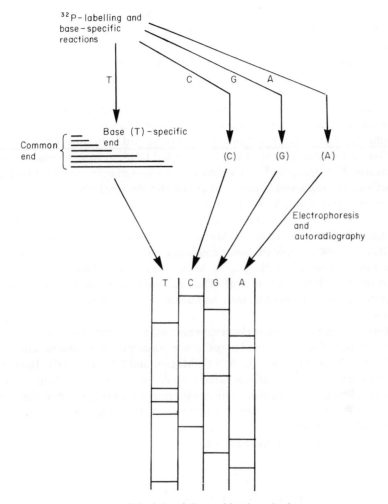

Fig. 1. Principle of the rapid gel methods.

chemical rather than enzymatic procedures in the base-specific reactions and they are well-annotated and carefully written protocol sheets. Although working procedures for this method are described in this chapter, time, space and the authors' experience preclude discussion of the method with the detail given by Maxam and Gilbert (1980).

In the chemical method, a single or double-stranded DNA fragment is radioactively end-labelled. There are two common methods for end-labelling DNA: using polynucleotide kinase and $[\gamma\text{-}^{32}P]ATP$ or $[\gamma\text{-}^{35}S]thioATP$ to label the 5' end; or, for double-stranded DNA only, using DNA polymerase and radiolabelled [α-^{32}P] or [α-^{35}S] deoxyribonucleoside triphosphate to label the 3' end. Other methods (e.g. labelling with terminal transferase) are available, but are rarely used. A double-stranded DNA fragment must then be manipulated to separate the labelled ends, either by strand separation or by secondary cleavage with a restriction enzyme and separation of the subfragments (Fig. 2). Whatever combination of labelling and separation methods is used, the result is a single DNA fragment labelled at one end and in one strand only. The labelled end constitutes the common end described in Section 1A.

The next step is to subject the end-labelled DNA to base-specific chemical cleavage. Typically, the DNA is divided into four aliquots and four chemical reactions are used which allow differentiation between the four bases. The reaction may be specific for more than one base, and involves three steps: modification of the base, removal of the modified base from the sugar-phosphate backbone and scission of the backbone at that site (Fig. 2). The first step must be base-specific and occurs under conditions which allow modification of only one base per strand in the region to be sequenced (which may be shorter than the complete DNA strand). The remaining steps must be quantitative, otherwise DNA fragments with anomalous electrophoretic mobilities may result.

Typically, the four sets of chemical reactions are specific for C, C+T, G and A+G, respectively. A number of other base-specific reactions are available (e.g. T, A > G, G > A, A > C; Maxam and Gilbert, 1980). In current protocols, the base-removal and strand-scission reactions are performed simultaneously by using piperidine. This causes removal of the modified base, then catalyses β-elimination of phosphates from the sugar. Piperidine is volatile and can easily be removed to leave a salt-free product which can be analysed on thin acrylamide gels (Section IV.A).

2. M13-cloning/chain-termination method

The chain-termination method (Sanger et al., 1977) was originally applied to the sequence analysis of single-stranded bacteriophage DNA (Sanger et al., 1978), and arose from the original 'plus and minus' method (Sanger and Coulson 1975). The method depends on the in vitro copying of a single-stranded DNA template by primed synthesis using DNA polymerase I (Fig. 3). Single-stranded template DNA can be obtained from almost any DNA by cloning double-stranded fragments of DNA into the replicative form (RF) DNA of bacteriophage M13. A series of bacteriophage M13 derivatives have been constructed specifically for this purpose (Table I; Fig. 4). Other single-stranded vectors are also available and are becoming more widely used, as are modi-

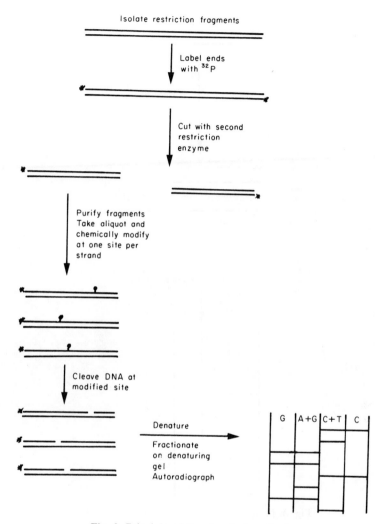

Fig. 2. Principle of the chemical method.

fications of the chain-termination method that can be used with double-stranded DNA (Section III.C.3).

Bacteriophage M13 (Ray, 1977) is a filamentous single-stranded DNA phage which infects only male (F⁺) *Escherichia coli* in a chronic non-lytic mode. There is no strict geometric constraint on the size of single-stranded DNA that can be packaged in the virion. The replicative form DNA is present in infected cells in high-copy number and can easily be isolated as a covalently-closed,

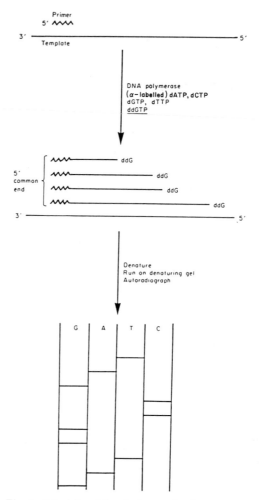

Fig. 3. Principle of the chain-termination method.

double-stranded DNA (RFI DNA). Although infected cells are viable, they grow more slowly than uninfected cells and continually extrude bacteriophage; M13 thus forms plaques of retarded growth when plated on a lawn of cells by the overlay method.

The bacteriophage M13mp series of vectors carry, in a non-essential region of the bacteriophage M13 genome, part of the *E. coli lac* operon: this is the regulatory region and the coding sequence for the first 145 amino acids of the β-galactosidase (the α-peptide). *In vitro* mutagenesis has been used to introduce

TABLE I
Bacteriophage M13mp vectors

Bacteriophage	Cloning sites	Insert fragments[a]	Reference
M13mp2	*Eco*RI	*Eco*RI	Gronenborn and Messing (1978)
M13mp7	*Eco*RI	*Eco*RI	Messing *et al.*
	*Pst*I	*Pst*I	(1981)
	*Bam*HI	*Bam*HI, *Bcl*I, *Bgl*II, *Sau*3AI, *Xho*II	
	*Sal*GI[b]	*Sal*GI, *Xho*I	
	*Acc*I[b]	*Acy*I, *Aha*II, *Asu*II, *Cla*I, *Hpa*II, *Mae*II, *Nar*I, *Sci*NI, *Taq*I	
	*Hinc*II[b]	any blunt-ended fragment (*Sma*I-cut M13mp8 is better)	
M13mp8[c]	All sites in M13mp7 plus		Messing and
M13mp9[c]			Vieira (1982)
	*Sma*I[b]	Any blunt-ended fragment	
	*Xma*I[b]	*Xma*I	
	*Hin*dIII	*Hin*dIII	
M13mp10[c]	All sites in M13mp8 and M13mp9 plus		Norrander *et al.*
M13mp11[c]			(1983)
	*Sac*I	*Sac*I,	
	*Xba*I	*Avr*II, *Nhe*I, *Spe*I, *Xba*I,	
M13mp18[c]	All sites in M13mp10 and M13mp11 plus		Yanisch-Perron
M13mp19[c]			*et al.* (1985)
	*Kpn*I/		
	*Asp*718I[d]	*Kpn*I/*Asp*718I[d]	
	*Sph*I	*Nla*III, *Nsp*(7524)I, *Sph*I	

[a]Enzymes recognizing related sequences (degenerate sites) of which only some give the desired termini are not included in the table (e.g. not all *Acc*I fragments can be cloned in the *Acc*I site).
[b]The sequence 5'-GTCGAC-3' is recognized by *Sal*GI, *Acc*I and *Hinc*II. The sequence 5'-CCCGGG-3' is recognized by *Sma*I and *Xma*I.
[c]M13mp8/M13mp9, M13mp10/M13mp11 and M13mp18/M13mp19 have inserts in opposite orientations.
[d]The site 5'-GGTACC-3' is cleaved by *Kpn*I to give a 3' tetranucleotide extension, and by *Asp*718I to give a 5' tetranucleotide extension; *Asp*718I is the better enzyme in our hands.

M13mp2
```
 1  2  3  4  5  6  7  8
 T  M  I  T  N  S  L  A
ATGACCATGATTACGAATTCACTGGCC
          EcoRI
```

M13mp7
```
 1  2  3  4  5  1' 2' 3' 4' 5' 6' 7' 8' 9'10'11'12'13'14' 6  7  8
 T  M  I  T  N  S  P  D  P  S  T  C  R  S  T  D  P  G  N  S  L  A
ATGACCATGATTACGAATTCCCCGGATCCGTCGACCTGCAGGTCGACGGATCCGGGGAATTCACTGGCC
          EcoRI      BamHI  AccI  PstI  AccI  BamHI    EcoRI
```

M13mp8
```
 1  2  3  4  5  6  1' 2' 3' 4' 5' 6' 7' 8' 9'10'11' 7  8
 T  M  I  T  N  S  R  G  S  V  D  L  Q  P  S  L  A  L  A
ATGACCATGATTACGAATTCCCGGGGATCCGTCGACCTGCAGCCAAGCTTGGCACTGGCC
          EcoRI SmaI BamHI AccI  PstI   HindIII
```

M13mp9
```
 1  2  3  4  1' 2' 3' 4' 5' 6' 7' 8' 9'10'11' 5  6  7  8
 T  M  I  T  P  S  L  A  A  G  R  R  I  P  G  N  S  L  A
ATGACCATGATTACGCCAAGCTTGGCTGCAGGTCGACGGATCCCCGGGAATTCACTGGCC
          HindIII  PstI   AccI BamHI SmaI EcoRI
```

M13mp18
```
 1  2  3  4  5  6  1' 2' 3' 4' 5' 6' 7' 8' 9'10'11'12'13'14'15'16'17'18' 7  8
 T  M  I  T  N  S  S  S  V  P  G  D  P  L  E  S  T  C  R  H  A  S  L  A  L  A
ATGACCATGATTACGAATTCGAGCTCGGTACCCGGGGATCCTCTAGAGTCGACCTGCAGGCATGCAAGCTTGGCACTGGCC
          EcoRI  SacI  KpnI     BamHI  XbaI  AccI  PstI  SphI HindIII
                           SmaI
```

M13mp19
```
 1  2  3  4  1' 2' 3' 4' 5' 6' 7' 8' 9'10'11'12'13'14'15'16'17'18' 5  6  7  8
 T  M  I  T  P  S  L  H  A  C  R  S  T  L  E  D  P  R  V  P  S  S  N  S  L  A
ATGACCATGATTACGCCAAGCTTGCATGCCTGCAGGTCGACTCTAGAGGATCCCCGGGTACCGAGCTCGAATTCACTGGCC
          HindIII SphI  PstI   AccI  XbaI  BamHI    KpnI  SacI  EcoRI
                                                SmaI
```

Fig. 4. DNA sequences at the cloning sites in some of the M13mp vector series (Messing et al., 1981; Messing and Vieira, 1982; Yanisch-Perron et al., 1985). The N-terminal sequence of the β-galactosidase α-peptide is shown in each case using the single-letter amino acid code, the additional amino acids due to the polylinker being numbered 1' etc. Note that the universal primer primes from right to left in this orientation.

unique restriction sites in the α-peptide coding region (Table I; Fig. 4; Gronenborn and Messing, 1978; Messing et al., 1981; Messing and Vieira, 1982; Norrander et al., 1983; Yanisch-Perron et al., 1985). All these M13 derivatives synthesize the α-peptide under the control of the lac regulatory region. In a suitable host (e.g. E. coli K12 JM101; Table II), which produces the corresponding C-terminal fragment of the lac β-galactosidase (the ω-peptide), α-complementation occurs to give an active β-galactosidase protein. The β-galactosidase activity can be detected in vivo by using the chromogenic substrate 5-bromo-4-chloro-3-indolyl-β-D-galactopyranoside. This colourless substrate, also known as X-gal, is hydrolysed to give a dark blue product. Figure 5 illustrates this process.

If DNA fragments are cloned into the unique restriction sites in the α-peptide coding sequence of M13mp derivatives, and the resulting recombinant molecules are used to transfect the indicator host, then the recombinant bacteriophage will form plaques on a lawn of indicator host. However, the infected cells will not contain active β-galactosidase due to the disruption of the α-peptide coding sequence, and plaques will be colourless on the chromogenic indicator, X-gal.

TABLE II
Some *E. coli* host strains suitable for bacteriophage M13mp vectors

Strain	Genotype	Reference and notes
71–18	K-12 *supE*44, *thi*, Δ(*lac*, *proAB*) (F′ *proAB*, *lacI*q, ZΔM15)	Messing *et al.* (1977)
JM101	K-12 *thi*, *supE*44, Δ(*lac-proAB*) (F′ *proAB*, *lacI*q, ZΔM15, *traD*36)	Messing (1979) Transfer-deficient F factor
JM107	K-12 *endA*1, *gyrA*96, *thi*, *supE*44, *hsdR*17, *relA*1, λ⁻, Δ(*lac-proAB*) (F′ *proAB*, *lacI*q, ZΔM15, *traD*36)	Yanisch-Perron *et al.* (1985) R. *Eco*K⁻
JM109	*recA*1 version of JM107	Yanisch-Perron *et al.* (1985)
TG1	*hsdR*⁻ version of JM101	Gibson (1984)
TG2	Tc^R, *recA*⁻ version of TG1	(T. Gibson, unpublished)

It is a simple procedure to grow 1 ml cultures of infected cells from each colourless plaque, removing the cells by centrifugation and precipitating the virions with polyethylene glycol. Sufficient single-stranded DNA for about ten sequence determinations can be made by phenol extraction of the bacteriophage pellet. In this way, many M13 recombinants can be processed at one time.

The M13-cloning procedure is most commonly used to produce random subclones of the DNA to be sequenced, such that the sequence is obtained in a random fashion and is gradually compiled. All the subclones produced have an identical (*lacZ* gene) sequence to the 3′ side of the inserted DNA. The 'universal primer' can be annealed to this common flanking sequence for primed-synthesis copying of the insert DNA in the chain-termination reactions. This primer is normally a synthetic oligonucleotide. The annealed primer-template mixture is divided into four aliquots, one for each specific base. These are incubated with the large fragment of proteolysed *E. coli* DNA polymerase

E. coli KI2 TGI (Δ(lac, pro), F'lacI^q ZΔMI5 pro)

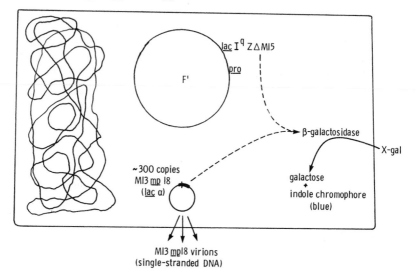

MI3 mpI8 virions
(single-stranded DNA)

Fig. 5. Diagrammatic representation of the M13mp selection system. The *lac* operon DNA in M13mp18 is shown as a thick black line; cloning a DNA fragment at the multiple cloning site, indicated by the vertical bar, prevents the production of β-galactosidase, but does not prevent virion production.

I (Klenow polymerase*), (α-labelled)dATP, dCTP, dTTP, dGTP and a chain-terminating compound specific for one base.

The DNA is internally labelled during primed synthesis; [α-^{35}S]dATP is normally used, but one of the other triphosphates can be used with minor variations in the subsequent protocol (Section III.B.2(b)). The specific chain-terminating compound used is usually the appropriate 2′,3′-dideoxyribo-nucleoside 5′-triphosphate. In the G-specific reaction, for example, a ratio of 2′,3′-dideoxyguanosine 5′-triphosphate (ddGTP) to 2′-deoxyguanosine 5′-triphosphate (dGTP) would be chosen such that a C residue in the template strand would direct the incorporation of a 2′-deoxyguanosine residue about 99% of the time and elongation of the radiolabelled DNA strand would

---

* 'Klenow polymerase' is the larger fragment of two peptide fragments produced by limited protease treatment and chromatography of the *E. coli* DNA polymerase I. This treatment removes the 5′ exonuclease activity normally present in DNA polymerase I, which would otherwise remove nucleotides from the 5′ end of the DNA primer fragment. If, as is usually the case, the DNA primer fragment is not removed by restriction endonuclease cleavage after primed synthesis (or if a short primer is used and removal attempted), then the prerequisite condition that all the DNA fragments have a common end as well as a base-specific end (Section I.B) will not be met if a 5′ exonuclease activity is present. A modified gene encoding only the large fragment of DNA polymerase I has been constructed and used to produce 'cloned Klenow polymerase'; this provides a better product more cheaply.

continue; the remaining 1% of the time a 2′,3′-dideoxyguanosine residue would be incorporated*. As the dideoxyguanosine residue has no 3′ hydroxyl group, it cannot participate further in phosphodiester bond formation, and chain elongation stops. The incorporated dideoxynucleoside residue is only very slowly removed by the 3′ exonuclease activity of DNA polymerase I (Atkinson *et al.*, 1969). The net result is a nested series of DNA fragments with identical 5′ ends, and variable base-specific 3′ ends.

3. Gel electrophoresis and autoradiography

Many of the problems preventing the rapid accumulation of good sequence data are encountered in obtaining well-resolved specific band patterns in auto-radiograms. In addition, the radioactive DNA products of the chemical method are chemically identical with those of the chain-termination method, except that in the former, they have an additional 3′-phosphate group. Therefore, most of the considerations of electrophoresis and autoradiography are common to both methods. With both methods, the DNA samples are denatured by heating in formamide before loading them onto the gel.

The gel system normally used is an acrylamide slab-gel with 1 in 20 cross-linking, in 7 M urea and 90 mM Tris-borate, pH 8.3, 2.5 mM EDTA (TBE) buffer (Peacock and Dingman, 1967). The total acrylamide concentration can be varied according to the experiment, from 20% for sequences close to the common end, to 5% for sequences much further from the common end. Urea is a denaturing agent. However it is a poor one for nucleic acids because 7 M urea reduces the melting temperature of DNA by only about 20°C. The high ionic strength of the buffer causes a high current to be drawn, and the gels run at a temperature of 60–70°C, thus minimizing the formation of intrastrand secondary structure in the DNA fragments. The simple practice of running the electrophoresis system for 30–60 min to heat up the gel before loading the sample, circumvents a number of artefacts that may cause problems in inter-preting the gel pattern.

The apparatus used for gel electrophoresis should give direct contact between gel and buffer, allowing high-voltage gradients to be used. Models similar to that described by Studier (1973) are commercially available. (Further comments on gel apparatus are made in Section IV.A.) Thin (0.35 mm) acrylamide gels (Sanger and Coulson, 1978) are normally run with the chain-termination procedure, and they can also be used with those reactions in the chemical method which give a salt-free product. Thin gels have the advantage that the bands on the autoradiogram are very sharp as a simple geometric consequence of reducing the distance between the film and the 'back' face of the gel. Thin

---

* The chemical ratio dGTP : ddGTP will be lower than 99 : 1, as dideoxynucleotides are poorer substrates than deoxynucleotides.

gels must be cast between plates of high quality float glass, otherwise variations in gel thickness cause abnormal mobilities of DNA fragments and cause localized variations in heating. 'Wedge gels' and 'buffer gradient gels' have been introduced to maximize the amount of sequence that can be read from a single gel. The use of ^{35}S-labelled nucleotides gives much sharper bands, which can be resolved further up the gel than with ^{32}P, but requires that the gels be dried down.

II. Procedures for the chemical method

Although the main procedures are outlined here, the reader is referred to the excellent treatise by Maxam and Gilbert (1980).

A. Preparation of DNA

Chapters 5 and 6 describe methods of preparing plasmid DNA and digesting it with restriction endonucleases. The preparation of DNA fragments for sequence analysis by the chemical method can be done by the methods described there. For many DNA sequencing projects, it is preferable to isolate large unlabelled restriction fragments in bulk in an initial step. This has the advantages that unnecessary DNA sequences (e.g. from a vector replicon) can be discarded before radioactive labelling, resolution of labelled DNA fragments in the area of interest is easier, experiments can be more readily designed to give the optimum distribution of restriction sites and problem areas can be more easily defined.

Some steps in the chemical method are inhibited or otherwise made less efficient by common contaminants of DNA. For example, the kinase reaction will also label oligodeoxyribonucleotides (fragments of chromosomal DNA) which are not efficiently removed by CsCl isopycnic centrifugation; some restriction endonucleases are stored in phosphate buffer, which will inhibit the phosphatase reaction; linear acrylamide will readily elute from low percentage acrylamide gels during extraction of DNA fragments and may inhibit DNA polymerase end-labelling. Care must be taken to remove such contaminants during the bulk preparations of DNA fragments.

Subsequent digestion of large DNA fragments with other restriction endonucleases can initially be done with small amounts of DNA to determine the best enzymes to use to get a good distribution of fragments. Enzymes with small (tetranucleotide) recognition sites giving DNA fragments with 5′ terminal extensions (for example HinfI, HpaII, Sau3AI, TaqI, and others) are preferable, due to the number of fragments produced and the greater efficiency and ease of labelling such fragments by 'end-repair'.

1. Bulk preparation of DNA fragments

Mix in a 1.5 ml centrifuge tube:
200 μg plasmid DNA.
30 μl 10 × restriction enzyme (RE) buffer (see supplier's recommendation).
Mix and add 40 units restriction endonuclease.
Mix and incubate for 10 h at recommended temperature.
(This provides a safety margin of two-fold overdigestion, which is sufficient for all except a few unstable enzymes.)
Add 200 μl phenol, vortex and centrifuge for 30 s.
Remove and discard lower phase.
Add 30 μl 3 M sodium acetate, vortex and add 750 μl ethanol.
Chill for 5–10 min in a dry ice-isopropanol bath.
Centrifuge for 15 min at 10 000 g and 4°C.
Remove supernatant and wash pellet by adding 95% ethanol, centrifuging for 1 min and discarding supernatant.
Dry under vacuum.
Resuspend DNA in 80 μl H$_2$O, 20 μl Native loading buffer (Section IV.A). Load into slot (2 cm × 0.15 cm) of 40 cm long slab-gel (5% acrylamide; 30:1 cross-linking), and run in TBE buffer at 300–400 volts for 16 h.
Exact running conditions may be varied, but lower voltages prevent heating and aid resolution.

To detect the unlabelled DNA fragments, remove one gel plate and either:
(i) Stain the gel in ethidium bromide (0.3 μg ml^{-1} in water for 5–10 min, and examine under U.V. light (300 nm or 340 nm). The ethidium can be removed later by phenol extraction, or,
(ii) Put the gel on transparent film (e.g. Saran Wrap) and place this on a TLC plate impregnated with fluorescent indicator. The DNA will absorb short-wave (260 nm) U.V. light and appear as dark bands against a fluorescent background. The short-wave U.V. light may cause some nicking of the DNA.

Each DNA fragment of interest should be excised in the minimum volume of gel and put in a tightly-stoppered tube (e.g. Sarstedt 55.516, 13 mm × 100 mm). Crush the gel slice to a lumpy paste (too fine crushing causes too much acrylamide to be released). Add 4 ml Gel Elution Buffer (Section VI.A), cap the tube and agitate at 40–50°C for 16 h.

Centrifuge and decant the supernatant and filter it, under pressure, through a Millipore HA 0.45 μm filter (or equivalent) into a 20 ml centrifuge tube. The filtration is best done using a syringe and filter holder rather than by negative pressure, as there is SDS in the buffer. Wash the gel fragments in a further 1 ml of Gel Elution Buffer, vortex and centrifuge; filter and pool the supernatant with the earlier filtrate.

Add 10 ml 2-butanol. Vortex to mix the phases; centrifuge briefly and discard the top phase. Repeat until the lower aqueous phase is 200–500 μl. (If it disappears completely — add water to regenerate it.) Transfer the concentrated DNA solution to a 1.5 ml microfuge tube, and precipitate the DNA with three volumes of ethanol at $-70°C$ for 5 min. Centrifuge and resuspend the pellet in 300 μl of 0.3 M sodium acetate, add 600 μl ethanol and chill at $-70°C$. Centrifuge, wash the pellet in 95% ethanol and dry under vacuum.

Resuspend the pellet in 200 μl TE buffer (Section VI.A) and store at $-20°C$. Ignoring losses, 1 μl of this solution contains fragment DNA from 1 μg of plasmid DNA.

2. Preparation of labelled DNA

Digestion of bulk-prepared DNA should be carried out on a small scale to determine which enzymes give the optimum number and location of cleavage sites. Enzymes generating fragments with 5' terminal extensions are ideal in that such fragments can be 5' or 3' end-labelled with good efficiency. The ideal pattern is one with fragments between 200 and 800 nucleotide pairs, with no comigrating fragments.

Mix:
   10 μl bulk-prepared fragment
   2.5 μl 10 × RE buffer (see supplier's recommendation)
   10 μl water
   2.5–5 units restriction enzyme (in 2.5 μl).
   Incubate for 3 h at the appropriate temperature.
   Add 70 μl water, 5 μl 0.25 M EDTA and 50 μl phenol.
   Vortex and centrifuge to separate the phases.
   Discard the bottom layer.
   Add 10 μl 3 M sodium acetate and 250 μl of ethanol.
   Chill, centrifuge and discard supernatant.
   Resuspend the precipitate in 50 μl 0.5 M sodium acetate and repeat the ethanol precipitation.
   Ethanol-wash and dry the pellet as before.

The DNA is now ready for 5' or 3' end-labelling. It may be useful to resuspend the DNA in water and dry it down in aliquots containing 10 pmol of fragment ends. For 5' end-labelling the number of pmole of fragments ends must be known. Simple formulae for this are:
   Number of pmol fragment ends

$$= \frac{2x.n}{y}$$

or

$$= \frac{3.4x.n}{z}$$

where x is the number of micrograms of original of plasmid used;
y is the size of the original plasmid in megadaltons;
z is the size of the original plasmid in kilobase pairs and
n is the number of fragments in the restriction digest to be labelled.

The calculations are approximate and ignore losses, giving a beneficial over-estimate of the number of pmol of ends to be labelled.

(a) 5' end-labelling (from Weaver and Weissmann, 1979). For an aliquot of 10 pmol of fragment ends:

Redissolve the DNA in 17 μl water.
Add: 3 μl 10 × phosphatase buffer (Section VI.B).
10 μl calf intestinal phosphatase (Molecular Biology Grade; diluted to forty units per microlitre).
Incubate at 37°C for 60 min.
Heat at 65°C for 60 min to inactivate the enzyme.
Add 5 μl 10 × kinase buffer (Section VI.B).
Use this mixture to dissolve 100–150 μCi [γ-³²P]ATP (5000 Ci mmol⁻¹; 20–30 pmol) recently dried from ethanol (do not leave the label in the dry state for more than 30 min).
Add 5–10 units T4 polynucleotide kinase and water to 50 μl.
Incubate at 37°C for 10–60 min.
Add 150 μl water and 10 μl 0.25 M EDTA.

(The specific radioactivity of the [γ-³²P]ATP used above is a good compromise between obtaining a high specific radioactivity of labelling and driving the reaction with a sufficiently high concentration of ATP to be above the K_m for the enzyme and to provide a sufficient excess molar ratio of ATP to ends. [γ-³⁵S]ATP can be used instead.)

Either, add 20 μl 3 M sodium acetate and 500 μl ethanol. Chill, centrifuge, discard the very radioactive supernatant, ethanol-wash the precipitate, and dry under vacuum.

Redissolve the DNA in 20 μl water.

Or, phenol-extract (100 μl phenol) to remove the kinase and ethanol precipitation twice if a second restriction cleavage is to be performed prior to separation of the labelled ends.

(b) 3' end-labelling. The most straightforward procedure for labelling the 3' ends of double-stranded DNA fragments is the end-repair of 5' terminal extensions with [α-³²P] or [α-³⁵S]dNTPs and Klenow polymerase. A number of enzymes generating DNA fragments with 5' terminal extensions now exist (Roberts, 1987; Kessler and Höltke, 1986). (Other methods are described by Maxam and Gilbert (1980) and Tu and Cohen (1980).) Klenow polymerase is normally used for these end-repair labelling reactions as it contains no 5' exonuclease activity. T4 DNA polymerase (from bacteriophage T4-infected E. coli) can be used for similar reasons, and has the advantage of a relatively higher 3' exonuclease activity which will convert 3' terminal extensions into fully base-paired termini. Fully base-paired termini can be 3' end-labelled using the combined 3' exonuclease and polymerase activities of T4 DNA polymerase to replace the terminal nucleotide with its a labelled equivalent. This turnover labelling gives much lower specific radioactivities than end-repair labelling. The choice of labelled nucleotides depends on the termini to be labelled, e.g. only dCTP and dGTP are required to fill in termini generated by HpaII or TaqI, and only dGTP is required to label a HaeIII digest by turnover. Some enzymes have degenerate recognition sites, and a subset of fragments may be labelled by using the correct labelled nucleotides.

For an aliquot of 10 pmol of fragment ends:

> Dry down 20 μCi of each α-labelled dNTP required (400–1000 Ci mmol⁻¹; 20–50 pmol).
> Resuspend the radioactive label in 45 μl water.
> Add this to the precipitated DNA.
> Redissolve the DNA fragments.
> Add 5 μl 10 × R buffer (Section VI.A) and 1 unit Klenow polymerase.
> Incubate at 12–16°C for 2 h.
> Add 150 μl water and 5 μl 0.25 M EDTA (to stop the enzyme and to minimize precipitation of unincorporated triphosphates).
> *Either*, add 20 μl 3 M sodium acetate and 500 μl of ethanol. Chill, centrifuge and discard the radioactive supernatant. Ethanol-wash and dry the pellet.
> *Or*, phenol-extract (100 μl phenol) to remove the polymerase and do two rounds of ethanol precipitation if a second restriction enzyme cleavage is to be performed prior to separation of the labelled ends.

(c) Isolation of DNA fragments labelled at one end. The first step in this procedure is often the isolation of single fragments labelled at both ends. This isolation is performed as described for the isolation of subfragments in Section (i) below. If the number of fragments is small and they are well resolved, this first step can be omitted.

The labelled ends of a fragment can be separated by secondary cleavage or strand separation. For smaller fragments (less than 500 bp), strand separation should be attempted, as it has the advantage that overlapping sequence from both strands can be obtained, and the majority of such strands can be separated satisfactorily. Strand separation will often work on longer fragments.

Secondary cleavage with restriction enzymes is best performed by splitting the labelled DNA into two aliquots and cutting these with different enzymes, each cleaving near one end. This obviates the problem that secondary cleavage may remove a small central region of the fragment, which would not be labelled and may not therefore be included in the composite sequence of the fragment; and it also allows overlapping sequence from both strands to be obtained.

(i) Isolation of subfragments after secondary cleavage (purification of single fragments)

> Redissolve labelled DNA in 20 μl water.
> Add 5 μl native loading buffer and mix well.
> Load the sample into one slot (1 cm × 0.15 cm) on a 40 cm long slab gel (5% acrylamide; 30:1 cross-linking) and run at 600 volts for 2–6 h in TBE buffer. (The conditions can be based on earlier analytical digests.)

Remove one glass plate, leaving the gel on the other, and cover it with Saran Wrap transparent film. Label the gel with either luminous paint or a radioactive ink marker sufficient to show in a short exposure time.

In the dark room put an X-ray film on the gel and leave under pressure for 2–20 min to expose (Section IV.B). A second exposure should be taken if a permanent record is required. Develop the film according to the manufacturer's instructions.

Identify the bands of interest and cut the photographic image from the film. Align the film with the luminous or radioactive marker and use it as a template to excise the DNA fragments in a small volume of gel. (Note: a much smaller amount of DNA is eluted here than in the bulk preparation of DNA fragments, and the protocol is therefore different.)

Lightly crush the gel slice with a glass rod and elute the DNA at 37°C overnight in 0.6 ml Gel Elution Buffer in a stoppered tube.

Vortex the tube, and decant into a 1 ml plastic pipette tip plugged with siliconized glass wool, resting in a 12 mm × 75 mm polycarbonate or siliconized glass tube.

Briefly re-extract the gel paste with 0.2 ml Gel Elution Buffer, and add to the pipette tip. Recover all the buffer by brief centrifugation. A hand Geiger Monitor can be used to determine the efficiency of recovery.

Extract the eluate with 2 ml of 2-butanol to reduce the volume, centrifuge and discard the top phase. Transfer the bottom phase to a 1.5 ml microfuge tube, add 20 μg tRNA and three volumes ethanol, chill and centrifuge to precipitate the DNA.

Unless the next step is to be strand separation of purified DNA fragments, the precipitate should be redissolved in 50 μl of 0.3 M sodium acetate and precipitated with three volumes ethanol, washed in 95% ethanol and dried. Strand separation can follow a single precipitation.

(ii) Strand separation of DNA fragments.
Redissolve labelled DNA in 25 μl TE buffer.
Add 25 μl 2 × Strand Separation Buffer (Section VI.B).
Heat at 90°C for 2 min.
Quickly chill to 0°C.
Without allowing the sample to warm up, load it into one slot (1 cm × 0.15 cm) in a 40 cm long strand separation gel (5% acrylamide, 0.083% bis-acrylamide in 0.55 × TBE; note that 60:1 cross-linking and 50 mM Tris-borate conditions are important).
Run gel at 200 volts (maximum) for 10–20 h.
The DNA is detected by autoradiography and eluted as described in Section (i) above. Two ethanol precipitations should be performed.

Multiple loadings are required to separate the strands of fragments of different sizes. (Remember to load the largest fragments first!) On these gels, xylene cyanol comigrates with strands approximately 200 nucleotides long.

The separate strands form intrastrand secondary structure under the conditions used, and separation is achieved according to small differences in secondary structure in each strand. Heating would denature the strands, and low voltage gradients and lower ionic strength buffer helps prevent this. Some renaturation of the double-stranded DNA may occur and be visible on the gel as a third band of different intensity to the other two. Note that the appearance of the sample entering the gel should not be cause for concern.

B. Modification and cleavage reactions

The following procedures are simple to perform and have many steps in common like the incubation temperatures. They give salt-free products and can be easily analysed on thin acrylamide gels. Other procedures are described in Maxam and Gilbert (1980). Details of buffers and solutions are given in Section VI.B.

Note that some of the reagents are dangerous and great care should be taken with them. All operations involving these reagents should be performed in a fume hood, and appropriate safety clothing should be worn.

Dimethyl sulphate (DMS) is mutagenic, causes burns and is poisonous. It can be inactivated by excess sodium hydroxide.

Hydrazine (HZ) is mutagenic, causes burns and is poisonous. It can be inactivated by saturation with ferric chloride.

Piperidine is poisonous and highly flammable.

1. Setting up the reactions

Resuspend the radioactively-labelled DNA in 40 µl water.
Label four 1.5 ml microfuge tubes 'G', 'A', 'T' and 'C'.
Keep the tubes at 0°C and add to them:
   G: 5 µl radioactive-DNA + 200 µl DMS buffer
   A: 10 µl radioactive-DNA
   T: 10 µl radioactive-DNA + 15 µl water
   C: 5 µl radioactive-DNA + 20 µl 4 M sodium acetate.
   (The unused 10 µl radioactive-DNA can be kept at −20°C, for use in case of accident!)
   Fresh aliquots of the following solutions should be prepared each day:
   1. HZ and DMS is removed from the stock bottles and kept in the fume hood.
   2. 88% formic acid is chilled to 0°C.
   3. tRNA is aded to the DMS Stop solution and to the HZ Stop solution (Section VI.B).

2. Base-specific modification reactions

These must be followed through without stopping. The reactions are identical from the first addition of ethanol.

Times marked (*) are approximate, and can be varied to suit the length of the required sequence and the efficacy of the reagents. Incubations at 20°C should be done in a water bath to ensure rapid temperature equilibration.

'G' reaction

   Add 1 µl of DMS to chilled tube 'G' from 1, using a mechanical pipetting device in a fume hood.
   Cap the tube, mix gently by flicking the tube, and incubate at 20°C for 6 min (*).

Return the tube to 0°C and add 50 μl cold DMS Stop solution.
Cap the tube and vortex.
Add 750 μl ethanol.
Cap the tube and invert several times to mix.
Chill at −70°C for 5 min.
Centrifuge for 5 min at 4°C in a microfuge.
Remove the supernatant (and carefully discard into 5 M sodium hydroxide to destroy mutagenic DMS).
Resuspend the pellet by vortexing in 200 μl 0.3 M sodium acetate.
Add 600 μl ethanol.
Chill at −70°C for 5 min or longer.
(If desired, the reaction can be put at −20°C at this stage.)

'A' reaction (Actually A + G reaction; Skryabin et al., 1978).

Add 25 μl chilled (0°C) 88% formic acid to chilled tube 'A' from 1.
Cap the tube, mix gently by inverting, and incubate at 20°C for 10 min (*).
Return the tube to 0°C and add 200 μl HZ Stop solution.
Cap the tube and vortex.
Add 750 μl ethanol.
Cap the tube and invert several times to mix.
Chill at −70°C for 5 min.
Centrifuge for 5 min at 4°C in a microfuge.
Remove the supernatant (and discard; nonmutagenic).
Resuspend the pellet by vortexing in 200 μl 0.3 M sodium acetate.
Add 600 μl ethanol.
Chill at −70°C for 5 min or longer.
(If desired, the reaction can be put at −20°C at this stage.)

'T' and 'C' reactions

These reactions (actually 'T + C' and 'C' reactions) differ only in the addition of sodium acetate (to 'C' in Section II.B.1) to suppress the reaction of T residues. Sodium chloride was used in earlier protocols but it tends to coprecipitate with the DNA.

Add 30 μl HZ to chilled tubes 'T' and 'C' from Section II.B.1. Use a mechanical pipetting device in a fume hood.
Cap the tube, mix gently by flicking and incubate at 20°C for 10 min (*).
Return the tube to 0°C and add 200 μl cold HZ Stop solution.
Cap the tube and vortex.
Add 750 μl ethanol.
Cap the tube and invert several times to mix.
Chill at −70°C for 5 min.
Centrifuge for 5 min at 4°C in a microfuge.
Remove the supernatant (and carefully discard into 3 M ferric chloride to destroy the mutagenic hydrazine).

Resuspend the pellet by vortexing in 200 μl 0.3 M sodium acetate.
Add 600 μl ethanol.
Chill at −70°C for 5 min or longer.
(If desired, the reaction can be put at −20°C at this stage.)

2. Cleavage reactions

All four base-specific modification reactions can now be treated together easily. A good vacuum pump is essential to the efficient removal of piperidine following these reactions, and a centrifugal drying system (e.g. Speed-Vac or UniVap) is preferable.

Centrifuge the tubes for 5 min at 4°C in a microfuge.
Carefully remove and discard the supernatant.
Gently add 0.5 ml ethanol without disturbing the pellet, centrifuge for 2 min and carefully remove and discard the supernatant.
Dry the pellet under vacuum.
Resuspend the DNA in 100 μl 1.0 M piperidine by vortexing.
Firmly cap each tube and almost fully immerse in a 90°C water bath.
Cover with a heavy weight (e.g. lead brick).
Incubate for 20 min.
Put the tubes at 0°C for 5 min.
Centrifuge briefly to collect condensate.
Poke holes in tube cap with heated needle (or cover with parafilm and puncture).
Freeze samples at −70°C.
Lyophilize samples to dryness.
Resuspend each sample in 100 μl water, vortex, freeze and lyophilize.
Resuspend each sample in 25 μl water, vortex, freeze and lyophilize.

The samples are now ready for suspension in the formamide-dye mix for electrophoresis. This is done immediately before heat denaturation of the sample and loading of the gel. If the samples are to be kept for a time before the gel is run, they should be stored dry at −20°C.

Conditions for forming, loading and running the sequencing gels are given in Section IV.

III. Procedures for the M13 cloning/chain-termination method

A. Cloning in bacteriophage M13 derivatives

1. Maintenance and growth of bacteria and bacteriophage

The standard method for M13-cloning/chain-termination uses bacteriophage M13mp derivatives which produce the α-peptide of β-galactosidase, and host E. coli (F⁺) host strains that produce the ω-peptide (Section I.B.3). The ω-peptide coding sequence is carried on an F′-plasmid which also carries the Pro marker; this plasmid can be selected and used in a $\Delta(pro,lac)$ host. The first steps in a series of M13-cloning experiments are the testing of the bacteriophage and bacterial strains.

The *E. coli* K12 strains (Table II) can be maintained as stab or lyophilized cultures. To select for cells containing the F'-episome, the strain is streaked onto minimal A-glucose plates, supplemented with vitamin B1. Cells can be maintained on minimal medium for one to two weeks. (Details of media, buffers, and other solutions are given in Section VI.C.)

The F'-episome is readily lost from the strains, and cannot be reacquired by mating in strains, where the episome is transfer-deficient. It is important that all liquid cultures are freshly inoculated with discrete colonies from the minimal medium.

All bacteriophage growth in liquid culture is done in $2 \times$ YT medium. However, for solid media, trypticase agar is preferred, as the plaques are more distinguishable. Top agar for plaque forming unit (p.f.u.) assays is used in 3 ml aliquots and the indicators for detecting the Lac$^+$ phenotype (IPTG and X-gal) are used at 30 μl of solution per 3 ml of top agar.

Bacteriophage M13 does not lyse its host and is therefore very easy to maintain and grow. Stocks can be kept for long periods as a suspension in $2 \times$ YT medium at 4°C. We normally remove cells and cell debris by centrifugation, and filter the lysate through a HAWP 0.45 μm membrane filter. Some workers freeze the M13 lysate at -20°C. The phage cannot be stored over chloroform. A stock of M13mp bacteriophage is easily obtained by taking one blue plaque from the *lac*-selective medium (IPTG + X-Gal), and incubating this plaque with shaking at 37°C for 6–8 h in 5 ml of $2 \times$ YT medium. The infected cells will grow and a phage titre of approximately 10^{12} ml^{-1} will be achieved. The culture is then briefly centrifuged and filtered into a sterile bottle.

In the following sections, the protocols will refer to the vector bacteriophage M13mp8 and host *E.coli* TG1, but other M13mp vectors and appropriate hosts can be used (see Fig. 4 and Table II).

2. Vector DNA

(a) Preparation of vector DNA. Take a single colony of *E. coli* TG1 from minimal medium and grow it up in a small volume of $2 \times$ YT medium at 37°C with shaking until $A_{550} = 0.6$.

Dilute bacteriophage M13mp8 stock to approximately 10^3 p.f.u. ml^{-1} in $2 \times$ YT medium.

To 3 ml molten BBL trypticase-top agar at 45°C, add 0.2 ml *E. coli* TG1 ($A_{550} = 0.6$), 0.1 ml diluted phage, 30 μl IPTG solution and 30 μl X-gal solution. Mix and pour onto a BBL trypticase-agar plate, agitate so that the plate is uniformly covered. Allow agar to set and incubate the plate inverted at 37°C until the colour has developed (about 9 h).

With a sterile toothpick, Pasteur pipette or similar sterile object, touch a blue plaque that is well-separated from the others and transfer it to 1 ml 2 × YT medium. Grow this at 37°C with shaking (300 r/min) for 6–8 h. Meanwhile, grow up *E. coli* TG1 (taken from a single colony on minimal A-glucose agar) in 10 ml 2 × YT, until $A_{550} = 0.5$ approximately. Add the *E.coli* TG1 culture and 1 ml of bacteriophage preparation to 500 ml of prewarmed 2 × YT medium in a 2 l Ehrlenmeyer flask. Shake this vigorously at 37°C for 4 h.

Collect the cells by centrifugation and prepare RFI-DNA by a cleared-lysis/dye-buoyant density centrifugation procedure suitable for small *E. coli* plasmids (Chapter 5). Two rounds of dye-buoyant density centrifugation are essential to remove all traces of contaminating chromosomal, single-stranded bacteriophage or other DNA. One litre of culture should yield about 500 μg M13mp8 RFI-DNA. This is kept frozen in small aliquots in TE buffer at 0.1–1.0 mg ml^{-1}.

(b) Cleavage and testing of vector DNA. Vector DNA must be tested prior to attempting to clone DNA fragments. The exact conditions for cleavage and testing depend on the vector and the restriction enzyme used. Table I lists some of the M13mp vectors so far available and the restriction enzymes that can be used to linearize the vector and to generate fragments for cloning.

Normally, sufficient linearized vector for a number of cloning experiments is prepared and tested. The vector can satisfactorily be stored frozen for several years. During each cloning experiment, the vector is tested to ensure that no spurious colourless plaques are generated in the ligation reactions.

(i) Sticky-end vector (e.g. using EcoRI)

Mix:
    2 μg M13mp8 RF-DNA
    2 μl 10 × *Eco*RI buffer
    Water to give a final volume of 20 μl
    3 units *Eco*RI.
    Incubate at 37°C for 1 h.
    Add 180 μl water (to bring to DNA concentration of 10 ng μl^{-1}).
    Heat at 70°C (this inactivates many enzymes).
    Extract with 100 μl phenol.
    Remove the aqueous layer to a new tube. Extract the phenol layer with 100 μl TE buffer, and pool with aqueous layer.
    Add 30 μl 3 M sodium acetate and 750 μl ethanol; keep at −20°C for 60 min.
    Collect the DNA by centrifugation in a microfuge (15 min at 4°C). Wash the pellet by centrifugation under 95% ethanol. Dry under vacuum.
    Resuspend pellet in 200 μl TE buffer, and store at 4°C (or −20°C for long-term storage).

Test this vector preparation by transformation of competent *E. coli* TG1 (Section III.A.4(c)) to ensure that it gives few plaques (less than 10/ng); and by religation (Section III.A4(a)) and transformation to ensure that it gives only blue plaques (300–400/ng). Religation of vector DNA can also be used to determine the correct amounts of T4 DNA ligase to use.

(ii) Blunt-end vector. *Sma*I-cut M13mp8 or M13mp9 is suitable for cloning blunt-end fragments. *Hind*II (or *Hinc*II) should not be used to generate a vector for blunt-ended fragments due to the high frequency (ca. 10%) of colourless plaques formed on religation. The cleavage and religation conditions are essentially the same as for sticky-end vector, but typically about five times the amount of DNA ligase is required.

(iii) Phosphatase treatment of vector can be carried out in order to increase the fraction of useful ligation events. (It is a good idea to save some of the cut vector before phosphatase treatment for control ligations.) Linearized M13mp vectors contain about 0.5 pmol ends per μg DNA.

Mix:
  1 μg cut vector DNA (sticky or blunt-ended)
  2 μl 10 × phosphatase buffer
  0.5 μl Calf Intestinal Phosphatase (40 units ml^{-1}).
  Water to 20 μl and incubate at 37°C for 30 min, then add another 0.5 μl C.I.P. and incubate for a further 30 min.
  Heat at 65°C for 10 min to inactivate the enzyme, extract with an equal volume of phenol, and precipitate the DNA with ethanol.

3. Preparation of insert DNA

The manner of preparation of the insert DNA depends on the sequencing objective. A very small or large region of DNA to be sequenced may demand different approaches. Large sequences may be cloned in a vector of known sequence (e.g. pUC18; Yanisch-Perron *et al.*, 1985), the whole chimeric plasmid subcloned in M13, and the subclones tested for homology with pUC18 by T-channel screening (Section III.B.3). Alternatively for smaller sequences, the DNA to be sequenced may be prepared in bulk (Section II.A.1).

DNA fragments can be prepared by restriction enzyme digestion, nonspecific nuclease digestion (Anderson, 1981) or sonication. Subsequently, the fragments may be fractionated by various methods like electrophoresis, chromatography or precipitation with polyethylene glycol, PEG (Lis, 1980). The number of possible permutations of methods is large. Some of the common methods are described below.

(a) Restriction fragments. Digest 1–3 μg of DNA according to supplier's recommendation. Usually the enzyme will not interfere with subsequent end-

repair and end-labelling reactions, but routinely the reaction is heated at 70°C for 10 min to inactivate any contaminating phosphatases, etc.

If the fragments are to be cloned directly, extract with an equal volume of phenol and then several times with ether.

Precipitate DNA by adding 0.1 volume 3 M sodium acetate, 2.5 volume ethanol, chill at −70°C for 5 min and centrifuge (15 min) in a microfuge. Wash the pellet by centrifugation under 95% ethanol and dry under vacuum. Resuspend the DNA in 20 μl TE buffer. Store at −20°C.

(b) Restriction fragments, end-repaired. Fragments with 5′ or 3′ terminal extensions which cannot be cloned directly into M13mp8 or its more recent relatives (e.g. *Hin*fI- or *Hha*I-generated fragments) can be converted into blunt-ended fragments.

To 20 μl digest (1–3 μg DNA fragments) from Section (a), add:
  2 μl dNTP mix (1.25 mM each deoxyribonucleoside triphosphate in water).
  0.2 unit T4 DNA polymerase (or Klenow polymerase).
  Incubate at room temperature for 10 min.
  Heat at 70°C for 10 min to inactivate the enzyme.
  Extract with phenol.
  Ethanol precipitate as described in Section (a).
  Resuspend in 20 μl 0.3 M sodium acetate and precipitate with ethanol.
  Wash the pellet by centrifugation under 95% ethanol and dry under vacuum. Resuspend the DNA in 20 μl water.
  Store at −20°C.

Ethanol precipitation will not remove all the deoxyribonucleoside triphosphates, and these may inhibit the subsequent ligation reaction. This inhibition is not usually sufficient to cause problems, but if difficulty with ligations is encountered, the triphosphates should be removed by chromatography on Sephadex (Brown, 1984) as a precaution.

End-repaired fragments can be cloned in *Sma*I-cut vectors. Care should be taken that two end-repaired fragments are not cloned together in the same vector molecule. These joins are not easily recognized and eliminated; for example, the sequence 5′-GATTACTC-3′ may result from two end-repaired *Hin*fI fragments, or may exist in the original DNA (and is not a *Hin*fI site). End-to-end cloning problems can be eliminated by size-fractionation of the DNA prior to cloning, or minimized by adding linkers or by choosing conditions in which only low frequencies of recombinant formation occurs.

(c) Restriction fragments plus linkers. More recent vectors and methods have displaced the addition of linkers as the routine method for generating insert DNA. However, this method may occasionally be useful, and the protocol is described by Sanger *et al.* (1980) or, in modified form, by Brown (1984). Linkers can now be purchased which are already 5′-phosphorylated.

(d) Sonicated DNA. Sonication offers advantages over restriction endonuclease digestion since the fragments are truly random. This is the preferred method for very large sequencing projects. If the DNA of interest can be cloned in a small fully-sequenced plasmid, this could be sonicated directly and the fragments cloned as described below. The vector sequences would then be eliminated from the compiled sequence, either by preliminary T-channel screening, or after sequencing. Sometimes the DNA has to be purified from vector before sonication.

A DNA fragment, typically 5–10 kb in size, is fractionated on 1% low-gelling temperature agarose in $1 \times$ TBE buffer, 0.3 μg ml^{-1} ethidium bromide. The fragment is excised from the gel, melted in four volumes of TE buffer at 67°C, then extracted four or more times with phenol until there is no detectable solid material at the phenol-buffer interface.

About 5 μg of the purified fragment is circularized by ligation at 15°C in 30 μl 50 mM Tris, pH 7.5, 10 mM MgCl$_2$, 10 mM Dithiothreitol (DTT) containing 20 units DNA ligase (where a unit is defined as the amount required to ligate 1 μg *Eco*RI-cut pBR322 completely in 1 h under the same conditions).

A cup-horn sonicator (e.g. the Heat Systems Ultrasonics model W-380) is ideal for sonicating DNA, as the sample can be placed in a sealed microfuge tube, and does not come into direct contact with the probe. A microtip probe can be used, but great care must be taken to prevent cross-contamination of samples. The sample should be kept cool, and the DNA should be sonicated until the average fragment size is 400–800 bp. Each sonicator must be calibrated. At maximum power in the above cup-horn sonicator, with the sample tube sitting about 1 mm above the probe in about 2.5 cm of cold water, 4 bursts of 40 s are normally used; the tube is centrifuged between each burst, and the water is replaced after two bursts.

The sonicated fragments are end-repaired by incubation at 15°C for 4 h with 2 μl dNTP mix (1.25 mM each deoxyribonucleoside triphosphate) and 20 units of T4 DNA polymerase. Klenow polymerase (20 units) can be used, but incubation should be overnight.

It is important that fragments are sized before cloning in M13, in order to prevent errors due to sequencing end-to-end cloned fragments. The fragments are mixed with native loading buffer and fractionated on a 1.5% agarose gel in $1 \times$ TBE, using a suitable plasmid DNA digest as size markers. The gel is run until the dye has moved about 2–3 cm, then the correct size range is eluted by cutting a slot in the gel at the lower size limit (350 bp), electrophoresing the DNA into the slot for about 45 s and removing the buffer from the slot into a tube. This is repeated until the upper size range is reached (600 bp). The fragments may be pooled in size ranges. Each pool is extracted with an equal volume of phenol, precipitated with 0.1 volume 3 M sodium acetate and 2.5 volume ethanol. The precipitate is collected, washed by centrifugation under 95% ethanol and dried under vacuum. Each size pool is redissolved in 25 μl TE for cloning in SmaI-cut, phosphatase-treated vector.

(e) Selection of specific fragments. For certain purposes, such as the end of a large sequencing project, it may be necessary to isolate and sequence a specific fragment. Sometimes this can be done by random cloning using suitable restriction enzymes and M13 vectors. Often the fragment must be purified from a gel. For large fragments, this can be done by phenol extraction of molten low-gelling temperature agarose, as described in Section (d); or by electroelution, as described by Maniatis *et al.* (1982). For small fragments (up to 400 bp), the purification can be done by separation on an 8% acrylamide gel and extraction of ethidium-visualized fragments by crushing and soaking, as described for the labelled fragments in the chemical method (Section II.A.2.(c).(i)). (Acrylamide concentrations below 6% should not be used, as significant amounts of linear polyacrylamide will coelute.)

4. Ligation and transformation

(a) The ligation reaction. If the concentration of DNA fragments is known, use a three-fold molar excess of fragments to vector. Otherwise try a range of different fragment concentrations.

Mix in a 400 μl reaction tube:
  1 μl linearized vector (10 ng; approximately 2 fmol)
  1–5 μl DNA fragments (approximately 6 fmol)
  0.5 μl 10 mM ATP
  1 μl 10 × C buffer
  0.5 μl autoclaved gelatin (1 mg ml⁻¹)
  1 μl T4 DNA ligase (diluted as determined empirically)
  Water to give a final volume of 10 μl.
  Incubate at 14°C for 3–12 h for sticky-end ligations, or for 24 h for blunt-end ligation.

A control ligation with no added DNA fragments must always be included in a set of ligation reactions to ensure that no spurious colourless plaques are formed. Occasionally, a control reaction with 10 ng of RFI vector DNA (CCC), and one with linearized vector but no ligase should be performed to test the efficiency of transformation, and to test the degree of linearization of the vector.

(b) Preparation of competent cells. Take a single colony of *E. coli* TG1 from minimal medium and grow up in 20 ml 2 × YT at 37°C with shaking. When A_{550} is approximately 0.3, centrifuge the cells at 4°C for 5 min at 6000 g in a sterile tube. Resuspend the pellet in 10 ml sterile 50 mM $CaCl_2$ at 0°C. Leave for 20 min at 0°C then pellet the cells again. Resuspend the cells in 2 ml sterile

50 mM CaCl$_2$. The cells can be used immediately or kept at 0°C for up to 24 h. If 20% glycerol is included in the final incubation, the cells may be frozen in 200 µl aliquots in liquid nitrogen, or ethanol-dry ice, and stored at -70°C. This is a simple protocol that gives cells of sufficient, rather than high, competence.

An alternative procedure is to grow up a larger culture of cells, and harvest when A$_{550}$ = 0.3. The pellet is resuspended in 2/5 volume TfbI (Section IV.C) and incubated at 0°C for 5 min. The cells are collected by centrifugation (5 min, 6000 g) and incubated in 1/25 volume TfbII at 0°C for 15 min. The cells can be frozen in 200 µl aliquots and stored at -70°C.

(c) Transformation and plaque assay. Either thaw sufficient pre-frozen 200 µl aliquots of competent cells at 0°C, or dispense 0.2 ml aliquots of fresh competent cells in sterile 1.5 ml reaction tubes and add part or all of the ligation mix. For blunt-end ligations use all the ligation mix, and for sticky-end ligations use about one-fifth of the ligation mix, depending on the DNA preparation used.

Mix and incubate at 0°C for 40 min.
Heat shock the cells at 42°C for 2 min.
To 3 ml molten BBL top agar at 45°C, add 30 µl IPTG, 30 µl X-gal solution, 0.2 ml exponentially growing TG1 and the transformed cells. Mix and pour on BBL trypticase agar plate. The agar surface should be well-dried and the plate kept at room temperature. Spread the top agar by gently tilting and shaking the plate, and allow to set (1–2 min). Invert the plate and incubate at 37°C for 9–24 h.

5. Preparation of templates

Templates should be prepared within one or two days of the plaque assay to prevent cross-contamination between plaques by diffusion. Twenty-four templates can be conveniently prepared at one time using a 12-place microfuge.

In early experiments, a template should also be prepared from a blue plaque for use as a control in the chain-termination sequencing.

(i) Add 250 µl of late exponential phase culture of TG1 to 25 ml 2 × YT. Dispense into 1 ml aliquots in culture tubes or universal bottles. Using a sterile toothpick, Pasteur pipette or similar object, transfer a different colourless plaque into each aliquot. Choose well-separated plaques and take care to touch only a single plaque. Shake (300 r/min) at 37°C for 5–6 h. Longer incubations under these conditions may give poor template preparations due to host cell lysis. (An alternative procedure is to place infected cells from the plaque directly to 1 ml of 2 × YT medium, with no added uninfected TG1. The 1 ml aliquots are kept at 4°C until late evening then shaken at 37°C overnight (maximum 16 h).

(ii) Transfer the 1 ml culture to a 1.5 ml microfuge tube and centrifuge for 5 min at 10 000 g.

Dispense approximately 0.8 ml supernatant into another microfuge tube and add 0.2 ml 2.5 M NaCl, 20% PEG 6000. Mix well and leave at room temperature for 15 min.

Centrifuge for 5 min at 10 000 g. Decant the supernatant and centrifuge the pellet for 1 min.

Remove all traces of the PEG supernatant using a drawn-out capillary or a piece of tissue. The PEG pellet should be visible, although it may be very small.

(iii) Dissolve the pellet in 100 μl TE buffer and add 50 μl buffer-saturated phenol. Vortex for 10 s, leave at room temperature for 10 min, vortex again and centrifuge for 1 min.

Transfer the upper aqueous layer to a new microfuge tube, using a drawn-out capillary or an automatic pipette.

Extract with 0.5 ml water-saturated diethyl ether to remove the phenol. Ethanol precipitate the DNA (10 μl 3 M sodium acetate, 250 μl ethanol); incubate at −70°C for 5 min; centrifuge for 15 min and ethanol-wash the pellet).

Dry the pellet under vacuum.

Redissolve in 50 μl TE buffer.

Store the template DNA at −20°C. The presence of single-stranded DNA can be tested by running 2.5 μl on a 1% agarose gel stained with ethidium bromide (do not use xylene cyanol FF marker dye).

B. Chain-termination sequencing

1. Preparation of primer DNA

(a) Several universal primers are available for sequencing in M13mp derivatives. The simplest to obtain are the synthetic primers available commercially (from suppliers in Section VI). These yield good results if they are carefully titrated against a typical template preparation. About 1 ng of a 20-nucleotide primer is used per hybridization. Too little primer gives a very weak autoradiogram making interpretation of the sequence difficult.

(b) Occasionally, internal primers, complementary to the DNA cloned into the M13 derivative, may be required. Although such primers can be prepared from other DNAs, a more common method is to synthesize suitable primer based on known nucleotide sequence. It is very important that the sequence of the proposed primer is checked for homology with other regions of the template (e.g. the M13 backbone), otherwise the priming may occur at additional sites.

2. The sequencing reactions

Several sequences can be determined simultaneously. There is a number of variations depending, for example, on whether [35]S- or ^{32}P-labelled nucleotides are used, and whether a centrifuge capable of taking microtitre plates is available. The various methods are similar, and that using microtitre plates will be described in most detail.

The volumes described have been chosen to make use of the Hamilton PB600-1 repetitive dispenser carrying a 100 μl gas-tight syringe (1710), which dispenses 50 × 2 μl aliquots.

(a) The microtitre plate method. Pipette 2 μl template (1) DNA into the first position of each of the top four rows of a microtitre plate, pipette 2 μl template (2) DNA into the second position of each of the top four rows, and continue in columns of four until all the templates have been set out.

For a number (N) of templates, mix (N + 1) times the following volumes:
  0.3 μl 100 mM Tris Cl, pH 8.0, 100 mM MgCl$_2$.
  0.3 μl primer solution (0.2 pmol μl^{-1}).
  1.4 μl T-mix.
Pipette 2 μl into the top well of each column of four.

Mix the same volumes of primer, buffer and nucleotide mix for 'C'-mix, and dispense 2 μl into the second well in each column of four. Do the same for 'G'-mix in the third row and 'A'-mix in the fourth row. Each template now has a single column of 4 wells, each with a different nucleotide mix. Cover the plate with parafilm and a lid and incubate at 55°C for 30 min to anneal the primer and template.

Then mix (N + 1) times the following volumes:
  0.8 μl 100 mM Tris Cl, pH 8.0.
  0.8 μl 100 mM DTT.
  0.5 μl [α-^{35}S]dATP (10 mCi ml^{-1}, ca. 600 Ci mmol^{-1}, in tricine).
  5.5 μl water.
  0.4 μl Klenow polymerase (2 units at 5 units μl^{-1}).
(The enzyme must be added immediately before the solution is used. If the label or the polymerase are at different concentrations, or if the label is dried down from 50% ethanol, the volume of water can be adjusted to a total of 8 μl per template.)
Add 2 μl of the enzyme-labelled mix to each well; centrifuge to mix; cover the plate and incubate at 37°C for 20 min.
Add 2 μl chase solution of 0.5 mM dNTPs (all four triphosphates) and incubate for a further 10 min.
Then either add 4 μl formamide-dye mix, centrifuge briefly, denature the DNA at 80°C for 15 min and run on a gel immediately (Section IV), or freeze the samples at −20°C prior to denaturation in formamide.

(b) Microfuge tube method. Three templates can be sequenced simultaneously in a standard 12-place microcentrifuge, and correspondingly larger numbers in higher capacity rotors (e.g. the 40-place Eppendorf 5413). For these smaller numbers of templates, the advantages of bulk-mixing primer and nucleotides are not so great, and a modified protocol is easier.

For N templates, mix (N + 1) times the following volumes:
    1.5 μl primer solution (0.2 pmol μl^{-1}).
    1.5 μl 100 mM Tris, pH 8.0, 100 mM MgCl$_2$.
    3.0 μl water.
Dispense into N microfuge tubes, and add 4 μl each template. Anneal at 67°C for 15 min (or 55°C for 30 min).
Dispense 4 × 2 μl aliquots of annealed primer-template into lidless microfuge tubes (e.g. Sarstedt 72.696 or cut off the lids).
Add 2 μl 'T'-mix to the first tube of each set of four, 2 μl 'C'-mix to the second, 2 μl 'G'-mix to the third, and 2 μl 'A'-mix to the fourth.
Then proceed as for the microtitre plate method from the preparation of the enzyme-labelled mix.

(3) Titration of dideoxynucleotide mixes

The length of the sequence that can be determined and the intensity of bands on the autoradiograph are functions of the ratio of 2′,3′-dideoxyribonucleoside triphosphates to the corresponding 2′-deoxyribonucleoside triphosphate. The correct ratios must be determined empirically. With the reagents used in the authors' laboratory, those concentrations given in Section VI work well. The ratio should be chosen such that bands are seen well beyond the limit of resolution of the gel, or occasional problems will be found due to spurious bands in the resolved region. Too low a dideoxynucleotide concentration will result in weak exposures on autoradiography. It is convenient to have two different sets of nucleotide mixes, a 'short' set for sequences close to the primer and a 'long' set for determining sequences much further away.

3. T-channel screening

If a random cloning strategy is used to generate templates for sequence analysis, the frequency of occurrence of templates containing new or useful sequence information decreases as the project progresses. Initial screening of the templates, using only the T-specific reaction, allows the elimination of poor templates, templates containing only a short inserted fragment, and templates containing known sequence. It is also useful for comparing several templates from a single fragment to identify both orientations. The T-pattern is compared with the known sequence (possibly using a computer program). The T-reaction is used, as the characteristic pattern of T residues in the vector distal to the cloning site is easily recognized on the autoradiograph (Fig. 6).

To screen N templates, mix (N + 1) times the following volumes:
    0.4 μl 100 mM Tris, pH 8.0, 100 mM MgCl$_2$.
    0.5 μl primer solution (0.2 pmol μl^{-1}).
    1.1 μl water.
(Note: this can be made in bulk and stored at −20°C.)

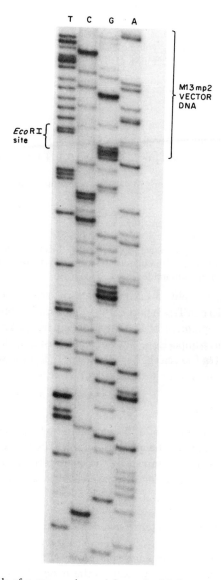

Fig. 6. Autoradiograph of a sequencing gel from the M13-cloning/chain-termination method. The sequence is a small fragment of Tn*501* DNA cloned in M13mp2. Note the sequence of the vector, particularly the characteristic pattern of T residues, beyond the insert.

Dispense 2 μl into the N wells of a microtitre plate, or into N capped microfuge tubes. Add 2 μl template. Centrifuge to mix, cover with parafilm (or cap tubes) and incubate at 55°C for 30 min.

Mix (N + 1) times the following volumes:
  0.1 μl 50 mM Tris Cl, pH 8.0, 1 mM EDTA.
  0.02 μl 0.5 mM dTTP.
  0.4 μl 0.5 mM dCTP.
  0.4 μl 0.5 mM dGTP.
  0.09 μl 10 mM ddTTP.
  1 μl water.
(Note: this can be made and stored in bulk at −20°C.)

To this is added 0.5(N + 1) μCi [α-^{35}S]dATP and 0.4(N + 1) units Klenow polymerase immediately before use.

Add 2 μl of nucleotide-enzyme mix to each primer-template preparation and centrifuge to mix.

Incubate at 37°C for 20 min.

Add 2 μl 0.5 mM dNTP mix. Centrifuge to mix.

Incubate for a further 10 min.

Then either add 4 μl formamide-dye mix, centrifuge briefly, denature the DNA at 80°C for 15 min and run on a gel immediately (Section IV) or freeze the samples at −20°C prior to denaturation in formamide.

After autoradiography, each T-pattern can be read (as e.g. TxTxxTTTxxxxT) in a T-rich part of the sequence and compared to the known sequence. The complement of the T-pattern or of the known sequence must also be used in the comparison. The sequence distal to the cloning site in the M13mp series (TTcgTaaTcaTggTageTgTTTccTgTgT) can be used to identify clones with small inserts, or with inserts of a specific size (see Fig. 6).

4. Reading a sequence

Following gel electrophoresis and autoradiography (Section IV), the DNA sequence can be read from the autoradiograph. The following points will assist in reading a sequence determined by the chain-termination method.

(a) Bands in each sequencing track are not of equal intensity. For doublets, the following rules apply:

(i) Upper C is always more intense than lower C;

(ii) Upper G is often more intense than lower G (and always so if preceded by a T);

(iii) Upper A is often less intense than lower A.

(b) Always read the spaces as well as the bands. Ensure that the proposed sequence fits into the available space on the gel (but see (d)).

(c) Occasionally, bands with the same mobility in all four channels are found on the gel. The known reasons for these include:

(i) End-repair of double-stranded primer DNA (this will be in one region of the sequence only and is rarely the case now that synthetic primers are used);

(ii) DNA polymerase stopping immediately after a long (>5) run of Gs; compressions (see below). Performing the sequencing reactions at higher temperatures may eliminate problems with polymerase stopping. Temperatures up to 50°C have been used in some laboratories, but it is advisable to add extra Klenow polymerase (in 10 mM DTT, 10 mM Tris, pH 8.0) with the chase solution. Alternatively, reverse transcriptase may be used instead of Klenow polymerase (e.g. Agellon and Chen, 1986).

(d) Bands on the autoradiograph may be clustered together. This is due to a persistent secondary structure in the radioactive DNA during electrophoresis. In an extreme case, the bands are not resolved. Usually a compression is followed by a slight expansion of band spacing above the compression. This is a common and serious problem with GC-rich sequences, and may result in an incorrect sequence being read, or the sequence being unreadable. This is now most easily remedied by the inclusion of 7-deaza-dGTP or dITP in place of

SYMBOL	MEANING			
1	Probably	C		
2	"	T		
3	"	A		
4	"	G		
D	Definitely	C	possibly	CC
V	"	T	"	TT
B	"	A	"	AA
H	"	G	"	GG
K	Definitely	C	possibly	CX
L	"	T	"	TX
M	"	A	"	AX
N	"	G	"	GX
R		A or G		
Y		C or T		
5		A or C		
6		G or T		
7		A or T		
8		G or C		
–		C or T or A or G		

Fig. 7. Uncertainty codes for DNA sequences for use with DBUTIL and related sequence compilation programs (Staden, 1980).

dGTP in the sequencing reactions (Mizusawa *et al.*, 1986), which reduces the stability of GC base pairs. Hot-formamide gels (see Brown, 1984) are rarely required.

(e) Problems in the interpretation of a sequencing gel can be allowed by the use of an uncertainty code, as shown in Fig. 7. This code covers all ambiguities that may arise on a sequencing gel. It is very useful in conjunction with programs such as DBUTIL (Staden, 1980) which allow a record of the primary gel-reading data to be kept. The above list covers most usual deviations from the ideal of equal-intensity bands separated by regularly decreasing spacing. Some unusual deviations or problems are covered in Section V.

C. Outline of some specialized techniques

In addition to those combinations of methods described in Sections II.A and III.B, the following protocols will be useful for certain sequencing problems.

1. Deaza-GTP substitution

One of the common difficulties in determining a DNA sequence is the presence of 'compressions' on the autoradiograph; these are due to fold-back intrastrand base pairing. This base-pairing is more stable in GC-rich regions than in AT-rich regions, and can be reduced by using 7-deaza-dGTP in place of dGTP in the chain-termination reactions (Mizusawa *et al.*, 1986). The 7-deaza derivative of dGTP is a better replacement for dGTP than is deoxyinosine triphosphate, which was similarly used (see Brown, 1984); 7-deaza-dGTP causes less premature termination and can be used at the same concentration as dGTP, with Klenow polymerase.

2. Kilobase sequencing

Several workers have described strategic modifications of the M13-cloning method which avoids random cloning and can be used to travel along the DNA sequence (e.g. Barnes *et al.*, 1983; Hong, 1982). A large fragment of DNA is cloned into an M13 derivative containing asymmetric restriction targets (e.g. M13mp9); a series of deletions are made from a fixed point in the vector, adjacent to the universal primer. The larger the deletion, the further into the original fragment the sequence is determined. By sequencing a series of deletions which have been separated according to size, the sequence of the large DNA fragment can be rapidly obtained.

The essential difference between these methods is in the vector used and in the manner of generating the variable (distal) end of the deletion. Barnes *et al.* (1983) use the vector M13mWB2344, or similar, and generate the distal end of the deletion using quantitative digestion with DNAase I in the presence of ethidium bromide to nick the DNA, exonuclease III to form a gap and nuclease S1 to cleave the other strand. Hong (1982) uses the vectors M13mp8 or M13mp9, and generates the distal end of the deletion using partial digestion with DNase I in the presence of Mn^{2+}. Misra (1984) uses Bal31 from a restriction site. The fixed end of the deletion is generated using a restriction site next to the primer. It is important that this site is unique on the molecule; thus, symmetrical vectors, such as M13mp7 are unsuitable for this method.

Detailed protocols describing these methods are given in the original papers, and commercial kits using related deletion methods are available.

3. Direct chain-termination sequencing of plasmid DNA

Protocols are now available for the direct sequence analysis of plasmids (Chen and Seeburg, 1985; Agellon and Chen, 1986). These may offer some advantages. For example, if the DNA of interest is already cloned in one of the pUC vectors (e.g. pUC18; Yanisch-Perron et al., 1985), it can be sequenced without further subcloning using the M13mp universal primer; and sequencing can be done in both directions on the same template. Newly-cloned regions may be sequenced using as primer the oligonucleotide with which they were identified.

The success or failure of double-stranded chain termination sequencing is very dependent on the primer and the template used, as well as the host strain from which the plasmid template is isolated. In our limited experience, sequencing in M13 derivatives is more reliable, although direct plasmid sequencing may be more convenient for some applications.

A booklet 'Guidelines for quick and simple Plasmid sequencing' (Heinrich, 1986), to which readers are referred, is available from Boehringer Mannheim GmbH; and this contains all the necessary recipes.

The template DNA is generated by alkali denaturation of the plasmid DNA, followed by neutralization and ethanol-precipitation. A large excess of primer oligonucleotide is annealed to the template and the chain termination reactions are carried out as with M13 derivatives.

4. Use of modified T7 DNA polymerase

Recently, Tabor and Richardson (1987) described the use of chemically-modified T7 DNA polymerase for chain-termination sequencing. This enzyme is highly processive, and efficiently incorporates nucleotide analogues such as 7-deaza-dGTP, dITP and ^{35}S-labelled deoxyribonucleoside triphosphates. There is less variation in band intensity than with Klenow enzyme or reverse transcriptase, and less background due to pausing at secondary structures. Commercial kits using this enzyme are available ('Sequenase' from United States Biochemical Corp.), and the enzyme can be bought separately. It cannot be directly substituted for the Klenow polymerase in the procedures given here because different ratios of ddNTP to dNTP are required. The method described by Tabor and Richardson (1987) involves a preliminary extension and labelling step without dideoxynucleotides, and a subsequent chain-termination reaction. We have not yet tested this system fully, but it does appear to offer some advantages.

IV. Procedures for gel electrophoresis and autoradiography

A. Standard electrophoresis

Chain-termination reactions and the salt-free reactions from the chemical method can be run on thin (0.3 mm) acrylamide-urea gels (Sanger and Coulson, 1978). Thin gels increase band resolution on the autoradiograph as a simple geometric consequence of having the radioactive source close to the film. The gel concentration used depends on the experiment being performed. A 6% acrylamide gel (with 1 in 20 cross-linking with bis-acrylamide) is satisfactory for most purposes, except for sequencing within the first 15 nucleotides from the common end (i.e. the labelled end in the chemical method). In this case, a 20% gel can be used. In all cases, the gels are heated prior to loading the sample, usually by pre-running the gel. This reduces problems caused by any persistence of secondary structure in the DNA fragments. Prerunning also removes electrolytes from 20% gels which would otherwise run with the smaller oligonucleotides and prevent the sequence being determined.

Making the gel

A variety of slab gel electrophoresis apparatus is available. Some commercial apparatus is unduly complex and may leak during use or be very difficult to set up and use. Readers contemplating buying a gel apparatus are strongly recommended to test commercial apparatus over several days prior to purchase. We use commercially-available apparatus of the Studier (1973) type. The gel plates must be very clean to prevent bubbles forming when the gel is poured. Each plate is cleaned with acetone, ethanol and distilled water when used. The back plate (with a cut-out to provide electrical contact with the cathode buffer reservoir) is siliconized on its inside face to facilitate its removal from the gel. Siliconization is done using a commercially-available solution of 2% dimethyl-dichlorosilane in 1,1,1-trichloroethane (e.g. Repelcote), and is repeated after the plate has been used several times. The front plate is soaked in detergent between uses.

Combs and spacers can be made from 0.35 mm thick Plastikard. New Plastikard is coated with a substance that inhibits the polymerization of the acrylamide, and this must be removed with neat detergent followed by thorough washing in water.

A recent innovation is the 'Shark's-tooth comb', in which the pillars of acrylamide which traditionally separate the channels on a gel are eliminated in favour of a flat surface separated into channels by sharp triangular 'teeth' of the comb. After the gel has been formed using the flat side of the comb, the comb is inverted and the teeth are pushed into the gel surface. Care must be taken to prevent cross-leakage between channels. The teeth are constructed such that there is room between them to load the gel and there is virtually no gap between the channels. Thus, it becomes much easier to determine the relative order of fragments on the autoradiograph, and thus to determine sequences from much higher up the gel. 'Shark's-tooth combs' have to be constructed very precisely, and are difficult to make without workshop facilities; they can be purchased commercially (e.g. from BRL).

(i) Clean the gel plates with acetone, ethanol and water. Clamp the plates together with Plastikard spacers and seal the sides and bottom with yellow vinyl tape (Sellotape 1607); ensuring that there are no wrinkles in the tape, otherwise the gel may leak during pouring.

(ii) Make the gel mix. For a 40 cm × 20 cm gel:

	6%	20%
Urea (ultra-pure)	21 g	21 g
10 × TBE	5 ml	5 ml
Deionized 40% acrylamide (38:2)	7.5 ml	25 ml
1.6% ammonium persulphate	1.6 ml	1.6 ml

With gentle warming and stirring make up to 50 ml with water. Cool and add 50 μl *N,N,N',N'-Tetramethylethylenediamine (TEMED)*.

Pour the gel immediately, using a large volume pipette with a mechanical pipetting device; with the plates at an angle of about 45°, run the solution down one of the side spacers until the gel mould is full. Lay the gel almost horizontal and insert the comb. Clamp the gel, or put weights over the wells to ensure that they form properly.

(iii) Leave 6% gels at least 1 h after setting before use; 20% gels are best left overnight, after layering buffer over the slot-former and covering with sealing film.

Running the gel

(i) Remove the tape from the bottom of the gel, and the comb from the top (or invert the 'shark's-tooth comb'). Clamp the gel in the electrophoresis apparatus and fill the apparatus with 1 × TBE buffer. An aluminium plate clamped to the front gel plate, and out of contact with the buffer chambers is a cheap and reasonably safe way to give good temperature equilibration and avoid a 'smile' across the gel. Prerun the gel for 30–60 min at 1.2 kilovolts for a 40 cm long 6% gel (this gives approximately 2 mA current per centimetre width of gel).

(ii) Add the formamide-dye mix to the DNA sequencing reactions, if this has not already been done, and denature as described earlier. Rinse out the wells in the gel using a Pasteur pipette — this removes the urea that will have leached out of the gel. Load 2–3 μl sample per slot.

(iii) Run the gel at 1.2 kilovolts. The duration of the run depends on the sequence to be determined. On a 6% gel fragments of about 24 nucleotides comigrate with the fast dye (bromophenol blue) and 110 nucleotide fragments run with the slow dye (xylene cyanol FF).

Other gel systems

Two stratagems have been adopted to reduce the spacing of the fragments in the lower part of the gel, and thus to maximize the amount of information that can be read from a single gel loading. Both alter the field in the gel so that fragments move more slowly as they proceed down the gel.

'Wedge gels' (see Olsson *et al.*, 1984; Heinrich, 1986) are constructed by placing a thicker spacer at the bottom of the gel, and the gel recipe is the same as for standard gels. If the gel is to be dried down (see next section), the bottom of the spacer must be thinner than 0.5 mm; 0.1 mm thick Plastikard can be used for the top of the spacer and for the comb.

'Buffer gradient gels' (Biggin *et al.*, 1983) are made with normal spacers, combs and plates, but the gel mixture is altered. It is easier to make the gel mixes in sufficient quantity to last about two weeks. There are two mixes:

	low TBE	high TBE
Sucrose	—	5 g
Bromophenol blue	—	5 mg
Urea (ultra-pure)	210 g	42 g
10 × TBE	25 ml	25 ml
Deionized 40% acrylamide (38:2)	75 ml	15 ml
Water to final volume of:	500 ml	100 ml

For each 40 cm × 20 cm gel, take
    30 ml Low TBE mix, add 60 μl 25% ammonium persulphate and 60 μl TEMED
    7 ml High TBE mix, add 14 μl 25% ammonium persulphate and 14 μl TEMED
(note polymerization will occur about 5 min after adding the TEMED)

Using a pipette controller (e.g. 'Propipette' or 'Pi-Pump') take up 4 ml of low TBE mix into a long 10 ml pipette, then take up the 6 ml high TBE mix. Draw up about 3 air bubbles through the solution to form a TBE gradient. Pour the mixture carefully down the edge of the gel plates, then top up with the remainder of the low TBE mix, as described for standard gels. (If gel plates wider than 20 cm are used, proportionally more low and high TBE mix must be used to generate the buffer gradient. If longer gels are used, the gradient must also be changed. If 50 cm long gel plates are used, rather than 40 cm, the high TBE mix should be $5 \times$ TBE and the relative volumes of the gradient are 6 ml of low TBE mix and 7 ml of high TBE; 45 ml of low TBE mix should be made, using 90 μl of ammonium persulphate and TEMED).

B. Standard autoradiography

1. Room temperature autoradiography

There are several ways of routinely autoradiographing gels. The method of choice depends on the sequencing technique that has been used, and the facilities available.

The simplest method is to remove the siliconized back plate, leaving the gel attached to the front plate. The DNA fragments are fixed in the gel by immersion for 5 min in 10% (v/v) acetic acid, 10% (v/v) methanol. The methanol removes the urea, and is only required if the gel is to be dried. The gel is rinsed well in water.

For ^{32}P autoradiography, the gel can be blotted dry with paper towels, covered with Saran Wrap, and marked with fluorescent paint or radioactive ink. The gel is then placed in a cardboard folder, a sheet of X-ray film (see below) is laid directly onto the gel, and a flat, heavy metal plate is placed on top to maintain close contact between the film and the gel. If the cardboard folder is attached to a 2 mm thick mild steel plate, then several gels can be stacked on top of one another. It is relatively simple to arrange these in a light-proof cupboard or box in a darkroom.

For ^{35}S autoradiography, the gel must first be dried down; this can be done by placing wet 3 MM chromatography paper on the gel and removing the gel from the glass plate to a gel drier. These thin gels will dry in about 20–30 min at 80°C. If a gel drier is not available, the front gel plate should be coated with a binding agent (e.g. Bind Silane) prepared according to the manufacturer's instructions. The gel will then be firmly attached to the glass plate, and can be dried at 80°C in an air-stream after fixing in acid-methanol. Sodium hydroxide solution can be used to remove the gel from the plate after autoradiography.

2. Low-temperature autoradiography

This method has the advantage that the signal-to-noise ratio for ^{32}P auto-radiography is higher at lower temperatures. A disadvantage is a potential loss of resolution, due to gel distortion when ice crystals form (Laskey, 1980). The gel is not acid fixed, but is immediately covered with Saran Wrap and marked with fluorescent paint or radioactive ink. The gel is placed in a metal cassette such as those used for medical X-rays. A sheet of X-ray film (see below) is placed directly on the gel. The cassette is closed and put at $-20°C$. Sequencing gels in metal cassettes can be stacked without problems from cross-exposure of films, although this is not always true of high-activity gels such as those used in strand separation in the chemical method.

3. Choice of autoradiography film

Several films are available which are suitable for autoradiography of sequencing gels. We have used Kodak X-omat AR, which is a high-sensitivity film with a clear base and a thin emulsion that can be developed in automatic processors. It can be used with blue-emitting intensifying screens for other purposes. Amersham's Hyperfilm is a film with a very clear base, and is suitable for most purposes. Usually a 16 h exposure is sufficient to visualize a DNA sequence, but longer exposures may be required. Indirect autoradiography (Section IV.C.2) is rarely required, and is disadvantageous due to its band-broadening effect, which reduces the resolution bands on the autoradiograph. However, sometimes it is necessary if a fragment being sequenced by the chemical method is poorly labelled or present in low yield. The details of the use of indirect autoradiography can be found in Laskey (1980).

V. Problems

A. Chemical method

Several different problems that may be encountered in the sequencing protocols are considered in detail by Maxam and Gilbert (1980) and are not dealt with here.

Probably the most common difficulty with the chemical method has been in 5'-end-labelling DNA fragments. Providing the chemical concentration of γ-labelled ATP is kept sufficiently high, and the protocols are followed exactly, little can go wrong at this step. With increasing availability of restriction endonucleases giving 5' extensions, procedures for 3'-end-labelling with DNA polymerase and α-labelled dNTPs are probably easiest for end-labelling DNA.

B. M13-cloning/chain-termination method

Some problems and their cures are given below. This list is not comprehensive, but may indicate possible solutions to other problems.

1. No bands or few bands show up on autoradiogram

(a) No primer DNA. Check on 20% acrylamide gel.

(b) No template DNA. Was there a small pellet on PEG precipitation of the phage? Check the template preparation by running on an agarose gel containing ethidium bromide. Try a new template preparation.

(c) Template DNA contains deletion of primer-complementary region. Some colourless plaques may be due to deletions in the *lacZ* region of the M13 vector. Bizarrely, these deletions may occur only with added fragments and not in the appropriate control. Try other templates from the same batch, and try making and sequencing template DNA from a blue plaque.

(d) Primer or template DNA is poisoned by impurities. Make fresh DNAs.

2. Several band patterns superimposed, and probably out of register

(a) Template is not pure. A mixed pattern of specific bands is due to two different template sequences being present. These can be repurified by transfection of competent cells with about 0.1–1 μl of template preparation, and making a fresh template preparation.

Alternatively, this effect may be due to lysis of cells during growth of the phage. This can be suppressed relative to the template sequence by increasing the ddNTP/dNTP ratio, or the template can be repurified. Shorter growth times should be used to avoid cell lysis. This effect is also seen if the template DNA is randomly nicked by nucleases; all reagents should be checked as being nuclease-free.

(b) Priming at more than one site. This may happen if a primer other than the universal primer is used. If there are two (or more) sites complementary to the primer another primer must be used.

3. Bands appear as doublets or triplets

(a) All bands in all channels. Possibly the primer has a heterogeneous 5′-end caused by exonuclease activity during its preparation or use. Check solutions for exonuclease activity or prepare new primer.

(b) Some bands or channels. Bands may appear due to chain elongation stopping before the nucleotide to be incorporated due to the relative low dNTP concentration. This is particularly noticeable if the chase step is omitted; and it occasionally occurs elsewhere (e.g. T-channel). Prepare fresh chase mix. Increasing the dNTP concentration of a problematic mix (with or without altering the ddNTP/dNTP ratio) may remove the artefact.

4. A band or bands across all the lanes in a particular region of the gel

(a) DNA polymerase cannot chain elongate through a particular sequence (e.g. at a run of Gs). This is sequence-specific, and may be circumvented by performing the elongation reactions at high temperature (up to 45°C), or by sequencing the other strand, or by using reverse transcriptase instead of DNA polymerase (e.g. see Agellon and Chen, 1986). If the DNA polymerase has been diluted and kept in the wrong buffer, it may give rise to a number of bands across the gel.

(b) A site-specific nuclease in low amounts contaminating one of the reagents would also give this effect. Change or repurify the reagent concerned. This can be a problem if the primer is removed by a restriction endonuclease which may be contaminated with, or intrinsically contain, a second activity (e.g. *Hind*III, contaminated with *Hind*II; or *Eco*RI and *Eco*RI* activity).

(c) A severe compression.

5. A sudden decrease in band spacing (sometimes to zero)

Secondary structure in single-stranded DNA during electrophoresis. Foldback structures at the end of the single-stranded DNA do not alter the hydrodynamic shape of the DNA in a progressive way and such structures tend to have similar mobilities and to run together on the gel. Running the gel at higher temperature (higher voltage) may remove small compressions. Sequencing the other strand will generally give rise to compressions in a different part of the sequence. Substitution of dGTP with 7-deaza-dGTP will also help reduce base pairing (Mizusawa *et al.*, 1986).

6. Failure to obtain a particular M13 template

(a) The cloning is lethal. The cloning could be attempted in a different member of the M13mp series to avoid in-frame polypeptide fusions; or without IPTG induction and X-gal selection to avoid expression of a lethal insert.

(b) The X-gal selection does not identify the recombinant, as the fragment contains an active promoter and N-terminal polypeptide sequence in fusion with the β-galactosidase. The cloning could be attempted in an M13 recombinant of opposite sense (e.g. M13mp18 instead of M13mp19).

(c) Large fragments may give very small, slow-growing plaques that may be overlooked. These are more easily identified by using fewer lawn cells when plating-out transfectants; and they have to be grown for a longer time prior to template preparation. Large fragments tend to suffer partial deletion.

VI. Buffers, reagents, equipment and suppliers

A. General-purpose buffers and reagents

1. *Restriction enzyme buffers (10 × RE buffer)*; unless otherwise stated, all digestions were done in accordance with manufacturers' recommendations. Where a restriction enzyme is used simultaneously with another enzyme, the buffer used is given.
2. *Phenol*; redistilled in air, and kept at −20°C in the dark. It is saturated with TE buffer (below) before use and the pH adjusted to 8.0. Larger quantities are kept at 4°C in the dark and 1 ml aliquots are kept frozen at −20°C.
3. *3 M sodium acetate*; pH adjusted to 6.5 before use.
4. *TE buffer*; 10 mM Tris-HCl, pH 7.5, 0.1 mM EDTA.
5. *10 × R buffer*; 500 mM NaCl, 100 mM Tris-HCl, pH 7.9, 100 mM 2-mercaptoethanol, 100 mM MgCl$_2$.
6. *10 × TBE buffer*; 900 mM Tris-borate, pH 8.3, 25 mM EDTA (108 g l^{-1} Trizma base, 55 g l^{-1} boric acid, 9.3 g l^{-1} Na$_2$ EDTA; pH should be correct without further adjustment).
7. *Acrylamide*; 40% (380 g l^{-1} acrylamide, 20 g l^{-1} NN'-methylenebisacrylamide) for sequencing gels. 30% (290 g l^{-1} acrylamide, 10 g l^{-1} NN'-methylenebisacrylamide) for native gels. The acrylamide solution is stirred with 20 g l^{-1} Amberlite MB-1 mixed-bed resin for approximately 30 min, then the resin is removed by filtration. The acrylamide is further filtered through nitrocellulose filters, degassed and stored at 4°C.
8. *Formamide*; deionized by stirring with 5 g l^{-1} Amberlite MB-1 or MB-3 mixed-bed resin. Filter through sintered glass. When making formamide gels, the formamide should be freshly deionized and should be filtered through nitrocellulose filters.
9. *Formamide-dye mix*; made by dissolving xylene-cyanol FF and bromophenol blue to a concentration of about 0.03% (w/v) in deionized formamide and adding 0.1 volume of 0.25 M EDTA. The optimal dye concentration should be determined empirically.
10. *Native loading buffer*; 15% (w/v) Ficoll 400, 25 mM EDTA, 0.03% (w/v) bromophenol blue.
11. *Gel elution buffer*; 500 mM ammonium acetate, 10 mM magnesium acetate, 0.1% (w/v) SDS, 0.1 mM EDTA.
12. *Radioactive ink*; standard stationers' ink with ^{35}S-sulphate added to about 0.1 mCi ml^{-1} will be sufficient for general-purpose use. ^{35}S is a good isotope to use as it has a longer half-life than ^{32}P, and it emits lower energy β-particles, giving sharper definition. Luminous ink should be used for very short exposures (e.g. preparative separation of radioactive DNA fragments by gel electrophoresis).

B. Chemical method

1. *10 × Phosphatase buffer*; 100 mM Tris-Cl, pH 9.5, 1 mM EDTA, 10 mM spermidine-HCl.
2. *Calf Intestinal Phosphatase*; Pharmacia, Molecular Biology Grade, diluted to 40 units ml^{-1} in 100 g ml^{-1} Bovine Serum Albumin; 10 mM Tris-HCl, pH 7.9, 1 mM ZnCl$_2$, 1 mM MgCl$_2$ 50% (v/v) glycerol.
3. *10 × Kinase buffer*; 500 mM Tris-HCl, pH 9.5, 100 mM MgCl$_2$, 50 mM DTT, 5 mM Spermidine-HCl, 0.5 mM EDTA.

4. *2 × Strand separation buffer*; 60% (v/v) DMSO, 1 mM EDTA, 0.03% (w/v) xylene cyanol, 0.03% (w/v) bromophenol blue.
5. *Strand separation gel*; 6 g acrylamide, 0.1 g bisacrylamide (N.B. 1 in 60 cross-linking). 6.7 ml 10 × TBE, 2 ml 1.6% (w/v) ammonium persulphate. Adjust volume to 120 ml with water; degas and start polymerization with 50 μl TEMED. Pour a 40 cm × 20 cm × 0.15 cm gel, using 1 cm wide sample wells.
6. *DMS buffer*; 50 mM sodium cacodylate, pH 8.0, 10 mM $MgCl_2$, 1 mM EDTA.
7. *Dimethyl sulphate*; store in a fume hood. With an automatic pipette dispense a small aliquot into stoppered bottle for each day's use. Dispose of small amount into 5 M NaOH.
8. *Hydrazine*; store in a tightly sealed bottle and dispense in a fume hood. With an automatic pipette, dispense a small aliquot into a stoppered bottle for each day's work. Dispose of small amounts into 3 M ferric chloride.
9. *DMS stop*; 1.5 M sodium acetate, 1 M 2-mercaptoethanol, 1 mM EDTA. Store at 4°C. To 0.5 ml add 20 μl of 10 mg ml^{-1} tRNA immediately before use (enough for ten fragments).
10. *HZ stop*; 0.4 M sodium acetate, 0.1 mM EDTA. Store at 4°C. To 5 ml, add 25 μl of 10 mg ml^{-1} tRNA immediately before use (enough for eight fragments).
11. *tRNA solutions*; dissolve in 1 mM Tris, pH 7.5, 0.1 mM EDTA. (Phenol-extract, ether-extract and dilute to about 10 mg ml^{-1} (A_{260} = 200). Store frozen.
12. *1.0 M piperidine*; make up fresh each time it is required. Store concentrated piperidine at 4°C. Dilute 0.2 ml into 1.8 ml chilled distilled water (do not use polycarbonate tubes), and rinse the piperidine out of the pipette several times. Mix well and store at 0°C.

C. M13-cloning/chain-termination method

1. *Bacterial media*; autoclaving at 15 p.s.i. for 20 min.
   a. *5 × A salts*: 5.25% (w/v) K_2HPO_4
         2.25% (w/v) KH_2PO_4
         0.5% (w/v) $(NH_4)_2SO_4$
         0.25% (w/v) Sodium citrate.2H$_2$O
         Autoclave.
   b. *Minimal A-glucose plates*: autoclave 10 g agar in 400 ml H_2O.
     Add 100 ml sterile 5 × A salts
         0.5 ml sterile 20% (w/v) $MgCl_2$.7H$_2$O
         5 ml sterile 20% (w/v) glucose.
   c. *2 × YT medium*: 1.6% (w/v) Bacto-tryptone
         1.0% (w/v) Yeast extract
         1.0% (w/v) NaCl
         Autoclave.
   d. *BBL agar*: 1.0% BBL Microbiology Systems trypticase, 0.5% NaCl, containing 1.5% agar for plates, or 0.5% (w/v) agar for top agar overlaps. Top-agar can be heated to dissolve the solids then dispensed into 3 ml aliquots before autoclaving.
   e. *X-gal solution*: weigh 20 mg on clean glassine paper. Dissolve in 1 ml dimethylformamide. Do not further sterilize. Store at −20°C.
   f. *IPTG solution*: weigh 24 mg on clean glassine paper. Dissolve in 1 ml autoclaved water. Do not further sterilize. Store at −20°C.

g. *Transformation buffers*: TfbI is 30 mM Potassium acetate, 100 mM RbCl, 10 mM $CaCl_2$, 50 mM $MnCl_2$, 15% glycerol, pH adjusted to 5.8 with 0.2 M acetic acid (do not overshoot and back-titrate). Filter sterilize.
TfBII is 10 mM MOPS (or PIPES), 75 mM $CaCl_2$, 10 mM RbCl, 15% glycerol, pH adjusted to 6.5 with KOH. Filter sterilize.

  h. *Other reagents* are sterilized by autoclaving (e.g. 50 mM $CaCl_2$ for making competent cells).

2. *DNA ligase dilution buffer*: 50% (v/v) glycerol, 50 mM KCl, 10 mM 2-mercaptoethanol, 100 g ml^{-1} autoclaved gelatin, 10 mM potassium phosphate, pH 7.4, store at −20°C.

3. *ACE electrophoresis buffer*: 40 mM Tris-acetate, pH 8.0, 5 mM sodium acetate, 1 mM EDTA, 1 g ml^{-1} ethidium bromide.

4. *10 × C buffer for ligations*: 500 mM NaCl, 100 mM Tris-HCl, pH 7.5, 100 mM $MgCl_2$, 10 mM DTT.

5. *Commercial primer solutions* should be used according to the manufacturer's instructions. Too little primer gives no sequence, too much primer can cause only short sequences to be obtained.

6. *Nucleotide mixes*
(a) T- C-, G- and A-mixes are made up from stock solutions kept in water at −70°C. The working concentrations are determined empirically, but initially try:

	T	C	G	A
0.5 mM dTTP	25	500	500	500
0.5 mM dCTP	500	25	500	500
0.5 mM dGTP	500	500	25	500
10 mM ddTTP	75	—	—	—
10 mM ddCTP	—	3.2	—	—
10 mM ddGTP	—	—	7.5	—
10 mM ddATP	—	—	—	3.2
TE buffer	1000	1000	1000	500

(Note that the dideoxynucleotides and deoxynucleotides can be kept separate, each with 500 μl TE buffer, until the optimum ratio is determined; and that a set of 'long' mixes will have less ddNTP).
(b) 0.5 mM dNTPs for the chase are made up in TE buffer.
7. *Klenow diluent*: 50% (v/v) glycerol. 50 mM potassium phosphate, pH 7.0.

D. Sources of chemicals, enzymes and equipment

The products listed are those in the authors' laboratory, or listed in other papers (e.g. Maxam and Gilbert, 1980). This does not imply that they are superior to products from other suppliers, merely that they are satisfactory for their purpose.

Reagents in addition to those listed below were obtained from a general chemical supply company (e.g. British Drug Houses Ltd), and are AnalaR grade where possible.

Bacterial media are obtained from Difco Laboratories Inc.
Radiochemicals are obtained from Amersham International plc, or New England Nuclear, GmbH.

Acrylamide, bisacrylamide (electrophoresis grade); B.D.H. Ltd.
Agarose (low electroendosmosis); Sigma Chemical Co.
Amberlite MB-1, MB-3: B.D.H. Ltd.
Dimethyl sulphate: Aldrich Chemical Co.; N.E.N., GmbH.
IPTG: Boehringer Mannheim GmbH.
Piperidine: Fischer Scientific: N.E.N., GmbH.
Sephadex: Pharmacia.
Spermidine hydrochloride: Sigma Chemical Co.
tRNA: Boehringer Corporation Ltd.
Urea (ultra-pure): Bethesda Research Laboratories Ltd.
X-gal (5-bromo-4-chloro-3 indolylgalactoside): Boehringer Mannheim GmbH.

Universal primers for the M13-cloning/chain-termination method are available from a number of sources, and can very economically be made by anyone with access to a DNA synthesizer.

Similarly, restriction endonucleases are available from many firms. Availability, purity and price are the factors most frequently determining choice of supplier; unfortunately there is great variation in all three factors.

The authors' laboratory normally obtains other enzymes as listed below:

Calf intestinal phosphatase: Boehringer Ltd (Molecular Biology Grade: 713023; although at the time of writing, a suitable quality enzyme has not been available for several months); Pharmacia Ltd.
Klenow polymerase: Amersham T.2140.
Polynucleotide kinase: Amersham T.2020.
T4 DNA polymerase: Amersham T.2040; P-L Biochemicals Ltd (0918).
T4 DNA ligase: Amersham T.2010; Bethesda Research Laboratories (5224); Boehringer Corporation Ltd (481220); P-L Biochemical Ltd (0870); New England Biolabs (202).

Some of the *E. coli* strains and bacteriophage strains required for the M13-cloning method are available from Bethesda Research Laboratories Ltd or from P-L Biochemicals Ltd.

A DNA sequencing kit for the chemical method is available for N.E.N. (NEK-010).

Most of the equipment used in the methods described in this chapter can be obtained from general and specialist laboratory suppliers. A few notes describing important features of particular pieces of equipment are included below.

Plastikard (thickness 0.015 inch) is from Slaters' Plastikard Ltd. Temple Road, Matlock Bath, Derbyshire, U.K.

Vinyl Tape (No. 1607; 25 mm wide) is from D.R.G. Sellotape Products Ltd. Theobald Street, Boreham Wood, Hertfordshire, U.K.

Gel apparatus for sequencing gels is available from several suppliers (Bethesda Research Laboratories, Scotlab Ltd. etc.). It is very important that the gel plates are of high quality float glass to minimize variation in thickness of the gel.

Saran Wrap (Dow Chemical Corporation) proves superior to other domestic transparent film that we have tested, as it does not adhere to autoradiography film and cause film-blackening due to static electrical discharge on removing the film.

Microfuge tubes (1.5 ml) must have a good seal, yet open with relative ease. Those manufactured by Sarstedt (No. 72.690), Treff (96.2494) or Eppendorf (2236411-1) prove satisfactory.

Watchmakers' luminous paint suitable for making location markers for autoradiography is available from Mahoney Associates, 58 Stapleton Road, Bristol BS5 0RD, U.K., or other watchmakers' supply companies.

VII. Computer methods

The assembly of DNA sequence data is now so rapid and its analysis so complex that computer assistance is required for both these functions. There are many packages available that will allow the compilation and analysis of DNA sequence data with relative ease. Some packages will run only on mainframe machines, others will run on personal computers. There are some good packages which are very cheap or free to academic users. Advice should be sought through the central facilities for molecular biologists provided on the academic networks, such as BIONET in the USA and SEQNET in the UK, or through specialized groups such as the Computer Club of the Society for General Microbiology. In this laboratory, we have used the STADEN package (from Dr R. Staden, MRC Laboratory of Molecular Biology, Hills Road, Cambridge, CB2 2QH, U.K.) for sequence compilation and analysis. A version of this package for the IBM-PC is available from Amersham International plc.

VIII. Concluding remarks

When the original version of this chapter first appeared (Brown, 1984), the question was posed 'Should I use the chemical method or the M13-cloning/chain-termination method?'. The reply was that this depended on the project and the expertise of the experimentalist. This is still the case, but now the question is also which variation of the method would give the optimum approach to a particular sequencing task. There is no easy answer, but some of the factors to consider include:

(i) the size of the project: a very large sequence may best be determined by the shotgun approach of M13 sequencing of sonicated DNA;

(ii) the DNA to be sequenced: very GC-rich DNA may be best sequenced by the chemical method;

(iii) associated experiments: if protein-binding is to be studied, then chemical sequencing has traditionally offered advantages;

(iv) the isolation of the DNA: if a large number of mutants across a small region of DNA are to be sequenced, the time invested in finding a suitable primer for direct sequencing of double-stranded plasmid DNA may be worthwhile.

We have omitted many important variations of techniques in this chapter, due to pressures of space and limitations in our expertise. For this we apologise; some of them are quite unusual, such as direct genomic sequencing (Church and Gilbert, 1984), and enable experiments to be considered that would otherwise be technically difficult. Others are individual preferences for strategy or reaction conditions. There are so many that we cannot include them all, and we apologize to those who do not find their favourite method in this chapter.

Acknowledgements

We thank those colleagues who have shared methods and problems with us over the years. In particular, A. Bankier, W. Barnes, B. Barrell, M. Biggin, M. Buttner, T. Gibson, J. Messing, R. Mullings, A. Perry, F. Sanger, R. Tizard, K. Weston. Work from the authors' laboratory was supported by the Medical Research Council and the Royal Society. NLB is a Royal Society EPA Cephalosporin Fund Senior Research Fellow.

References

Agellon, L. B. and Chen, T. T. (1986). *Gene Analysis Tech.* **3**, 86–89.

Anderson, S. (1981). *Nucleic Acids Res.* **9**, 3015–3027.

Atkinson, M. R., Deutscher, M. P., Kornberg, A., Russell, A. F. and Moffat, J. G. (1969). *Biochemistry* **8**, 4879–4904.

Barnes, W. M., Bevan, M. and Son, P. M. (1983). *In* "Methods in Enzymology" (R. Wu, Ed.), Vol. 101, pp. 98–122. Academic Press, New York and London.

Biggin, M. D., Gibson, T. J. and Hong, G. F. (1983). *Proc. Natl. Acad. Sci. U.S.A.* **80**, 3963–3965.

Brown, N. L. (1984). *In* "Methods in Microbiology" (P. M. Bennett and J. Grinsted, Eds), Vol. 17, pp. 259–313. Academic Press, New York and London.

Chen, E. J. and Seeburg, P. H. (1985). *DNA* **4**, 165–170.

Church, G. M. and Gilbert, W. (1984). *Proc. Natl. Acad. Sci. U.S.A.* **81**, 1991–1995.

Gibson, T. J. (1984). Ph.D. Thesis, University of Cambridge.

Gronenborn, B. and Messing, J. (1978). *Nature (London)* **272**, 375–377.

Heinrich, P. (1986). "Guidelines for quick and simple Plasmid sequencing." Boehringer Mannheim GmbH, Mannheim.

Hong, G. F. (1982). *J. Mol. Biol.* **158**, 539–549.

Kessler, C. and Höltke, H. J. (1986). *Gene* **47**, 1–153.

Laskey, R. A. (1980). *In* "Methods in Enzymology" (L. Grossman and K. Moldave, Eds), Vol. 65, pp. 363–371. Academic Press, New York and London.

Lis, J. T. (1980). *In* "Methods in Enzymology" (L. Grossman and K. Moldave, Eds), Vol. 65, pp. 347–353. Academic Press, New York and London.

Maniatis, T., Fritsch, E. F. and Sambrook, J. (1982). "Molecular Cloning. A Laboratory Manual". Cold Spring Harbor, New York.

Maxam, A. and Gilbert, W. (1977). *Proc. Natl. Acad. Sci. U.S.A.* **74**, 560–564.

Maxam, A. and Gilbert, W. (1980). *In* "Methods in Enzymology" (L. Grossman and K. Moldave, Eds), Vol. 65, pp. 499–560. Academic Press, New York and London.

Messing, J. (1979). *Recombinant DNA Technical Bulletin* **2**, 43–48.

Messing, J. and Vieira, J. (1982). *Gene* **19**, 269–276.

Messing, J., Crea, P. and Seeburg, P. (1981). *Nucleic Acids Res.* **9**, 309–321.

Messing, J., Gronenborn, B., Müller-Hill, B. and Hofschneider, P. H. (1977). *Proc. Natl. Acad. Sci. U.S.A.* **74**, 3642–3646.

Misra, T. K. (1984). *Gene* **34**, 263–268.

Mizusawa, S., Nishimura, S. and Seela, F. (1986). *Nucleic Acids Res.* **14**, 1319–1324.

Norrander, J., Kempe, T. and Messing, J. (1983). *Gene* **26**, 101–106.

Olsson, A., Moks, T., Uhlen, M. and Gaal, A. B. (1984). *J. Biochem. Biophys. Methods* **10**, 83–90.

Peacock, A. C. and Dingman, C. W. (1967). *Biochemistry* **6**, 1818–1827.

Ray, D. S. (1977). *In* "Comprehensive Virology" (H. Fraenkel-Conrat and R. R. Wagner, Eds), Vol. 7, pp. 105–178. Plenum, New York and London.

Roberts, R. J. (1987). *Nucleic Acids Res.* **15**, 189–217.

Sanger, F. and Coulson, A. R. (1975). *J. Mol. Biol.* **94**, 441–448.

Sanger, F. and Coulson, A. R. (1978). *FEBS Lett.* **87**, 107–110.

Sanger, F., Nicklen, S. and Coulson, A. R. (1977). *Proc. Natl. Acad. Sci. U.S.A.* **74**, 5463–5467.

Sanger, F., Coulson, A. R., Barrell, B. G., Smith, A. J. H. and Roe, B. A. (1980). *J. Mol. Biol.* **143**, 161–178.

Sanger, F., Coulson, A. R., Friedmann, T., Air, G. M., Barrell, B. G., Brown, N. L., Fiddes, J. C., Hutchison, III, C. A., Slocombe, P. M. and Smith, M. (1978). *J. Mol. Biol.* **125**, 225–246.

Skryabin, A. G., Zakharyev, V. M. and Bayev, A. A. (1978). *Doklady Akad. Nauk. U.S.S.R.* **241**, 488–492.

Smith, A. J. H. (1980). *In* "Methods in Enzymology" (L. Grossman and K. Moldave, Eds), Vol. 65, pp. 560–580. Academic Press, New York and London.

Staden, R. (1980). *Nucleic Acids Res.* **4**, 4037–4051.

Studier, F. W. (1973). *J. Mol. Biol.* **79**, 237–248.

Tabor, S. and Richardson, C. C. (1987). *Proc. Natl. Acad. Sci. U.S.A.* **84**, 4767–4771.

Tu, C.-P. D. and Cohen, S. N. (1980). *Gene* **10**, 177–183.

Weaver, R. F. and Weissmann, C. (1979). *Nucleic Acids Res.* **7**, 1175–1193.

Yanisch-Perron, C., Vieira, J. and Messing, J. (1985). *Gene* **33**, 103–119.

Note added in proof: A simple technique giving results similar to poured buffer gradient gels is to make the bottom buffer chamber, 0.33 M sodium acetate in TBE, at the end of the pre-run; pre-run the gel for a further 10 min before loading. A constant current power pack should be used.

Index

310

Contents of published volumes